Modeling and Simulation in Thermal and Fluids Engineering

This textbook comprehensively covers the fundamentals behind mathematical modeling of engineering problems to obtain the required solution.

It comprehensively discusses modeling concepts through conservation principles with a proper blending of mathematical expressions. The text discusses the basics of governing equations in algebraic and differential forms and examines the importance of mathematics as a tool in modeling. It covers important topics including modeling of heat transfer problems, modeling of flow problems, modeling advection-diffusion problems and Navier-Stokes equations in depth. Pedagogical features including solved problems and unsolved exercises are interspersed throughout the text for better understanding.

The textbook is primarily written for senior undergraduate and graduate students in the field of mechanical engineering for courses on modeling and simulation. The textbook will be accompanied by teaching resources including a solution manual for the instructors.

Modeling and Simulation in Thermal and Fluids Engineering

Krishnan Murugesan

CRC Press is an imprint of the
Taylor & Francis Group, an **informa** business

First edition published 2023
by CRC Press
6000 Broken Sound Parkway NW, Suite 300, Boca Raton, FL 33487-2742

and by CRC Press
4 Park Square, Milton Park, Abingdon, Oxon, OX14 4RN

CRC Press is an imprint of Taylor & Francis Group, LLC

ISBN: 978-0-367-56046-1 (hbk)
ISBN: 978-1-032-16322-2 (pbk)
ISBN: 978-1-003-24807-1 (ebk)

DOI: 10.1201/9781003248071

Typeset in Times
by Deanta Global Publishing Services, Chennai, India

Access the Support Material: www.routledge.com/9780367560461

Contents

Preface xiii
Author Biography xvii

Chapter 1 Introduction 1

1.1 Modeling 3
1.1.1 Physical Models 3
1.1.2 Mathematical Models 4
1.1.2.1 Perfect Gas Equation 8
1.1.2.2 Hooke's Law 8
1.1.2.3 Deflection of Beam under Load 9
1.1.2.4 Fluid Mechanics 10
1.1.2.5 Heat Transfer 11
1.2 Simulation 12
1.3 Conservation Principles 13
1.3.1 Mass Conservation 14
1.3.2 Momentum Conservation 15
1.3.3 Energy Conservation 17
1.3.4 Species Conservation 18
1.4 Types of Physical Problems 20
1.4.1 Equilibrium Problems 21
1.4.2 Eigen Value Problems 23
1.4.3 Propagation Problems 23
1.5 Models in Engineering Analysis 25
1.5.1 Lumped Parameter Model 25
1.5.2 Continuum Based Model 28
1.6 Solution of Differential Equations 29
1.6.1 Analytical Techniques 30
1.6.2 Numerical Techniques 32
1.6.3 Computing Techniques 35
References 36
Exercise Problems 36
Quiz Questions 38

Chapter 2 Conservation Equations 41

2.1 Solid Medium 42
2.1.1 Energy Transport in Unsteady State Conditions 43
2.1.1.1 Generalized Conduction Equation in Cartesian Coordinates 46

2.1.1.2 Generalized Conduction Equation in Cylindrical Coordinates....48
2.1.1.3 Generalized Conduction Equation in Spherical Coordinates....50
2.1.1.4 Initial and Boundary Conditions....50
2.1.1.5 Initial Condition....51
2.1.1.6 Boundary Conditions....51
2.1.2 Energy Transport in Steady State Condition....53
2.1.2.1 Steady State Heat Conduction in Plane Wall....54
2.1.2.2 Steady State Heat Conduction in Cylinder....56
2.1.2.3 Steady State Heat Conduction in Sphere....58
2.2 Fluid Medium....59
2.2.1 Mass Conservation....61
2.2.1.1 Material Derivative Form....65
2.2.1.2 Incompressible Fluid Flow....66
2.2.2 Momentum Conservation....66
2.2.2.1 Relation between Stress and Viscosity....72
2.2.2.2 Momentum Balance Equations for Incompressible Flow (μ=constant)....73
2.2.3 Energy Conservation....75
2.2.3.1 Energy Balance....76
2.2.3.2 Rate of Change of Energy in CV....78
2.2.3.3 Net Efflux of Energy from CV....78
2.2.3.4 Rate of Work Done by Surface Forces....78
2.2.3.5 Work Done by Body Forces....79
2.2.3.6 Net Addition of Heat due to Conduction and Radiation Heat Transfer....80
2.2.3.7 Heat Generation within Control Volume....81
2.2.4 Species Conservation....87
References....88
Exercise Problems....88
Quiz Questions....90

Chapter 3 Finite Difference and Finite Volume Methods....93

3.1 Finite Difference Method....95
3.1.1 One-Dimensional Conduction....95
3.1.2 Taylor's Series Principle....97
3.1.3 Polynomial Method....100
3.1.4 Application to Ordinary Differential Equations....102
3.1.4.1 Equations for the Boundary Nodes 1 and M....104

3.1.5 Application to Partial Differential Equations 107
3.1.5.1 Two-Dimensional Conduction Equation...... 107
3.1.5.2 Difference Equations for Boundary Conditions 111
3.1.5.3 Corner Nodes 114
3.1.5.4 Boundary Nodes 119
3.1.5.5 Comparison of Two-Dimensional Conduction Results with Analytical Solution 120
3.2 Finite Volume Method 123
3.2.1 Heat Flux Boundary Condition at M (x=L) 127
3.2.2 Convective Boundary Condition at Node M (x=L) 128
3.2.3 Example Problem for Finite Volume Method – Fin 128
3.2.4 One-Dimensional and Two-Dimensional Applications 128
3.2.4.1 One-Dimensional Application 128
3.2.4.2 Two-Dimensional Application 131
3.2.4.3 Boundary Conditions 133
3.2.4.4 Corner Nodes 136
3.2.5 Complex Geometry and Variable Property 138
3.2.5.1 Complex Geometry 139
3.2.5.2 Variable Property 140
3.2.5.3 Variable Area 141
References 143
Exercise Problems 144
Quiz Questions 146

Chapter 4 Finite Element Method 147

4.1 Galerkin's Weighted Residual Method 148
4.1.1 Integration of Shape Functions 154
4.1.2 Boundary Conditions 158
4.1.2.1 Convective Boundary Condition 158
4.1.2.2 Dirichlet Boundary Condition 159
4.1.3 Example Problem: Fin (Computer Code fin_FEM.for) 159
4.2 Domain Discretization and Isoparametric Formulation 162
4.2.1 Domain Discretization 162
4.2.2 Isoparametric Formulation 163
4.3 Discretization of One-Dimensional Domain 168
4.4 Discretization of Two-Dimensional Domain 170
4.4.1 Rectangular and Quadrilateral Elements 172
4.5 Discretization of Three-Dimensional Domain 176

4.6 Mesh Generation 180
4.7 Transfinite Interpolation Technique (TFI) 182
4.7.1 Multi-Block TFI Grid Generation 188
4.7.2 Three-Dimensional TFI Meshing 188
4.8 Time-Dependent Problems 194
4.8.1 Stability Conditions 199
4.8.1.1 Explicit Scheme 200
4.8.1.2 Implicit Scheme 201
4.8.1.3 Semi-Implicit Scheme (Crank-Nicholson Scheme) 202
4.8.1.4 Significance of Fourier Number 202
4.8.1.5 Alternate Direction Implicit (ADI) Method 204
References 205
Exercise Problems 205
Quiz Questions 207

Chapter 5 Modeling of Heat Transfer Problems 209

5.1 Heat Transfer Problem – One-Dimensional Conduction with Heat Generation 210
5.1.1 Derivation of Energy Conservation Equation 211
5.1.2 Identification of Boundary Conditions 213
5.1.3 Solution Using Finite Element Method 213
5.1.4 Incorporation of Boundary Condition 215
5.1.5 Computational Algorithm 217
5.1.6 Computer Programming 217
5.1.7 Mesh Sensitivity and Validation Results 220
5.1.8 Simulation Parameters and Results 221
5.2 Two-Dimensional Problem – Heat and Mass Transfer through Soil: Landmine Detection 225
5.2.1 Derivation of Conservation Equations 226
5.2.1.1 Conservation Equation for Heat and Moisture Transport within the Soil Medium 227
5.2.2 Initial and Boundary Conditions 231
5.2.2.1 Initial Conditions 231
5.2.2.2 Boundary Conditions – Soil Medium 232
5.2.2.3 Top Side 232
5.2.2.4 Bottom Side 233
5.2.2.5 Vertical Sides 233
5.2.2.6 Mine-Soil Interface Boundary Condition 233
5.2.3 Solution Using Finite Element Method 234

5.2.4 Inclusion of Convective Boundary Condition on the Top Surface of the Soil Medium 235
5.2.5 Solution of Energy Equation for the Landmine 239
5.2.6 Computational Algorithm 240
5.2.7 Computer Programming 242
5.2.8 Simulation Parameters and Results 243
5.2.9 Discussion of Simulation Results 244
5.2.9.1 Mesh Sensitivity and Validation Results 244
5.2.9.2 Simulation Results 247
References 253
Exercise Problems 254
Quiz Questions 255

Chapter 6 Modeling of Flow Problems 257

6.1 Fluid Mechanics – Filling of Water Tank 257
6.1.1 Derivation of Mass and Momentum Conservation Equations 258
6.1.2 Boundary Conditions and Initial Conditions 259
6.1.3 Solution Using Analytical Method 260
6.1.4 Computational Algorithm and Computer Program 260
6.1.5 Simulation Parameters and Discussion of Results 261
6.2 Two-Dimensional Flow Problems – Stokes Flow 262
6.2.1 Description of Problem 262
6.2.2 Mathematical Modeling 263
6.3 Three-Dimensional Stokes Flow 267
6.3.1 Governing Equations for Three-Dimensional Stokes Flow 267
6.3.2 Finite Element Solution Procedure 270
6.3.3 Enforcement of Dirichlet Boundary Conditions in Finite Element Solution Procedure 275
6.3.3.1 Computational Steps to Incorporate Dirichlet Boundary Conditions 276
6.3.4 Global Matrix-Free Finite Element Algorithm 278
6.3.4.1 Matrix Storage Schemes for Large Size Problems and Solvers 278
6.3.4.2 BICGSTAB and Element-by-Element Scheme for Parallel Computing 278
6.3.4.3 Procedure to Implement Global Matrix-Free Finite Element Algorithm 279
6.4 Results for Three-Dimensional Stokes Flow 284
6.4.1 Comparison of Memory Storage of GMFFE Algorithm with Column Format Scheme 286

6.4.2 Flow Results for Three-Dimensional Stokes Flow Using 51^3 Mesh 286
6.4.2.1 Mesh Sensitivity and Validation Results 286
6.4.2.2 Velocity Vectors Distribution 288
References 290
Exercise Problems 290
Quiz Questions 292

Chapter 7 Navier-Stokes Equations 293

7.1 Momentum Balance of Fluid in a System 293
7.1.1 Fluid Dynamics 295
7.2 Navier-Stokes Equations in Primitive Variables Form 296
7.2.1 Navier-Stokes Equations 297
7.2.2 Application of Predictor-Corrector Method 298
7.2.3 Finite Element Solution Procedure 300
7.2.4 Computational Algorithm 309
7.2.4.1 Computer Program – Subroutines 310
7.3 Navier-Stokes Equations in Velocity-Vorticity Form 311
7.3.1 Derivation of Velocity-Vorticity Equations as Generalized Formulation 313
7.3.2 Computation of Vorticity Boundary Conditions 318
7.3.2.1 Node i on Side AB – For Wall Normal Parallel to Positive y-Axis 320
7.3.2.2 Node j on Side CD – For Wall Normal Parallel to Negative y-Axis 320
7.3.2.3 Node k on Side DA – For Wall Normal Parallel to Positive x-Axis 321
7.3.2.4 Node m on Side BC – For Wall Normal Parallel to Negative x-Axis 321
7.3.3 Solution Using Finite Element Method 321
7.3.4 Finite Element Formulation of Vorticity Transport Equation 323
7.3.5 Finite Element Solution Procedure for Velocity Poisson Equations 326
7.3.6 Computational Algorithm 327
7.3.7 Simulation of Lid-Driven Square Cavity Flow Problem 329
7.3.8 Simulation Results 329
7.3.9 Simulation of Natural Convection in a Square Cavity 336
7.3.9.1 Finite Element Solution Procedure 338

7.3.9.2 Simulation Results for Natural Convection in a Differentially Heated Square Cavity....340
References....343
Exercise Problems....344
Quiz Questions....345

Index....347

Preface

Engineers and scientists have been making numerous attempts to understand the behavior of nature and the environment in which the systems they design have to operate for the wellbeing of mankind. The discoveries and inventions made by the creative power of many of these intellectuals have helped human beings to lead a civilized life over centuries. The laws and hypotheses proposed by scientists have been well documented over the decades and have been put in use to design and analyze many scientific and engineering systems used by society. All the scientific hypotheses used by the current scientific community are either based on an experimental approach or on theoretical propositions. Design and performance analysis of engineering problems either employ experimental techniques or numerical approaches. In the recent past, both experimental and numerical methods have been used in combination to obtain more realistic results and data to improve the performance of systems and to design very efficient machines.

The study of thermal and fluids engineering problems finds wide engineering applications such as in internal combustion engines, aero-space applications, heat exchangers, atmospheric science, nuclear engineering, electronic cooling etc. to mention a few. Though many of these engineering applications have been well established, there is a continuous attempt to improve the systems to make them both energy efficient and environment friendly. Analysis of engineering problems employ theoretical, numerical, and experimental techniques to obtain the data required for performance study and design improvements. Experimental methods are age-old methods that have provided reliable data for engineers and scientists. However, it requires huge investment to set up experimental and laboratory facilities to investigate the mechanisms of many man-made machines. Though the experimental approach is very popular in many engineering applications, it is not always possible to obtain all the required data for design and analysis only from experimental methods. For example, problems such as heat and moisture transport through unsaturated porous media, turbulent flow problems, radiation problems etc. cannot be fully understood using only experimental methods. The study of unsaturated porous solids involves multi-scale transport phenomena, turbulent flow problems are yet to be fully understood and it becomes challenging to analyze non-linear radiation heat transfer phenomenon in many real-life applications. Atmospheric science is another important field of research, which helps to predict weather patterns for the welfare of mankind. Many attempts have been made by scientists to measure wind patterns and other weather-related parameters during critical weather conditions; however, such efforts provide only limited data for analysis to understand fully the weather pattern. Design and performance analysis of nuclear engineering equipment poses health hazards for the operating personnel during the time of experimentation, thus limiting this approach in this field of science. Hence, it can be appreciated that though experimental techniques provide valuable data for analysis, their applications are limited in many real-life problems.

Mathematical modeling is one of the alternate methods for experimental techniques. Modeling is used to study the underlying physics in many applied engineering problems and has been in use in science and engineering analysis for centuries. The definition of the perfect gas equation based on Boyle's law and Charles' law, material stiffness, spring constant etc. are some of the mathematical models that are in use. As understood from physics, any physical phenomenon is governed by certain conservation laws or constitutive equations, and when this is expressed using a mathematical expression, then it is called a mathematical model. It has to be understood that a mathematical model does not simply represent some differential equations or some complex mathematical expressions. The physics behind those mathematical equations has to be well defined before these expressions are said to represent a mathematical model. Without clearly affirming the underlying physics in these mathematical equations, the said equations do not convey any meaning about the phenomenon under investigation. Thus, in a sense, a mathematical model can be viewed as the combination of mathematics and physics used to represent the behaviors or phenomena that take place in a system. Mathematical models also can be viewed as an extension of a theoretical approach that had been widely employed to study the behavior of nature, and such approach is built on strong hypotheses established either through natural observations or experimentations. For example, the laws postulated in thermodynamics do not have any mathematical proof, however, they are accepted and applied in many analyses because the consequences of these laws could not be disproved by any method up to now. With the help of mathematical models, engineers and scientists can predict the behavior of a system without fabricating the system, and this procedure is called simulation program.

The invention of high-speed computing machines has led to a revolution in mathematical modeling because the governing equations used in a mathematical model can be solved easily using numerical methods with the help of computers, even though analytical solutions may exist for those equations sometimes. Nowadays, many computer programs in the form of commercial software are available and are handy to solve many industrial related problems. Recent advancements in numerical methods, availability of new computer coding languages and high-speed computing machines have made possible the simulation of many complex problems. Initially these kinds of commercial software were developed as general computer programs catering to find the solution to variety of problems; however, now due to widespread specific requirements in some fields, custom-based software has been developed, focusing on some special applied problems. As the need of development of new software for engineering problems has increased over the decades, there is a widespread academic interest from many universities across the globe in mathematical modeling. Though mathematical modeling was known originally only for some engineering problems, such as in aircraft industries, automation, computational fluid dynamics etc., this method has occupied its own place in other fields of science such as chemistry, physics, bio-science, bio-engineering, biotechnology, economics, social science etc. Hence, now there is a need for training the undergraduate and post-graduate students of science and engineering in mathematical modeling and simulation programs. Many universities and institutes have introduced courses on modeling and simulation at various levels to meet the demand by industries and society at large. A word of caution is felt necessary at this juncture when the importance of modeling and simulation programs is being realized – the final usage of the simulation programs depends on the validation data either obtained from experimental techniques or other well-established methods; these are used to validate the mathematical models.

The main objective of this textbook is to motivate students to understand the physics behind many engineering and scientific problems and develop their own computer programs for the mathematical models developed for specific problems. The textbook deals with four major topics, conservation principles, numerical methods, application of modeling and simulation of thermal and flow problems, and finally case study problems, and it is organized in 12 chapters with a specific focus in each chapter. An introduction to modeling and simulation programs are discussed in detail in Chapter 1, giving some examples for mathematical models that are known to readers from different perspectives. Discussion on the importance of conservation principles, different type of models used in engineering analysis, types of equations employed in models and the solution techniques are given as an introduction to this textbook. This makes a good beginning for students new to modeling and simulation subjects. Chapter 2 is completely devoted to deliberations on different conservation principles relevant to thermal and flow problems and their respective equations related to thermal and flow problems. In order to deal with conduction heat transfer problems, the initial section has been allotted to conservation equations for the solid medium. The later part of the chapter is reserved for conservation principles related to convective heat transfer and flow problems.

Numerical methods are important tools for modeling and simulation programs. Hence, the basic numerical methods, finite difference method, finite volume method and finite element method are explained focusing on their fundamental principles. Finite difference and finite volume methods are explained in Chapter 3 whereas the finite element method is discussed in Chapter 4. The methodology for the application of these numerical methods for problems in thermal and fluids engineering is demonstrated by solving some example problems. In order to motivate readers to develop their own computer programs, computational algorithms have been explained wherever possible. Chapter 5 deals with modeling of heat transfer problems in which the basic principle of mathematical modeling and simulation programs is elaborated in detail in step-by-step procedures using a one-dimensional heat conduction problem. The discussion has been further expanded for a two-dimensional heat and moisture transport problem. Computational algorithms also have been explained in detail. Modeling of flow problems is elaborated in Chapter 6, starting with a simple practical problem of the filling of a water tank. All the steps involved in modeling and simulation of this problem are demonstrated in detail in this chapter. Two-dimensional and three-dimensional Stokes flow problems are also discussed by introducing the global matrix-free finite element algorithm.

Chapter 7 is devoted to the discussion of the numerical solution procedure of Navier-Stokes equations. Initially, the primitive variables form of Navier-Stokes equations were discussed by implementing the solution procedure using the

Eulerian-velocity correction method. One of the forms of Navier-Stokes equations without the pressure term, called the velocity-vorticity form, is also derived and solved using the finite element method. Finally, a problem on natural convection in a square cavity is explained at the end of the chapter.

Chapters 8 to 12 are available online at www.routledge.com/9780367560461.

Author Biography

Professor Krishnan Murugesan is currently teaching at the Indian Institute of Technology Roorkee, Roorkee, India in the Mechanical and Industrial Engineering Department. He is specialized in broad research areas such as computational fluid dynamics, modeling of heat and moisture transport through porous solid, fuel cells, cooling towers and ground source heat pump systems. The modeling and simulation subject is his favorite to teach to master level students, which he has done for more than a decade in the department. Experience gained during teaching this course for master level thermal engineering students, interaction with the students and conducting practical programming classes for the course has motivated him to write this textbook. It is the opinion of the author that it is very difficult for students to understand the underlying principles of modeling and simulation without developing a computer code. Apart from teaching the course on modeling and simulation, the author has also guided many dissertation works, at undergraduate, post-graduate and doctoral levels, on modeling different types of problems ranging from modeling fuel cells, hydrogen storage, nuclear waste disposal facilities, cooling of fuels cells, landmine detection, concrete exposed to fire, computational fluid dynamics problems using nanofluid as working fluid, different shapes of nanofluids, computational fluid dynamics in the presence of a magnetic field etc. Professor Krishnan Murugesan has been in the field of modeling and simulation since his master program at the Indian Institute of Technology Bombay, Mumbai; and doctoral program from the Indian Institute of Technology Madras, Chennai. During his post-doctoral program at National Taiwan University, Taipei, Taiwan and a number of academic visits to Geoenvironmental Research Centre (GRC), Cardiff University, Cardiff, UK, the author worked on a variety of problems related to computational fluid dynamics and heat and moisture transport through porous solids using modeling and simulation techniques with his personal computer codes.

1 Introduction

Over centuries, the human race has strived to develop different kind of systems and understand natural phenomena for better lifestyles and civilization. Scientists have tried to describe nature using many laws of physics with the help of experiments and theories. Comprehending the intricacy of matter was a breakthrough by scientists which enabled characterization of materials using material chemistry. It is worth quoting Albert Einstein, who said that 'Physics is essentially an intuitive and concrete science. Mathematics is only a means for expressing the laws that govern phenomena'. This can be interpreted as that combined physical science and mathematical knowledge enabled engineers to evolve systems and processes for comfortable living. The physical mechanisms that are observed in many man-made systems, devices, machines etc. are well described using conservative laws. Many natural processes and incidents are explained using certain conservation principles which are not yet contradicted even with advanced science. All observable events should be described using measurable material, space and time variables so that they can be controlled. Apart from scientific principles and mathematics, the power of the human imagination has also played a vital role in the renewal of existing systems. For example, the cars that were manufactured 50 years back were very primitive in design and function. However, these cars have undergone revolutionary transformation over the years, with increased human comfort and functionality. There is no doubt that they will continue to improve in the future and will have very advanced features. Human imagination does not have any limits to construe new ideas for upgrading current systems and processes. When products developed by intuitive ideas satisfy physical laws and produce sustainable performance, such products can withstand the test of time. Scientists and engineers attempt to develop viable structures, systems, products, processes etc. for better living of humankind. Their understanding of physical laws, material behavior and mathematical tools help them to promote technological advancements to the next level.

Mathematical modeling has become a powerful tool for engineers and scientists to understand feasible design methodologies and efficient system development. One should realize that mathematical modeling does not merely represent some complicated mathematical equations with hi-fi data hidden in it: a mathematical model simply represents system behavior after satisfying all the underlying physical laws. Mathematical representation in equation form enables the variables to take any value under the specified bounds for a given problem. As understood in basic mathematics, variables can be classified as dependent or independent variables and are related for simplicity, as a function. This relation between variables may be two or more, can yield some algebraic equations or some differential equations. What is more important is the underlying physics that connects the relation between these variables. This forms the basic core of mathematical modeling principles. Without grasping the

DOI: 10.1201/9781003248071-1

elemental physical principles that control the expected behavior of a system, a mathematical model will not serve its purpose. For this reason, a mathematical model can be defined as *the functional relations between variables under observation governed by physical laws*, simply a combination of physics and mathematics. These physical laws generally consist of conservation principles and material constitutive equations. A system is defined by its constituent substance, called the working medium or the material of which the system is made. The systems undergo changes under the given conditions according to certain conservative propositions. Understanding these conservation principles and relating them with the useful variables becomes the first step in mathematical modeling. This step is also called derivation of the *governing equations*: those equations representing the conservation principles that govern the behavior of the system. With the knowledge of mathematics, the nature of these governing equations can be identified and classified. The constraints under which a system has to function are termed *boundary conditions* of these governing equations. Without these data, the equations cannot be solved because boundary conditions are the data in real numbers that are common on the boundary separating the system and surroundings. The terminology *solution* means the real numerical values of the variables under consideration, obtained by solving the governing equations. A well-formulated mathematical model will definitely give rise to a unique solution for the system. Depending on the nature of the governing equations, either analytical techniques or numerical methods can be employed to solve the equations to obtain the final solutions.

With the advent of affordable and user friendly computers, the solution of governing equations are obtained using computer programming. Developing computer programs using languages such as C++, Fortran etc., even for analytical solutions, is useful because these programs reduce repeated manual calculations for change of certain variables. The major advantage of modeling is that, with the help of computer programs, a repeated number of solutions for the variables can be obtained easily. However, experimental exploration is very difficult due to the amount of effort and money required for performing experiments under different combinations of variables. The important feature of mathematical modeling is its ability to predict the solution for the variables under various possible combinations of boundary conditions. Such an exercise is called *simulation*, which is the far-reaching benefit of mathematical modeling. A computer program for the solution of the governing equations is developed based on a well-set logical sequence of steps, called a *computational algorithm*. Like the governing equations, a computer program is written using notations for the different variables that evolve in the solution procedure, by following a *syntax* as specified by the chosen computer programming language. The steps followed while computing the variables in the *computer code* follow a certain pre-set logic, which will indirectly satisfy the underlying physical laws, laid down as conservation principles. Among the many notations used in the computer code, there are certain variables whose numerical values have to be provided as *input*, otherwise it is not possible to get the solution. These are the input numbers to the computer code that finally produces the required solution in numbers. When these input numbers are changed, then there will be corresponding changes in the output also. Once again, the relation between these changes in the input and output numbers should also indirectly satisfy the underlying physical laws. Hence, the final aim in mathematical modeling and simulation is not simply to get some output in numbers or graphs but the expected numbers that satisfy the governing principle with which the governing equations were obtained originally. For this reason, before getting into detailed simulation analysis, initially the computer code developed is tested for its validity against the governing principles. This exercise is called *validation*, for which some known and reliable or established results for the same problem are considered. For the purpose of validation, a problem with established results is tested with the computer code developed. The minimum error of comparison as possible should be achieved, so that the computer code can be accepted for further simulation of results.

1.1 MODELING

In the context of engineering applications, a model can be construed as a representative system of something that is being proposed to be built in the future. Therefore it may be imaginative of something which does not exist at present. Ideas about this model may come up because of past experience, observations, expectations or be purely theoretical. However, the functions of what this particular model is supposed to perform might have been decided or be under consideration. It is very obvious that one cannot see this right now because it may not exist. The process of creating a model is called modeling. Model studies have been part of engineering analysis for decades. The functions and objectives of a model are clearly defined. The very purpose of a model has to be characterized before building the model. Most importantly the outcome of the model and the expected performance of a model are decided well before the model building. A proposed model may be unique in its functions and utility, created only for a specific purpose. As a model is built based on previous observations or conceptualization, or based on an existing system, certain assumptions have to be made to define the model. Some of these assumptions may be certain constraints or limitations of the model. Hence, the tools or propositions to be used to analyze the model are expected to work under these assumptions. If provisions for improvement of a model is built in at the very beginning of modeling, then some of the constraints could be overcome later by improvising the model. The person or the team involved in model development should consider this arrangement while planning a model. The feasibility of building the model and workability of the model under different operating conditions should be taken into account in advance. In a sense, complete control of the behavior of the model has to be established while designing the model. These models can be broadly classified as physical models and mathematical models.

1.1.1 PHYSICAL MODELS

The study of physical models in megastructures has played a critical role in engineering analysis. Construction of multistory buildings, cruise ships, commercial aircrafts, long bridges, water reservoirs, chimneys and cooling towers in thermal power plants etc. involve initial studies using models. Most of these studies focus on understanding the effect of solid-fluid interaction that will result in physical damage to these structures. A specialized field in fluid mechanics, called similitude analysis, is employed for such analysis [1]. There are well-defined methodologies in this field of research that have been established for decades. Figure 1.1 shows some actual structures along with their scaled models to explain the principle of physical models.

A wind tunnel is a facility in which these scaled physical models can be tested by allowing air to flow over the structures at different flow conditions depicting the actual working environments. Figure 1.2 demonstrates the wind tunnel testing of an aircraft model and other testing models for bridges and chimneys.

The experimental data obtained from wind tunnel testing will be transformed to actual data corresponding to the real structures using similitude analysis. Sometimes scaled models are used in the esthetic design of building architecture and designing consumer products made of different materials. Plaster of Paris (POP) is a commonly used material for making models in the esthetic design of consumer and electronic components. The cost of the final value-added product can be kept beneficial by making different models of the same product to meet the required functions, finish as well as the demands of a competitive market. Prototype models are also well known in manufacturing engineering industries.

1.1.2 Mathematical Models

A conceptual representation of an existing or non-existing system or process on whose behavior the interest is focused is called a mathematical model. Unlike physical models, which are real but scaled, mathematical models are not real, at least at the time of description. The system behavior and attributes can be predicted using mathematical models. They simply imitate the system, if at all to be built, which may be or may not be present at that moment. It is the outcome of imagination of the model developer based on his/her observation, vision or expectation. Its strength lies in the fact that before being fabricated or designed, it can produce all the expected results or outcomes for which it is intended. A mathematical model is also flexible enough for any change or modification so that the expected results can be obtained accurately. In the recent decades, mathematical models have occupied almost all fields of science, engineering and humanities and still provide opportunity for further expansion in new fields in the future. Mathematical models are highly recommended for their contribution to weather forecasting, space science, bio-medical engineering etc. Mathematical models have influenced medical engineering to such an extent that their predictions can precisely guide a physician to make an incision only at the required spot on a human body to minimize blood loss [2]. Accurate weather forecast models have helped communities from natural calamities by giving warnings before damage happens. Economic models have helped systematic planning by policy makers to consider all areas of society in order to make improvements to people's lives. The application of mathematical models in biological fields has uncovered new dimensions to the analysis of cell

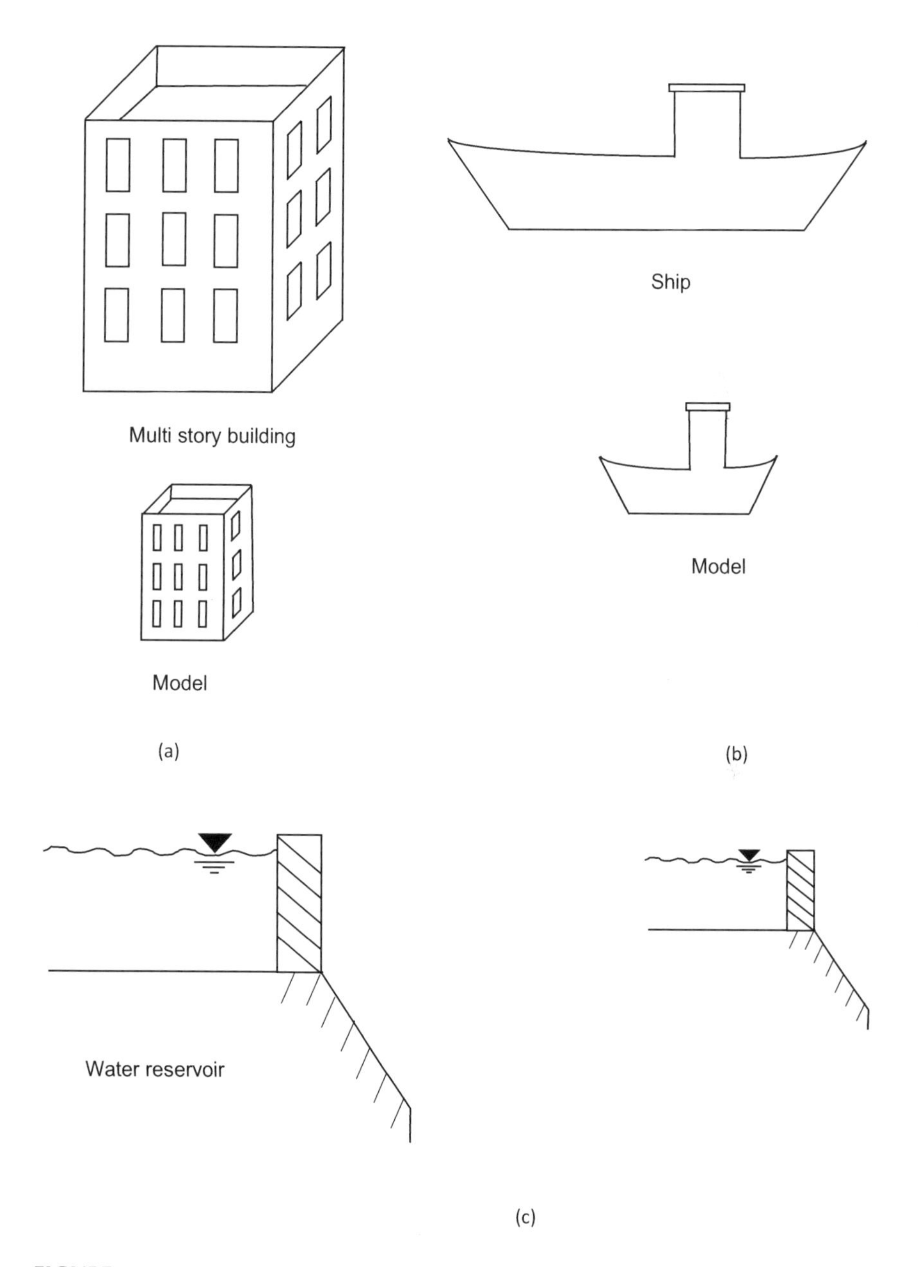

FIGURE 1.1 Physical models.

structure. In-depth knowledge of material characteristics has been made possible with the advent of mathematical models in material science. Prediction of damage of high performance concrete structures such as skyscrapers, nuclear reactors etc. due to unexpected fire has supported engineers to predict the time available to save valuable human lives. Computational fluid dynamic analysis of aerodynamic

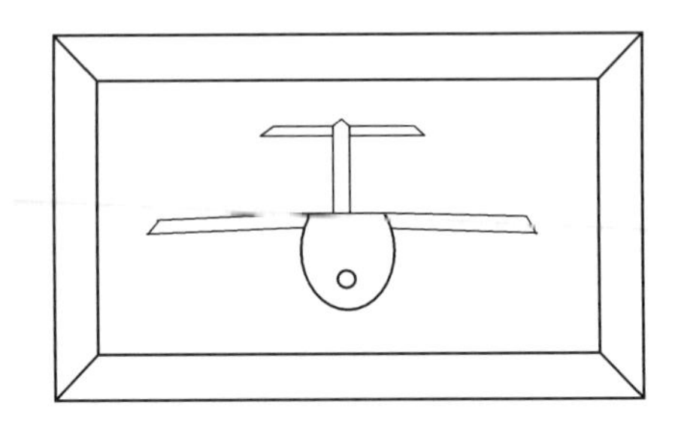

Wind tunnel testing of aircraft model

(a)

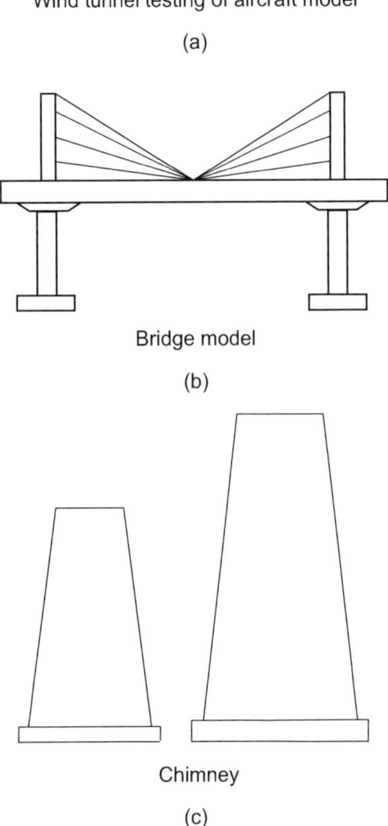

FIGURE 1.2 Testing of models in wind tunnel.

structures is of huge benefit to engineers to achieve efficient design with minimum frictional losses, thus reducing operational costs.

The conceptualization of a mathematical model is explained in Figure 1.3.

Intuitive and imaginary mathematical models are based on certain proven physical principles, laws and mathematics. In most of engineering analysis problems, some of the variables or attributes in the systems considered are continuous in nature

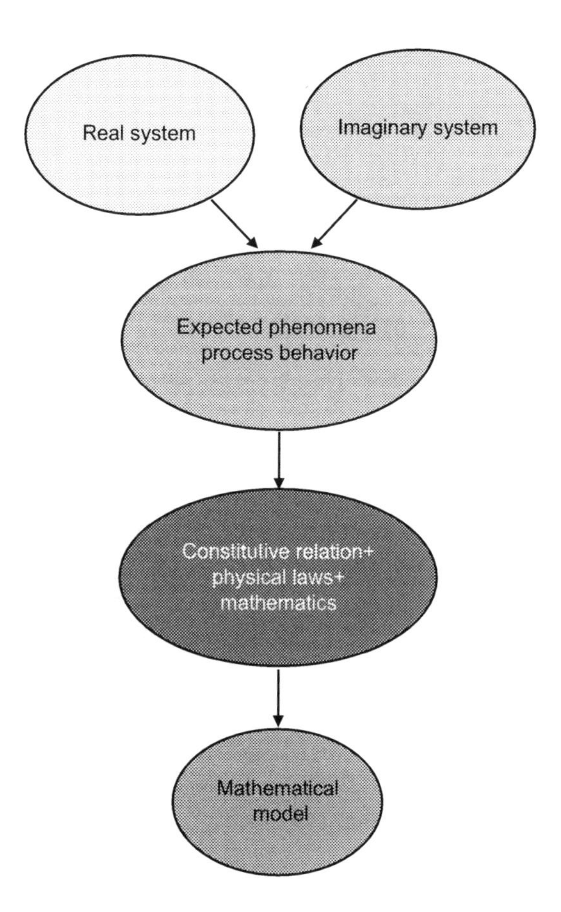

FIGURE 1.3 Conceptualization of mathematical model.

and others are not. For example, when water is heated in a vessel, the water at the top layer also gets heated up after some time. That means some attributes are continuous in the water medium, causing the heat to reach everywhere. Air flowing over a car body offers some resistance for the movement, due to variation of some parameters from the mainstream to the exact boundary of the car body. When a stone is thrown into water in a pond, the water ripples created travel some distance in a particular pattern. That means there are many situations where some characteristics of the system vary continuously. Unless the nature and type of variation of these variables are predicted or estimated, then the heat transfer within water, fluid friction on a car body and the distance to which the water ripples could travel cannot be estimated. Mathematics is a powerful tool to represent the continuous variation of attributes with space and time. With the assumption of continuum in a system, mathematics can reproduce continuous variations with space and time. This principle is called differential formulation, where differential equations are developed after satisfying certain physical principles. The physical laws proposed by many scientists in the fields of science and engineering do not depend on the size of a system. Hence, these principles are applied initially to an infinitesimal element of space or time, giving rise to differential equations. Later these equations are solved for a given geometry with some dimension, using either analytical techniques or numerical methods. There is another set of problems, wherein instead of continuous variation, the change of attributes only at certain locations becomes important. For example, in suspension bridges, engineers are more interested in estimating displacements due to the load on the bridge at specified locations than continuous variation of these displacements. The physical laws employed for such analysis of linkages are concerned with the collective effect of such displacements due to the presence of many other members taking the load. Such analysis will give rise only to algebraic equations. Modeling principles can be well understood by examining the examples shown in Figure 1.4.

1.1.2.1 Perfect Gas Equation

The idea of perfect gas equations is learned in high school science (Figure 1.4(a)), as the combination of Boyle's law and Charles's law for any particular gas as

$$pv = RT \tag{1.1}$$

In this, R is called the characteristic constant for a particular gas, which is also related to universal gas constant and molecular weight. The above equation can be considered as a mathematical model relating pressure (Pa), specific volume (m^3/kg) and temperature (K) for a particular gas. This is a well-known mathematical model that has been widely used in science and engineering calculations. However, it has to be noted that this equation is obtained as a result of Boyle's law and Charles's law with certain constraints or assumptions. This equation is applicable at certain state points and can be extended between two state points connected by constant pressure, volume and temperature processes. Its validity depends on the temperature and pressure of a given gaseous medium. However, the above equation can be modified further to ensure its validity for wide range of pressures and temperatures. Such equations are called van der Waals equations and virial equations of state.

1.1.2.2 Hooke's Law

This is represented in solid mechanics that within the elastic limit of a material, the strain and applied force can be related (Figure 1.4(b)) as

$$F = kx \tag{1.2}$$

where F is the applied force (N), x is the displacement (m) and k is the spring constant (N/m). This is another example of a simple mathematical model that relates applied force and displacement, of course with a constraint of elastic limit of the material. This formula becomes very useful to predict the displacement for any given applied force F, once the value of k is determined by performing some simple experiments. The above principle can be extended for different combination of forces acting on different machine elements. It is worth noting that this simple formula is obtained from the principle of Hooke's law.

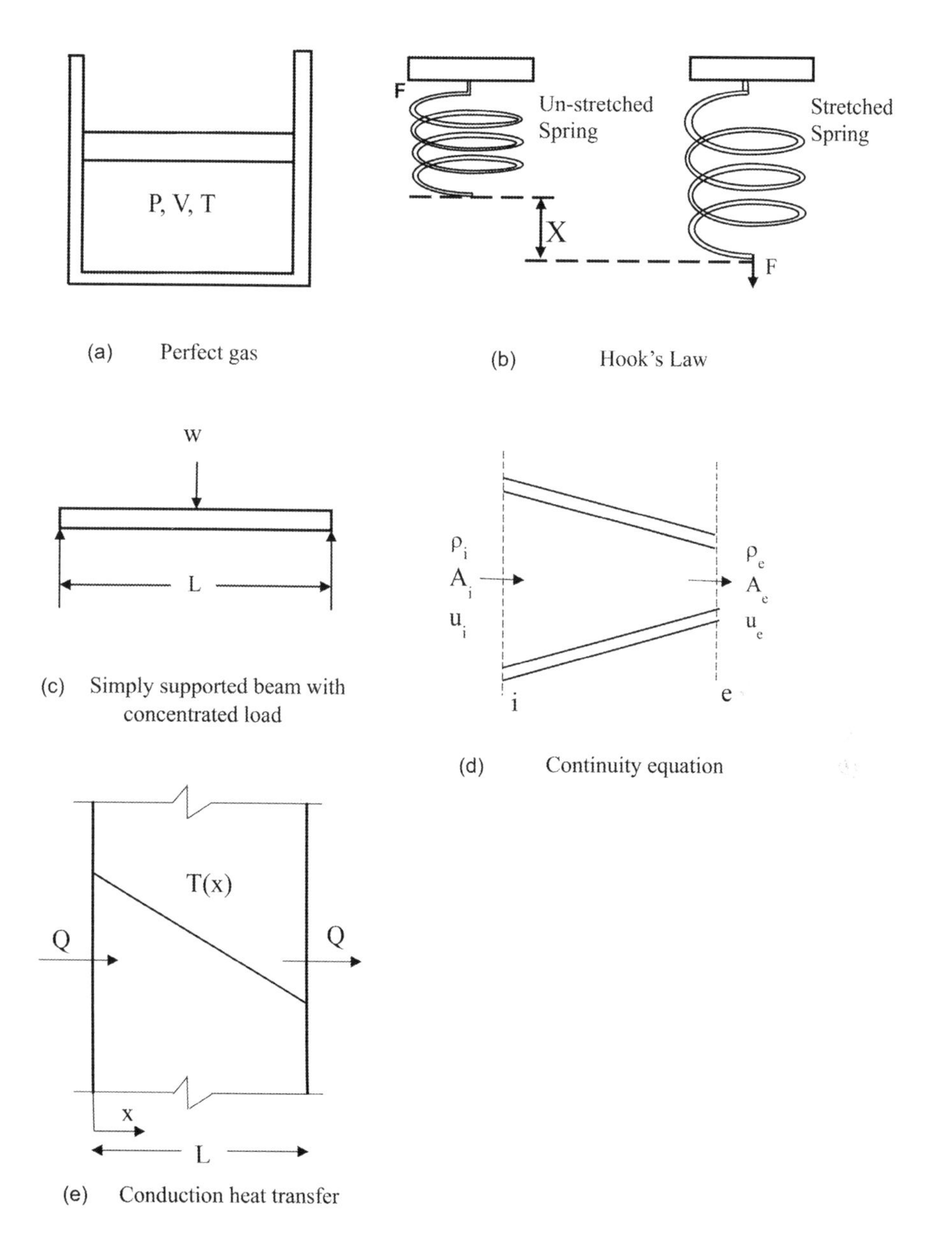

FIGURE 1.4 Examples for simple mathematical models.

1.1.2.3 Deflection of Beam under Load

Accurate prediction of deflection of beams under the action of force is required in many applied engineering problems such as machine elements, building structures and vibrations of stationary and flight structures. For a simply supported beam with a concentrated load w at the center of a beam of span L, the standard formula for maximum deflection, δ at the center (Figure 1.4(c)) is given as

$$\delta = \frac{1}{48}\frac{wL^3}{EI} \tag{1.3}$$

where EI represents the resistance of the beam for deflection. The above equation is obtained from theory of beams, which assumes that the deflections are small in relation to span length of the beam. This formula also serves as a simple mathematical model to estimate deflection of a beam of given length and applied load. The value of EI varies depending upon the material and its cross section. The maximum deflection at the center of the span of the beam can be computed for varying value of the concentrated load w. The above expression can be developed further to compute deflection under different loading conditions.

1.1.2.4 Fluid Mechanics

In fluid flow problems mass conservation is an important condition that has to be satisfied before balancing momentum or energy in the flow field. This is because any increase or decrease in mass within the control volume will lead to corresponding increase or decrease of momentum and energy. When mass continuity is not satisfied, then any further analysis will always give rise to misleading results. A simple first order differential equation for mass balance within a control volume for unsteady flow can be written as follows.

$$\frac{dM_{CV}}{dt} = \dot{m}_i - \dot{m}_e \tag{1.4}$$

where M_{CV} is the mass of fluid in the control volume, $\dot{m}_i$is the rate of mass flow entering the control volume and $\dot{m}_e$is the rate of mass flow leaving the control volume. This is a simple mathematical model to represent mass conservation in internal combustion engines where the mass of fuel and air entering the engine may be different from the mass of exhaust gases coming out of the engine. During small intervals of time, the flow of gases may be considered as unsteady, wherein the mass of gases equal to the difference between the air and fuel supplied and exhaust gases may accumulate within the engine cylinder. It has to be noted that the left side of Equation (1.4) cannot be negative; this means the mass of fluid leaving the control volume cannot be greater than that of the fluid entering the control volume. These two mass flows may be equal under steady state conditions (Figure 1.4(d)) as

$$\dot{m}_i = \dot{m}_e \tag{1.5}$$

For fluid with varying density between one inlet and one outlet, the above equation can be rewritten as

$$\rho_i A_i U_i = \rho_e A_e U_e \tag{1.6}$$

where ρ, A and U indicate fluid density, area of cross section and velocity and 'i' and 'e' represent the inlet and outlet of the control volume respectively. When the density of fluid remains constant like in the case of nozzles, orifice meters etc. then the equation becomes reduced to

$$A_i U_i = A_e U_e \tag{1.7}$$

The above equation is widely used in the analysis of fluid mechanics problems.

1.1.2.5 Heat Transfer

It is known in basic physics that the rate of heat energy can be transported by three modes of heat transfer, namely conduction, convection and radiation. Heat and work are quantities called energy, measured in J or kJ. Energy interactions in the form of heat and work are dealt with in detail by laws of thermodynamics for systems that are considered in thermodynamic equilibrium. However, the rate of energy interaction between a system and its surroundings is explained with transport equations. A conduction heat transfer equation is one such transport equation proposed by Fourier as the Fourier law of heat conduction, which was postulated based on experimental observations for one-dimensional heat flow by conduction. Heat conduction through a material is measured in terms of change in temperature across the thickness of the material and thermal conductivity. The mathematical model for one-dimensional conduction heat transfer across a material of constant thermal conductivity under steady state conditions can be expressed (Figure 1.4(e)) as

$$\frac{d^2T}{dx^2} = 0 \tag{1.8}$$

The solution of the above equation will provide an expression for temperature variation as a function of 'x'. As it is a second order differential equation in 'x', it requires two boundary conditions along the 'x' direction to get the solution. Integrating twice Equation (1.8), one can obtain

$$T = C_1 x + C_2 \tag{1.9}$$

Equation (1.9) demonstrates that the temperature variation along the length of the solid is linear for the material with constant thermal conductivity under steady state conditions. Once the boundary conditions at both ends of the solid, $x=0$ and $x=L$ are known, then the temperature at any location 'x' along the length of the solid can be computed. The mathematical models discussed in this section are some basic models that have been used widely, without identifying them as mathematical models. However, these examples are simple in nature, and the real benefit of mathematical models can be appreciated with application to problems which otherwise cannot be analyzed easily, either with experimental or other methods. The main advantage of a mathematical model is its flexibility to predict results for the system subjected to conditions that are not present at that moment but that the system may be subjected to in the future; such an analysis is called *simulation*. Forecasting the behavior of the system or process conditions under new environmental conditions, or *mimicking* the behavior of the system using non-existing conditions, is called simulation.

1.2 SIMULATION

Mathematical modeling is always followed by simulation for any given problem. In other words, the very purpose of developing a mathematical model is to carry out simulation analysis for the given system. It is well established that experimental work will provide reliable data for design and analysis. However, it may not be possible to explore many physical phenomena using experimental techniques. Any installed experimental facility is designed to operate or provide data under limited environmental conditions. The amount of effort and time required in experimentation is also enormous when long-term experiments are conducted to analyze the performance of systems. During experimentation, there are a number of parameters that can be varied, and hence a scientific scheme for experimental trials have to be prepared before the start of the experiments. Design of experiments (DoE) is one such scientific approach that is widely followed to minimize the number of experiments without compromising the required data. The limitation of experimental technique leaves the scope for modeling and simulation. Wind tunnel testing of aircraft models provides valuable information for the simulation of behavior of actual aircraft during flight. Simulation of weather forecast data is another field of research where recent accurate modeling, considering many physical phenomena, has enabled exact prediction of weather conditions. Parameters that cannot be varied in laboratory scale studies can be easily varied using model studies. For example, study of buildings exposed to accidental fire has attracted design engineers recently. Research on structures exposed to fire has limitations for laboratory studies due to the risk involved during experimentations. However, modeling building structures for fire environmental conditions will provide useful simulation results. The mathematical model developed can be subjected to any kind of severe high temperature exposure to predict valuable information on the strength of building elements at high temperatures and types of failure. The simulation results of such studies will provide useful data for design engineers. Of course, the outcome of these model studies are as good as the assumptions made while developing the models.

Modeling and simulation attracted many researchers after the advent of affordable high speed and high performance computers. A mathematical model may have analytical solutions for the given geometry and boundary conditions, or it needs to be solved using some numerical techniques. In most of the modeling exercises, computer programming is an important element, which makes the simulation process simple. Depending upon the background of programming knowledge of the user, program tools such as MATLAB, FLUENT etc. can be employed for simulation purposes. A mathematical model gives rise to a set of equations that may be linear or non-linear, differential or partial differential, or coupled or non-coupled differential. The solution of the mathematical model has to be obtained either using analytical tools if an analytical solution exists for the equations obtained from the mathematical model. Otherwise a suitable numerical technique has to be selected. The type of numerical technique to be used depends on the geometry of the system, constant or varying properties of the working medium, ease of programming etc. Generally, numerical techniques such as *finite difference*, *finite volume* and *finite element* are employed to convert the differential equations into algebraic equations. The number of algebraic equations depends on the type of *discretization* of the *system domain*. Finally, the solution of a set of algebraic equations has to be achieved using equations solvers. Then the values of the variables are determined at the required locations in the system domain. This is the first stage in the simulation analysis. After the implementation of the numerical technique, a computer program is developed to carry out the computations in different stages through logical sequence of calculations, called computational algorithms. Once the computer program is executed to obtain the basic solutions for the problem, simulation results can be obtained. Validation of the mathematical model is an important part of simulation. Initially, the results are compared with the computational predictions of a known problem or what is available in the literature. Then a systematic scheme is formulated for simulation, in that a strategy for varying different input variables for the problem is prepared. The blueprint for simulation analysis is drawn up by keeping in mind the expected outcome of the simulation.

1.3 CONSERVATION PRINCIPLES

All the systems and processes by which we are affected or influenced in day-to-day life follow certain physical principles. These principles may be postulated in science from physics, chemistry or mathematics, and most of them are postulated in the form of laws which are taught in high school science. It could be said that these physical principles have existed since the inception of the world and the human race, and it has taken time for humans to understand and conceive of these natural laws in their own way. With time, the understanding of nature by scientists and engineers has gone from deep into deeper knowledge and will continue to grow. Interpretation of this knowledge through observation has resulted in the discovery and invention of new systems and processes for the betterment of humankind. Mastering these scientific facts has made possible the development of advanced technology which we relish today. A learned scientist or engineer will never forget the basics of physical laws that govern the system or process he/she is investigating. The failure of many engineering systems or processes is mainly due to a lack of understanding of the underlying principles while designing the systems. Physical laws are very well laid down in the form of conservation principles in many fields of science and engineering. In this textbook, the main focus of modeling and simulation is on problems related to thermal and fluids engineering. Hence, discussion in this chapter and in future chapters will focus only on conservation principles related to thermal and fluids engineering. Systems and processes considered in thermal and fluids engineering involve certain kinds of working media or substances, and there may be one or more working fluids in the system, interacting with another system through energy mode in the form of heat or work. In the case of combustion of fuels, more than one gaseous species will be produced, and their influence on other processes need to be understood. Hence, conservation principles connected with mass, momentum, energy and species need to be understood. In mathematical modeling, this is the first step to be followed to recognize the basic principles of the working of the system.

1.3.1 Mass Conservation

In basic science, it is taught that mass can neither be created nor be destroyed. In engineering applications, generally there are two types of systems, the first one with fixed mass wherein the same fluid gets circulated within the system through different components. Air-conditioners and refrigerators (Figure 1.5(a)) are the best examples of such closed systems, in which the working fluid called the refrigerant is always confined within different components connected with one another that produce the required cooling effect.

Fluid in shock absorbers and coolant circulated in cooling system of automobile vehicles are other examples of closed systems. In these systems, the mass or the quantity of the working substance remains the same without having an entry or exit to and from the system, however, delivering the required functions. Mass is always conserved within such systems. The second kind of systems are called open systems, in which there is some entry and exit of fluid to and from the system during its operation. In most cases, the fluid entering will be equal to the fluid leaving the system with some changes in its properties. For example, in a household pump (Figure 1.5(b)), water enters the pump through the suction pipe and comes out at the delivery side at relatively higher pressure compared to the inlet. In the absence of any leakage, the same quantity of water that enters the pump leaves the pump. A similar phenomenon is observed in the case of a compressor wherein atmospheric air is sucked into the compressor and delivered at a higher pressure after compression. Here again the amount of air leaving the compressor remains the same as that of the air that enters the compressor. When the exit quantity becomes equal to the quantity that enters the system, it is called steady flow system. It has to be noted that the number of entries and exits need not always be one, it may be more than one, like in the case of municipal water supply pipe line circuits designed for city water supply. Most of the engineering flow systems come under the category of open systems, where there is definite entry and exit of fluid to and from the system. There are situations wherein the fluid leaving the system will not be the same as that of the fluid entering the system, in which some amount of fluid gets accumulated within the system. The best example is the filling of gas cylinders, in which the main gas supply line supplies gas to different gas cylinders and the remaining leaves the main line. Such flow systems are classified as unsteady flow systems. The terminology 'unsteady' is used to indicate the accumulation of fluid within the system changes with respect to time. The rate of change of mass in the control volume is equal to the difference between the mass entering and leaving the system as expressed by Equation (1.4). Whenever fluid momentum and energy conservation are considered in modeling, then there must be provision to ensure mass conservation within the system. Without satisfying mass conservation, other equations cannot be solved and correct solutions cannot be obtained. In the case of fluid momentum represented by

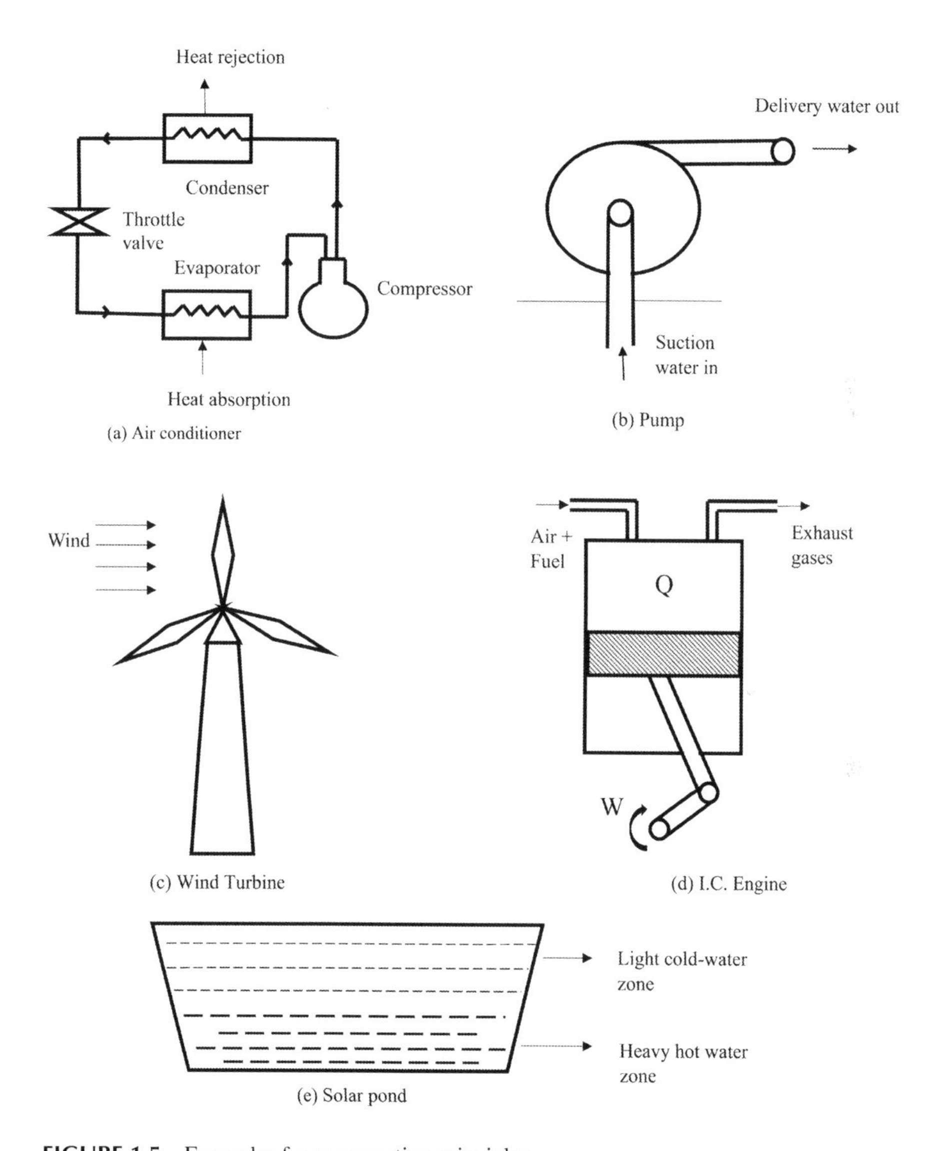

FIGURE 1.5 Examples for conservation principles.

the well-known Navier-Stokes equations, when any new methodology is proposed to solve these equations, the first necessary condition to be satisfied is conservation of mass. In the solution methods, the way in which mass conservation will be satisfied has to be explicitly elaborated in order to make the method acceptable for use.

1.3.2 Momentum Conservation

In applied engineering problems, mass flow of a particular fluid is always accompanied by momentum transport. Water flowing through a pump acquires momentum from the rotating impellers to gain kinetic energy and an increase in pressure. When water at high level flows over impellers fixed on rotating water wheels, momentum is transported to the wheels, which in turn run an electric generator. The momentum imparted by wind on wind turbines (Figure 1.5(c)), is used to produce electricity. In an internal combustion engine, the combustion of fuel with air produces gaseous products at a high temperature and pressure, which impart momentum to the piston head for turning the engine shaft. The fluid momentum imparted by compressed air in a pneumatic spanner turns the nuts firmly and saves human effort, especially for large machinery. Rockets are able to travel at speed more than sound because of the fluid momentum of the gases coming out of the engine. That means that whenever fluid flow takes place through a system, there will be a change in momentum of the fluid which will affect the overall force balance. It is known from basic physics that a change of momentum will give rise to force. Everyone observes or experiences changes of momentum in many kinds of equipment used in day-to-day life. In household water pumps, one can sense the vibration near the pump which is stationed on a well-anchored platform. The base of the pump needs to be bolted to the base in order to balance the force due to changes in the momentum of the water. Other than the vibrations caused by the rotating shaft in the pump, there is a change in momentum of the water inside the pump casing. When water flows through the rotating impellers it gains momentum because of changes in the velocity achieved by the shape of the impeller. Now the force due to the momentum change of water and force due to the rotating impeller at the shaft have to be balanced so that the pump is in equilibrium at the desired position. Pipes carrying fluids with high velocity have to be properly fastened with support because the change in momentum of the fluid in bends and valves gives rise to force. That means for systems with fluid flow, momentum conservation has to be considered to achieve force balance of the system. Problems in fluid mechanics for flow through nozzles, orifices etc., the flow domain is treated as one-dimensional flow. That means the variation in velocity is considered significant only in the direction of flow, generally the axial flow. However, flow through pipe or ducts is three-dimensional in a real sense. Based on the directions in which significant variation in velocity is observed, the flow is treated as one-dimensional, two-dimensional or three-dimensional. Flow over wings or blades in a rotor is treated as two-dimensional. In one-dimensional flow situations, there may be change in velocities in other directions also, but the magnitude may be very small such that their contribution for change in momentum is negligibly small. Engineers and scientists need to categorize the given flow problem based on the significant variation of velocities in specific directions. Such assumptions make the modeling simple and reduce the computational cost and effort. For example, flow through a pipe can be analyzed by treating it as three-dimensional. But ultimately the difference between the results obtained using one-dimensional and three-dimensional approach will be very small.

It is very important to satisfy momentum balance while modeling systems that deal with fluids at high velocity. In the case of nozzles used along with water or steam turbine, the change in velocity is very high, hence for the given mass flow rate, the change in velocity will give rise to a force of great magnitude. In such situations, proper mechanical balancing of the reaction forces has to be taken into account, otherwise there may be damage to the nozzle assembly. Needless to say, a type of reactive force is expected from a jet engine due to the exit of high velocity gases. While analyzing momentum conservation of fluid medium, inertial force, viscous force and force due to changes in pressure play a major role. In some applications, buoyancy force and magnetic force also become important. Hence when modeling a fluid system, the pattern of expected velocity variation and contribution of other forces have to be considered so that a complete force balance is achieved. Forces that are not considered in modeling may cause damage to the actual system when designed based on the simulation results of the mathematical model.

1.3.3 Energy Conservation

In this textbook, energy conservation is focused only on thermal energy conservation, specifically heat energy conservation. In basic thermodynamics, it is known that energy can neither be created nor be destroyed. However, energy takes different forms using energy conversion devices. In the thermal engineering field of study, many systems involve energy transport in the form of heat. Work is also a form of energy in a thermodynamic sense. However, the conservation principle is stated only in general form as energy conservation. For example, in an internal combustion engine, fuel is burnt with air inside the cylinder, which gives rise to gases at high temperatures and pressures. These gases expand in the cylinder-piston arrangement to produce mechanical work as output (Figure 1.5(d)). When the overall energy conversion is considered in this example, one can appreciate that heat energy available in the fuel is converted into mechanical work. Of course, the same heat can be produced by burning coal or other fuels, and such engines are called external combustion engines. Now in the internal combustion engine, the total heat supplied by burning the fuel has to be accounted for in different forms. When part of the heat is available in the form of mechanical shaft work after overcoming friction in different machine components, some amount is rejected from the engine through exhaust gases; a fraction of the total heat is lost to the cooling water circuit used to cool the engine wall from overheating; and some heat is lost to the atmosphere through radiation and convection losses from the surface of the engine.

The energy conservation principle has to be satisfied while balancing different forms of energy in a system. In the internal combustion engine example, one can appreciate the mass conservation principle that can be understood by appreciating the fact that the mass of gases leaving the engine cylinder in the form of exhaust gases must be equal to the total mass of fuel and air supplied to the engine. When the combustion products are produced during combustion inside the engine cylinder, there will be a significant change in velocity of gases during expansion and compression, which will give rise to change in momentum of the gases. Finally, the force due to the change in momentum of these gases only exerts the force to move the piston, doing mechanical work on the engine shaft. In this particular example, all the three conservation principles – mass, momentum and energy – are involved in the analysis. In a household geyser, the temperature of water is increased by heating it through an electric heater. The total heat gained by the water must be equal to the heat supplied by the electric heater, and both these quantities can be estimated by measuring water flow rate, and changes in temperature, electric current and voltage. In this example, only mass and energy conservation are involved; this means that whatever water enters the geyser comes out of it, and whatever heat supplied by the electric heater manifests as an increase in the temperature of the water. A solar water heater is another example in which only mass and energy conservation are applicable in the analysis. Heat energy in the form of solar radiation is absorbed by a collector which in turn heats the water being circulated through the collector. A clothes dryer is an example that deals with only energy conservation. In this system, the water present in the clothes is removed by evaporation using heat from a heater provided in the dryer. When this system is modeled, it may only be necessary to estimate the amount of heat required for evaporating different levels of moisture in the clothes. The initial moisture present in the clothes is important, and the moisture evaporated is not accounted for unless the residual moisture in the clothes is measured after drying, which would be complex. Hence in this example, only energy conservation is considered in the analysis.

It is worth mentioning here that energy conservation is considered in situations where there is a significant change in the temperature of the working fluid. Heat is measured as a quantity in terms of change in temperature that is caused by the heat interaction. However, in energy balance, all types of energy will be accounted for. For example, in a steam turbine analysis, the enthalpy of steam at the inlet and outlet of turbine; work output from the turbine; change in kinetic energy in the form of change in velocity; and any heat loss will be accounted in the energy balance. In the case of a water pump, the momentum conservation principle is generally employed because there is no significant change in temperature of water during pumping. The work input to the pump in the form of motor work is considered while calculating its efficiency. Hence while modeling a system, the relevant conservation principles have to be identified based on the working of the system and objectives of the analysis.

1.3.4 Species Conservation

Conservation of species concentration is with regard to mass transfer processes involving one or more than one species transport in a medium. This has to be distinguished from mass conservation principle discussed earlier, which is regarding the conservation of the whole working medium. However, species conservation is with regard to the transport of different substances within the working medium itself. For example, let us consider convective heat transfer within a building enclosure during a space cooling application. There may be a definite quantity of air entering and leaving the enclosure with a change in temperature due to the cooling temperature; naturally cold air enters and hot air leaves the building. Mass, momentum and energy conservation principles have to be accommodated while modeling the building. Now, consider spraying a deodorant inside the building to maintain a pleasant odor in the ambience. The transport of deodorant vapor into the air medium inside the room needs to be studied using a species conservation equation. This means the account of dispersion of the deodorant vapor into the air medium has to be analyzed using this conservation principle. Such analysis finds wide engineering applications in the fields of solar ponds, chemical vapor deposition, drying processes, nuclear waste disposal facilities etc. Energy and concentration conservation principles are based on transport processes; that is, in a more generalized unsteady situation, the rate of change of energy or species in a control volume is due to the diffusive and convective transport of energy or species concentration. These conservation principles differ from mass conservation wherein the mass of working medium enters and leaves a control volume by virtue of its flow velocities and gets accumulated inside the control volume. However, energy and species concentration transport are controlled by certain mechanisms, known as diffusion and convection. Diffusion takes place due to the difference in concentration of species in species transport, whereas in heat transport it is due to the temperature difference. Convection is a phenomenon that is governed by the bulk movement of heat or species from one region to another region.

In solar pond applications (Figure 1.5(e)), the solar heat energy absorbed during the daytime can be stored in a solar pond by maintaining the density gradient of salt due to temperature difference. Heat absorption and extraction to and from the pond takes place by virtue of density gradients controlled by temperature gradients. Transport of salt in a water medium is governed by species transfer due to the concentration gradient of salt. Here the salt is treated as a species. Chemical vapor deposition is a technique used to deposit certain material on the surface of another material surface for many engineering applications. The material to be deposited is known as a species and has to be transported from its source to close proximity of the surface to be coated by means of a convection mechanism. The surface characteristics of the material to be coated and the type of species to be deposited are the crucial parameters in this process. In such application, the transport of species towards the material surface needs to be modeled with the principle of species concentration. In the design of air-conditioning systems, study of moisture transport plays a vital role in deciding the design parameters. Thermal comfort in a room depends on both temperature and humidity of air. In normal conditions, the change in humidity is purely due to the perspiration effect of occupants present in the room. However, sometimes moisture transport is very significant to such an extent that its transport mechanism has to be modeled for efficient operation of the air-conditioning system. For example, in rooms in which cooking takes place or rooms closer to swimming pools or water bodies, there will be continuous transport of moisture into the room other than the nominal water vapor added due to human perspiration effect. While modeling such environments, sufficient care must be taken to consider the additional water vapor transport as species transport so that its effect on room humidity is taken into account in all the calculations. Hence, the team or person involved in modeling should consider suitable species concentration equation in the formulation.

Drying is another engineering application that is associated with intense moisture transport within the environment. Clay products such as brick, tile etc. contain large quantities of initial moisture. In order to achieve a bone dry condition, a suitable drying strategy has to be implemented so that all the moisture is removed without causing damage to the products. As the drying proceeds, that is, as the moisture is removed from the products by the warm air, there is certain moisture transport mechanism set up inside the solids being dried. Depending on the amount of moisture being removed from the solid and carried by the convective air, the replenishment of air has to be achieved so that drying continues to take place. Now the moisture enters the convective air as species transport which needs to be modeled accordingly so that the simulation results become relevant to meet the objectives. Thus it is important to list out the final outcome of modeling and the expected simulation results. In thermal and fluids engineering related problems, the basic physics behind the working of any given system has to be understood beforehand. It has to be remembered that modeling does not simply represent some simple or complicated equations but that the governing equations are representative of the system behavior. Hence, all the physical mechanisms involved in the system operation have to be grasped before choosing the relevant conservation equations. The conservation equations available in different sources such as textbooks, research papers or reports are not the final form, they are only general form. The definite form of the equations needs to be derived for every problem, in other words, the governing equations are unique for a given system. Any misconception in understanding and correlating the physical phenomena in a system at the beginning of modeling will lead to erroneous results and conclusions.

1.4 TYPES OF PHYSICAL PROBLEMS

In thermal and fluids engineering applications, different type of entities are encountered for analysis. Initially any system under study has to be described or the system has to be identified under certain spheres of definition. For example, in a thermodynamic sense, systems are classified as closed systems in which only energy interactions are permissible, whereas in open systems, apart from energy interactions, mass interaction also takes place. The main objective of mathematical modeling will be to understand or predict the behavior of a system under certain conditions. Hence, the definition of closed and open systems do not exactly represent the actual physical processes in the system. As has been stressed in many sections in this chapter, modeling is based on certain physical principles or conservation principles that govern the functioning of the system. Before proceeding to portray a system, we should understand what the outcome of mathematical modeling and simulation is. Predicting or mimicking the behavior of an entity under varying circumstances is called simulation. In a sense, before the simulation results are obtained, the modeler should have some expectations or intuitions about the outcome of the mathematical model. That shows how well the modeler has understood the guiding principles of the system behavior under given conditions. Hence, the classification demands the knowledge of the controlling physical laws of the system. The system attributes may vary depending on different types of problems in thermal and fluid sciences. It has to be remembered that the strength and limitations of applications of many methods of analysis have been understood with past experience in real life situations. These practices have helped us to understand system behavior as equilibrium type, eigen value and propagation type. Of course, it may be very difficult to distinguish these behaviors exactly by definition. According to the expected function of a system, such differentiations are required, and they are very convenient to model real systems as well.

1.4.1 Equilibrium Problems

Generally, equilibrium is said to be achieved in a system when the attributes of the system are balanced. Equilibrium problems are also called steady state problems whenever time variation is concerned. If the attributes of a system remain constant with respect to time during any interaction with another system, such a process is called steady state process (Figure 1.6(a)).

The equilibrium may be static or dynamic. For example, in hydrostatic analysis, the system is said to be equilibrium when there exists a force balance. Pressure measurement using a manometer is an example for static equilibrium in fluid mechanics. Pascal's law is satisfied at the continuous expanse of a single fluid by the balance of the forces due to the heights of manometric fluids with their relative densities (Figure 1.6(b)). A bridge is said to be in equilibrium as long as it does not fail during

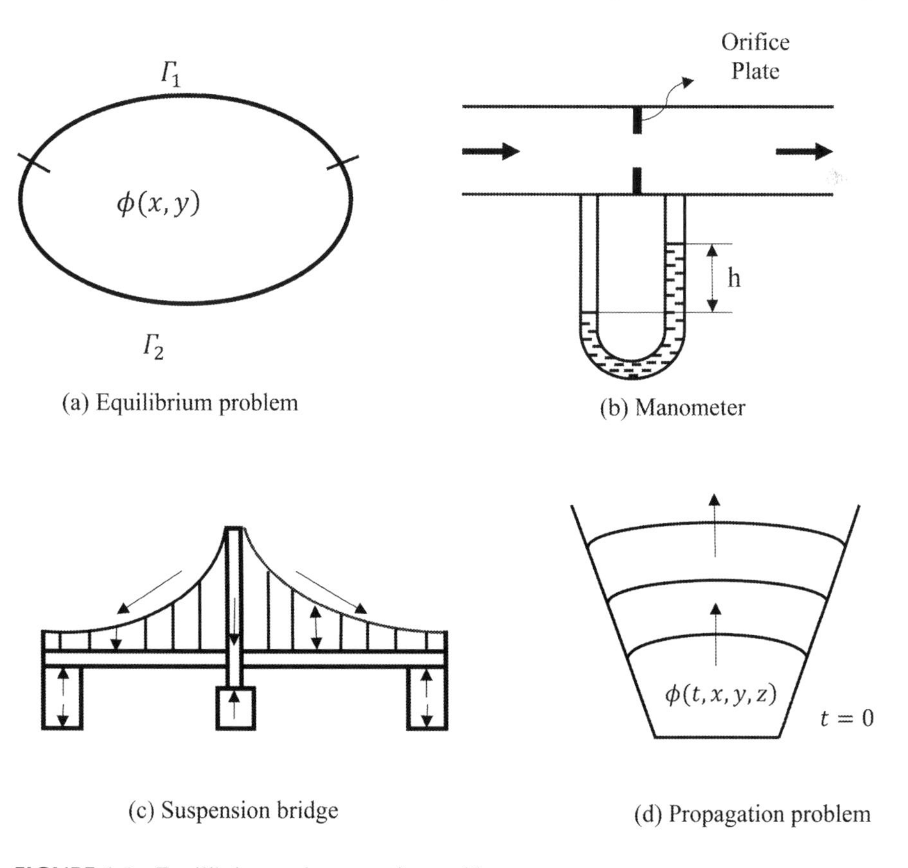

FIGURE 1.6 Equilibrium and propagation problems.

its use under the influence of different types of forces acting on the bridge. Dynamic equilibrium is said to exist when force balance is achieved during the movement of the system. In the same bridge example, when the bridge is subjected to maximum load, then there may be some swing noticed. This can be observed in the case of suspension bridges (Figure 1.6(c)). When water is pumped by a household pump, there are different dynamic forces acting on water during the continuous flow of water from inlet to outlet. As long as the discharge and pressure of the exit water remains constant, then the pump is said to be in dynamic equilibrium. In a geyser, heat is supplied to the water in the storage tank. When the hot water pipe in a geyser is shut, one can easily understand that heat transport to the water has reached equilibrium condition when the thermostat switches off the power supply. That means for the water stored in the tank, the quantity of heat supplied is enough to heat it to the desired temperature set by the thermostat. Once the hot water pipe is opened and hot water is being used continuously at a fixed flow rate, then it takes some time to reach an equilibrium condition at which the amount of heat drained by the water is just equal to the amount of heat supplied by the heater assuming no heat loss. This condition is called dynamic equilibrium. In a static equilibrium condition, the desired attributes of the entity remain constant whereas in dynamic equilibrium these attributes may change but reach a balanced condition. This balancing give the characteristic of equilibrium. While modeling, the type of behavior of a system has to be established so that the suitable physical laws or conservation principles representing the exact physics can be implemented. In thermal and fluid science applications, most of the systems undergo dynamic equilibrium conditions. An internal combustion engine working at constant speed developing constant work output can be designated as an equilibrium system, though the temperature, pressure and specific volume of the combustion products continuously vary. However, when the output from the engine is measured in terms of power output, which is rate of energy generated with time, certain parameter variables that contribute to work output have to be modeled as time variant characteristics of the system. Hence, the situation may change and create a problem – for this reason, the modeler has to decide at the beginning of modeling what the basic objectives of the final simulation are.

Equilibrium conditions demand balancing of either the static or dynamic attributes of the system. Fluid mechanics problems usually involve momentum balance of the fluid medium in the system, and in the case of heat transfer problems, it is always energy balance. Once the system is expected to produce simulation results corresponding to equilibrium conditions, then the required balance equations have to be identified. Other than the use of standard generalized momentum balance equations, the modeler has to decide the type of other contributing forces or energy for the changes expected in the system. Many times, doubt arises about including certain factors into modeling and how to determine the factors that are most influential and less influential. To overcome this situation, one can refer to available literature on similar work or carry out some parametric studies to choose the most influencing parameters. In thermal related problems, variation of properties of the working medium with pressure or temperature or both needs to be clearly understood. In problems involving significant changes in temperature or pressure, the variation of all the properties of the medium have to be taken into account to obtain accurate predictions.

1.4.2 Eigen Value Problems

Eigen value problems are also equilibrium problems but with certain characteristic functions or parameters. These have a distinct form of equation sets that have many applications involving linear matrix equations. Many problems encountered in dynamic analysis in structural mechanics and vibrations are well represented by these specific form of linear matrix equations. Depending on the nature of systems and expected variations in the attributes, these set of equations may be obtained from a set of algebraic equations or a set of ordinary differential equations. Sometimes, partial differential equations can be solved with the concept of eigen value problem in mathematical analysis of certain kind of problems encountered in fluid dynamics and heat transfer. Structural dynamic analysis is mostly based on eigen value problems which are specified by certain characteristics, for example the natural frequency of the structural element subjected to some disturbance. Analytical methods are well established to solve many kinds of eigen value problems, though numerical techniques also have to be implemented in some cases. In fluid dynamic problems involving shock waves, the mathematical analysis of Euler equations gives rise to eigen value problems. Sometimes attempts are made to convert the resulting partial differential equations obtained during modeling for the purpose of obtaining solutions that satisfy specific characteristics. Certain kind of heat diffusion equations can also be solved as eigen value functions. The basic objective of this textbook is to guide the modeler to develop mathematical models in thermal and fluids engineering, solve them using numerical techniques, evolve a computational algorithm and develop a computer code to obtain the required simulation results. Hence, methodology and procedure will be demonstrated for the common type of problems in thermal and fluids engineering. In this context, eigen value problems will not be focused on in detail except in this brief introduction.

1.4.3 Propagation Problems

In thermal and fluids engineering, propagation problems refer to situations in which either fluid particles or heat energy propagate through space and time. The variations in properties of a system under dynamic conditions refer to propagation problems. Fluid in a static condition cannot be called a propagation problem. Heat conduction in a solid under a steady state condition can be viewed only as an equilibrium problem, not a propagation problem. As a convention, a time varying phenomenon is called propagation problem. It can be considered as an open end problem started at some initial condition and spreads the effect of changes in properties to other region of the system with the passage of time (Figure 1.6(d)). During convective heat transfer, thermal energy is spread from hot regions to cold regions. Whenever the velocity of a fluid medium changes with time, fluid momentum is transmitted from one region to another region. Physical entities can undergo changes where the conditions do not remain constant with time, and changes continue to take place with the advancement of time. Sometimes propagation problems can be called unsteady problems, and most of the steady state problems can be treated as equilibrium problems. In the case of propagation problems also, either fluid momentum balance or energy balance have to be satisfied, the only difference being that the attributes under investigation will change with time. When we consider the working of an internal combustion engine, which is running at constant speed and constant load conditions, based on the time scale with which the attributes are measured the problem can be treated both as steady state or unsteady state problem. In this case, if the main interest is the study of the variation of temperature and pressure of combustion products inside the engine cylinder, then these variations can be plotted with respect to the crank angle, which in turn is a function of time. However, if the focus is only on the net output of the engine and its performance at different loading conditions, then the transients can be ignored.

In thermal and fluid flow propagation problems, only first order time derivative situations are considered. These propagation problems always have some initial conditions. That means the values of the attributes of interest are known at the beginning of the process or changes taking place in the system due to its interaction with another system. Initial conditions are generally represented as conditions at time, t=0. In reality there is no such time as t=0. In a propagation analysis, the system variables are predicted or simulated under certain conditions at different intervals of time. The total time duration of simulation is divided into a number of time steps, and the size of time step has to be decided based on a stability analysis depending on the problem. Sometimes, the size of time step is determined by trial and error after comparing the final results with standard results if regular methods cannot be implemented. The larger the time step, the lower the computational time and vice versa, however, the main objective of getting a stable solution has to be kept in mind while deciding the time step. When two-dimensional transient heat conduction through a solid is considered, the governing equations are second order partial derivatives in space and single derivative in time. The spatial boundary conditions have to be provided to compute the spatial variation of temperature, and these are called boundary conditions. For a second order partial differential equation in two-dimensional domain, four boundary conditions, two in each direction, need to be specified before the solutions can be attempted.

The presence of non-linear terms in the derivatives of the variables or variation in properties of material make the equations non-linear. Of course, it is assumed that these non-linear equations are obtained while applying the conservation equation to the system. Consider a simple case of temperature-dependent thermal conductivity of materials in conduction problems. Unless the temperature is known, the thermal conductivity of the material cannot be determined, which is an a priori requirement for solving the differential equations. During modeling, the effect of variation of properties with respect to temperature and pressure have to be considered if these variations are going to contribute to the final simulation results. The thermo-physical properties of most of the refrigerants are strong functions of temperature, hence while modeling, these variations have to be considered for accurate prediction of results. There are engineering problems wherein all the variables being computed are dependent on each other. For example, in drying of porous materials, the amount of moisture evaporated depends on heat transport or temperature, and temperature depends on the amount of water evaporated, thus both water vapor and temperature are coupled to each other. The governing equations are generally non-linear and coupled because some of the mass transfer and heat transfer properties related to the solid depend both on moisture and temperature. Concrete structures exposed to the high temperature of fire are another interesting example where temperature, amount of water vapor and pore pressure are coupled to each other. All the diffusion and transport properties employed in the resulting governing equations are highly non-linear in nature. These coupled equations are always transient in nature, because evaporation of water vapor and temperature vary significantly with time. At every time step the required balancing of mass and energy have to be achieved using the suitable conservation principles. However, the non-linearity poses the challenge in calculating the attributes of the problem. Hence, an iterative solution procedure is adopted to overcome the difficulty of decoupling the variables. The computations are carried out with certain initial assumptions for the variables, then a convergent solution is approached by means of iteration until the required conditions based on conservation principles are satisfied. At every time step such iterative calculations are performed, and after convergence the solution for the next time step is found. As has been insisted earlier, the modeler should have some knowledge, idea or forecast how the simulated results will look for the given conditions. These expectations are strengthened with better understanding of the physics of the problem under analysis and employing the appropriate conservation principles.

1.5 MODELS IN ENGINEERING ANALYSIS

System behavior and description are two important steps in mathematical modeling. The underlying physics of a system decides the type of behavior of the system attributes, whereas system description focuses on how it can be viewed from the point of view of the variations of the attributes. For example, thermal and fluids systems behave in a manner such that all the attributes vary continuously in the domain considered. It is assumed that the fluid domain responds to any system variable by satisfying the continuum principle. The continuum concept has already been discussed in thermal and fluids engineering. Any attribute of a system can be differentiated if the system satisfies the continuum principle. There are systems which can be represented as a single unit having certain unique characteristics. . Such systems can be called lumped systems and do not satisfy any continuum principle. In mathematical modeling, a clear distinction between lumped and continuum systems has to be made so that suitable conservation principles can be implemented. For the purpose of mathematical modeling in thermal and fluids engineering, a given system can be classified as either a lumped system or continuum.

1.5.1 Lumped Parameter Model

In the literature, there is mention of the lumped system, discrete system and distributed system. However, the definition of these classifications varies based on the type of problem being analyzed. We try to distinguish between lumped and continuum models with respect to thermal and fluids engineering from a mathematical modeling point of view. Many engineering systems have been successfully modeled using these classifications. A system is called a lumped one when it can be defined using its own characteristics in the presence of other systems. The system under consideration may be connected to some other systems as well. In a way, lumped parameter analysis is a simplification for continuum problems. In certain situations, these simplifications are helpful to come up with simplified governing equations, instead of treating it as continuum, which otherwise might not have given better results. In heat transfer application, when transient heat conduction in a solid is considered, in a normal course, it will give rise to a second order differential equation in space with a first order time derivative, resulting in partial differential equation. In a convective heat transfer environment, if the conduction resistance for heat transfer within the solid is ignored, then the equation simply reduces to a first order ordinary differential equation in time only. Thus a partial differential equation is simplified to a first order ordinary differential equation, which is easy to solve. Such an assumption is valid under certain conditions when the convective resistance at the surface is significant compared to the thermal resistance within the solid. In reality, there will be temperature variation within the solid, however, for the application considered, if this variation is negligible then the system can be described using the lumped parameter model. It is well understood that any engineering analysis involves some approximations within the permissible limits. Again, the modeler has to take a judicial decision about under what circumstances the given system can be modeled as a lumped parameter model without sacrificing the main objectives of simulation.

In fluid mechanics applications, lumped parameter modeling is applied to represent a fluid element which can be characterized by a characteristic equation. For example, fluid elements and actuators can be modeled as lumped parameter models. It is common to define fluid elements such as resistance and capacitance by making use of the electrical analogy, in the field of hydraulic and pneumatic system analysis (Figure 1.7(a)). Fluid resistance is defined as the resistance for the discharge of a certain quantity of fluid across a given pressure difference. Of course, the relationship between these variables has to be established as either linear or non-linear depending on the type of problem. Now the defined resistance element can be employed during the analysis of the layout of the system of pumps or pneumatic devices. This is very similar to the thermal resistance concept defined using an electrical analogy while computing heat transfer across a temperature difference. It is essential to identify what the group of properties of the system are that can be combined together to give rise to a characteristic to be used repeatedly. This will very much simplify the final equations to be solved.

In structural mechanics, many types of lumped parameters are employed to carry out a stability analysis of structures. 'Distributed' and 'discrete' models are commonly used terminology in structural mechanics. The lumped system model is treated as a simplified version of the distributed system which is considered as a number of discrete entities. One of the discrete entities possessing certain unique characteristic is called the lumped parameter model. For example, a building can

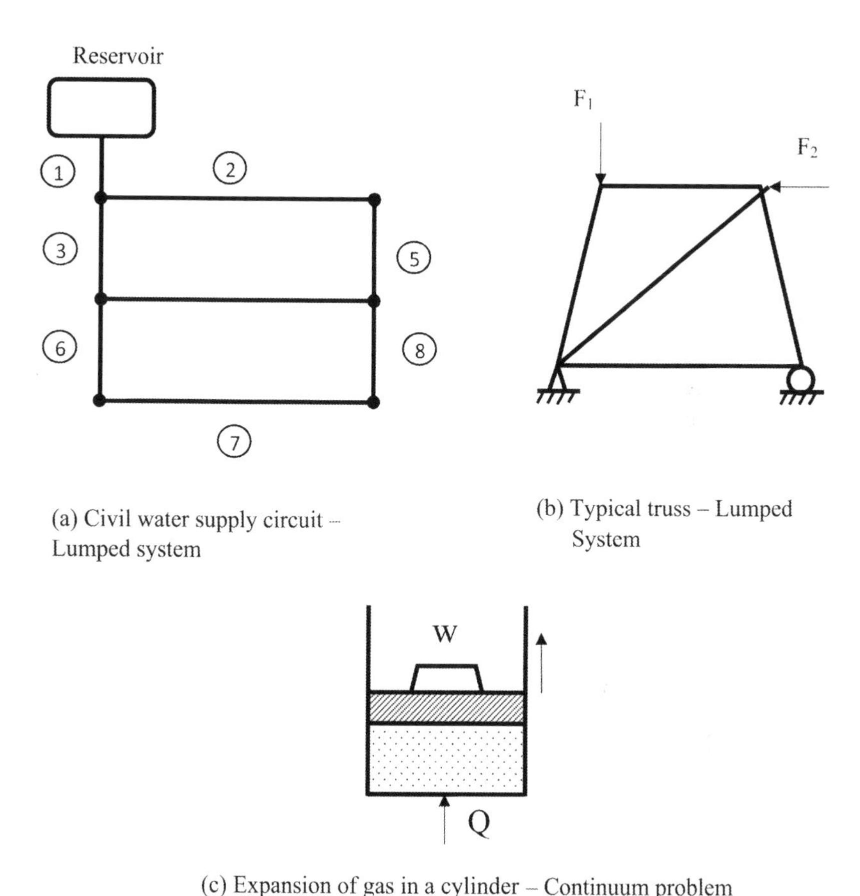

(a) Civil water supply circuit – Lumped system

(b) Typical truss – Lumped System

(c) Expansion of gas in a cylinder – Continuum problem

FIGURE 1.7 Lumped and continuum models.

be considered as a distributed system consisting of a number of discrete elements such as beams and columns and other structural elements (Figure 1.7(b)). A beam has certain characteristics which can be uniquely defined under loading conditions. Similarly, a column also can be treated as a discrete entity having its own qualities. Assemblage of all such structural elements will give rise to a final structure, well connected by these elements, using the required laws of mechanics. Hence, some transient heat transfer problems can be treated as a lumped system whenever the thermal resistance is small compared to other resistances in the given situation. By defining fluid elements in the form of lumped parameters, the analysis of fluidic systems has become much easier. Use of a lumped parameter model may be to reduce partial differential equations to simple ordinary differential equations in heat transfer analysis. In the case of fluid mechanics applications, use of lumped system approximation gives rise to simple algebraic equations. However, the incompressible and compressible nature of fluid in the system has to be taken into account while defining the lumped parameters.

1.5.2 Continuum Based Model

The continuum principle is the basis for the analysis of many field problems in thermal and fluid sciences. The definition of a continuum is well defined in fluid mechanics and related subjects in undergraduate classes. Any system or control volume is said to satisfy the continuum principle if the field variables such as temperature, pressure, velocities etc. can be treated as continuous functions. That means the medium of substance or fluid in a system allows continuous variation of the defined attributes under interactions with other systems or under the action of certain imposed boundary conditions (Figure 1.7(c)). In thermal and fluid sciences, it is required to compute the continuous variation of the variables of interest as a function of spatial coordinates and time coordinates. For example, $T(x,y,z,t)$ indicates a function that is a function of spatial coordinates x, y, z and time t. Differential operators are employed to represent the variation of these field variables resulting in ordinary and partial differential equations. These differential equations are obtained as a result of the application of certain conservation principles during mathematical modeling of a system. If we look at the reason why and how differential equations are derived from the statement of some conservation principles, one can easily understand that these principles are applied to a differential element. Conservation laws stop only at stating the physical principles for a given system, and not all the conservation laws are directly stated in differential form. The differential element considered is assumed to be part of the continuum established by the medium of substance or fluid in the system. The application of limiting theorem states that the variation of a given variable with respect to another variable that tends to zero provides the first order differential. As far as mathematics is concerned, this limiting value of Δx is always assumed to tend to 0. However, in engineering analysis, there is nothing like Δx becoming 0, there is always some discrete distance between which the variation of the attribute has to be computed. While comparing the simulation results with experimental results, this becomes important because in experiments one has to use a sensor, which has a finite size and dimension. In real life situations, the differential operator becomes meaningful with certain approximations because of the definite size. Numerical results predict the variation of the variables at selected nodal points separated by a small distance, and how small the distance is effects the accuracy of prediction for the given size of the problem. The differential operator may refer to some value of the limiting function in the case of a strain gauge, which may be very small compared to the limiting function considered for atmospheric modeling. In a strain gauge sensor, it may be in millimeters, whereas in atmospheric modeling it may vary from a few meters to a few kilometers. In both problems, the principle of continuum is satisfied where the variation of the required attributes can be represented as differential functions.

Differential equations are solved to obtain a solution as a function of spatial and time coordinates. These field variables should be differentiable at least to the first order in thermal and fluid sciences because many parameters are computed using the first order derivative. For example, for steady state heat conduction through a solid, when temperature is considered as the field variable, then a second order derivative in temperature is obtained which has the solution in the form of a straight line, indicating that the first order derivative is a constant with constant thermal conductivity of the material. For the same problem, if heat quantity is considered as the variable, its first order derivative itself becomes zero because the heat quantity is constant under steady state condition. The first order derivative of solution for temperature becomes a constant, which finally satisfies the principle of the Fourier law of heat conduction for one-dimensional steady heat conduction, that for constant thermal conductivity and constant heat transfer, the temperature gradient is a constant. In fluid mechanics, the fluid stress is defined as a product of fluid dynamic viscosity and velocity gradient. Computation of velocity gradients in a flow field over structural elements or automobiles is an important field of study. Wind load acting on huge building structures like power plant chimneys, cooling towers and skyscrapers is simulated to estimate the stability of these structures under severe wind conditions. During re-entry of space vehicles into the atmosphere closer to the earth, the surface of the space vehicles will be subjected to shear stress, leading to heating of surfaces, a phenomenon called ablation. The fluid stress over the surfaces has to be simulated accurately in order to take care of thermal stress of the material. Calculation of velocity gradients in such problems becomes crucial for accurate simulation results useful for design. Fluid flow through pipes or conducts is governed by the pressure gradients that exist in the direction of flow. In a two-dimensional flow domain, fluid flow is controlled by pressure gradients in both directions and hence the velocity distribution in the domain is influenced by the pressure gradients. Fluid shear stress on the surface of the domain is determined by the distribution of velocity gradient for fluid with constant viscosity. The type of distribution of these shear stresses depends on how the boundaries of the domain are controlled by enforced variables. While developing the mathematical model for such problems, the fluid momentum and continuity equations are derived; that means the fluid momentum balance is ensured in the domain. Once the originating physical principles of the conservation equations are realized, then the nature of the final simulation results can be interpreted technically, so that the insight into the problem is achieved. This is what is expected as the outcome of mathematical modeling. With the help of simulation results, the modeler will be able to include the physical phenomena that would have been present if the system existed and was subjected to the assumed conditions.

1.6 SOLUTION OF DIFFERENTIAL EQUATIONS

Mathematical modeling of thermal and fluid science problems gives rise to differential equations as governing equations. These equations may be either partial differential equations or ordinary differential equations. Governing equations in the form of differential equations represent the governing physical laws complying the conservation principles for the given system. In any differential operator, one variable is dependent and the other one is independent. When a dependent variable is a function of only one independent variable, then the resulting equation is an ordinary differential equation. If the dependent variable is a function of more than one independent variable, it gives rise to partial differential equation. The governing equation may be a single differential equation or it may be a set of differential equations depending on the number of conservation principles that have to be satisfied in the model. During modeling the selection of dependent variables is very important because most of the simulation results will be obtained directly in terms of those variables. Sometimes, derived quantities will be obtained using these dependent variables computed as a solution during simulation. Once all the required physical laws and conservation principles are employed during modeling, the required number of final differential equations will be obtained. Here, most of the discussion is focused mainly on differential equations though there may be mathematical models which give rise to directly linear or non-linear algebraic equations. All the continuum problems produce differential equations as a result of modeling. The terminology solution is used to represent the final results of the dependent variables as a function of independent variables in a given domain. The domain can be classified as spatial or temporal/time. Spatial domain is represented by Cartesian, cylindrical or spherical coordinates. A very cautious move taken in this textbook is that only Eulerian coordinate problems will be discussed in the upcoming chapters. After obtaining the differential equations from modeling, the next step is to get the solution for the variables.

1.6.1 Analytical Techniques

Analytical methods are well-established techniques to solve differential equations. With continuous focused research on finding new analytical solutions for different kinds of differential equations, now a stage has been reached where most of the engineering problems can be solved using analytical techniques. The most challenging part in dealing with the solution of differential equations using analytical techniques is the limitation on the boundary conditions and geometrical shapes. It is surprising to note that many users prefer numerical techniques or a software platform to get solutions of differential equations which otherwise could have been solved by analytical techniques. The reason may be that it is easy to get solutions using software than resorting to complex analytical tools for a given set of differential equations. However, one should not forget that an analytical solution gives a compact expression for the dependent variable in terms of the independent variables. The great advantage of such a solution is that the dependent variables can be computed at any number of spatial coordinates as represented by the independent variables. The knowledge of analytical solutions for differential equations educates the modeler about the kind of solution for the type of differential equations considered. For example, when partial differential equations are classified as elliptic, parabolic and hyperbolic, the nature of the solution of these equations give the modeler the physical interpretation of the expected solution. Elliptic equations are also called Laplace equations and represent the variation of a given property at a given location in terms of variation of the same property in the neighboring locations. This is a very useful form of equation and represents many physical phenomena in heat transfer and fluid mechanics science. Steady state heat conduction in three-dimension is described by a Laplace equation in temperature, T. Similarly, in fluid mechanics, the velocity potential written for the continuity equation gives rise to Laplace equation. It is easy to obtain an analytical solution for Laplace equations. The problems represented by elliptic equations are enclosed with a definite boundary; like an ellipse it is confined and the solution is explored within the domain defined by the boundary. An incompressible fluid with zero friction gives rise to a potential problem when the flow field is irrotational. For example, consider the fluid friction estimation over the streamlined body of an automobile moving at a particular velocity. The fluid very close to the boundary of the vehicle body will only experience the effect of fluid friction or shear stress; however, at a far-off distance from the boundary, there will not be any effect of fluid friction. Such a region is easily modelled by treating the fluid as potential flow, which is modeled by Laplace equations. The modeler has to identify clearly where the permissible approximations, such as inviscid fluid behavior and irrotational nature of fluid, can be implemented. In an automobile body, the fluid near the boundary that is responsible for the skin friction and the fluid that is far away from the boundary are the same; that is the atmospheric air. However, due to the specific characteristic of the flow field, one near the boundary and the other away from the boundary, the type of governing equations used are entirely different. This is where the importance of understanding the physical phenomena comes into picture. A modeler is not simply a person with thorough mathematical and numerical knowledge to produce some numbers, he must also be a person with good imagination. In a way, imagination plays a major role in mathematical modeling. One has to conceptualize the system in consideration, its behavior under the assumed conditions and the expected results. Some modeling exercises may require many brainstorming sessions to understand and imagine the expected outcome of simulation. Useful discussion and deliberation is possible in such sessions only when all the team members look at and analyze the situation with good understanding of the underlying physics.

Parabolic differential equations indicate system behavior when the expected attribute of the system varies in one open-ended direction. That means there is a starting point for the enclosed system with a boundary; however, the solution is expected to vary in one direction endlessly. Of course, in engineering analysis, engineers are interested in studying problems which have solutions for a definite period of time, and even in exponential variation cases, the time should be predictable within the reach of engineering design. Transient heat conduction problems are represented using parabolic equations. The condition of temperature at time, $t=0$, is known at all the boundaries of the domain as well as in the computational domain. Now the solution of the equation is obtained for temperature variation within the domain for all the time greater than zero, that is, $t>0$. The question arises of what the duration of time t is, up to which the solution has to be obtained. Even in the exponential decay of nuclear waste material during disposal of it, a finite time duration is predicted so that a suitable design can be proposed. An analytical solution of parabolic equations provides the solution decaying in time. This is an interesting proposition that the modeler has to consider while obtaining the analytical solution. Using separation of variables technique, the parabolic heat equation will be reduced to an eigen value problem with suitable manipulations. While selecting the eigen value, λ, there comes the choice of $+ve$ or $-ve$ value for the eigen value. Until this step, an analytical technique is helpful; however, while selecting the sign of eigen value, λ, the modeler has to apply his knowledge or experience of real life problems. The positive value for λ will give rise to solutions that will increase exponentially with an increase in time, which may be difficult to achieve in real situations. When a building structure is exposed to fire and where the temperature increases exponentially with time, the temperature of the building will definitely keep increasing at the beginning of exposure. However, there is a definite period beyond which the temperature rise will cease to go further because by that time all the heat dumped by the fire might have been consumed by the material of the building being damaged, burned or become ash. The only difference will be this time duration; it may be a few minutes, a few hours or a few days, like forest fire. This time will never be infinite and has a tendency to increase the given attribute continuously. With knowledge of this physics for any given problem, the modeler will choose the *-ve* sign for λ to obtain the solution. Similar examples can be found in fluid mechanics also. No engineer will come across a problem or will need to design a system wherein either velocity or pressure of the fluid medium will keep increasing with time. That is why a decaying solution is assumed in parabolic problems. In thermal and fluids science, parabolic differential equations play a major role for defining many real life problems. The hyperbolic differential equations well represent the wave propagation in both heat transfer and flow problems. Thus the basic characteristics of the differential equations are well utilized to model real life problems. An analytical solution of differential equations involves a series solution, which in turn can be explained using trigonometric functions. The nature of trigonometric functions provides insight into the expected solution of the differential equations during modeling. Hence, in modeling, analytical solutions are also employed wherever possible to reduce the computational time and effort.

1.6.2 Numerical Techniques

Mathematical modeling and simulation concepts have attained popularity after the accessibility of many computational commercial packages became easily affordable and cost effective computing machines became available. The power of analytical methods for the solution of even complicated equations like Navier-Stokes equations were well established decades back. In a way, the absence of high computing machines made people consider novel methods to solve such coupled equations. The concept of similarity parameters proposed in the solution of Navier-Stokes equations and related energy and concentration equations was one of the innovative techniques born out of imagination derived from the power of analytical tools. However, when systems with complicated geometries such as aircraft wings, structure of bridges etc. have to be analyzed, the limitations of analytical tools became explicit. In the meantime, the breakthrough in powerful programming languages and computers have made the modeler feel very comfortable with numerical techniques. Now a stage has been reached such that even problems that have analytical solutions are being solved using numerical techniques or software packages. That is how the numerical simulation has occupied many fields of science and engineering with ease of accomplishing the required objectives. The author would like to suggest that numerical techniques are powerful tools to solve many real problems. The modeler initially has to look at the possibility of solving the final governing equations using analytical techniques. The reason for this is that the analytical solution approach allows the modeler to identify the physics of the expected simulation results in advance, by interpreting the nature of solution that could be obtained at the end of the simulation. The geometry of the system may demand the use of numerical techniques as a solution methodology. Getting an insight into the expected outcome of the simulation results makes the modeling and simulation exercises more sensible and meaningful. It is repeated once again; simulation results are not sheer numbers but they possess a lot of physical meaning in the context of analysis of the specific problem being modeled.

Numerical techniques are approximate solution methods employed to obtain solutions of differential equations. The main advantage of a numerical technique is its ability to handle any type of complex geometry along with different kind of boundary conditions, which otherwise would have required complicated solution formulation with analytical methods. Finite difference, finite volume and finite element methods are some of the most popular numerical methods well established in modeling and simulation for problems encountered in structural mechanics and thermal and fluid mechanics fields of research. Though these techniques are approximate in nature, their accuracy can be improved with many efficient algorithms. The implementation of all the numerical techniques for the solution of differential equations, ordinary or partial, reduces the governing equations to simultaneous algebraic equations, which can be solved using an equations solver. In the numerical approach, the solution means obtaining the variables or attributes by solving the equations at a definite number of nodal points defined or designated by some discretization approach. Every numerical technique follows certain principles that retain the conservation principles based on which the differential equations are derived. Unlike the analytical method where the solution of the variables can be obtained at an infinite number of space or time coordinates, in numerical methods the solution can be computed only at definite number of points. That is why these methods always have a prefix, *finite*. The differential operators in differential equations are transformed into variables represented over discrete domains. A real problem is configured into an imaginary problem by means of enclosing the real system using boundaries suitable for defining the system. The sphere of space of the system is known as the *computational domain*; this means the domain over which the required solution for the variables are estimated. A computational domain of even 1 mm by 1 mm may consist of infinite number of points. In numerical techniques the required number of points in the domain, over which the solution has to be achieved, is obtained by a technique called *domain discretization*. In simulation terminology, this is also called *meshing* or grid *generation*. A grid is nothing but a set of lines drawn within the computational domain along the coordinate directions, in order to discretize the domain to obtain a certain number of node points at which a solution can be obtained. These discretized sub-domains may be regular geometry or may be irregular depending on the type of mesh generation technique employed. It has to be remembered that the specific numerical technique should possess the capability to satisfy the given governing equation at each and every sub-domain created by the grid lines.

The differential equation representing a conservation principle is also satisfied at every sub-domain, maintaining the continuity between the variation of the variables between the sub-domains. This is the unique characteristic of a given numerical technique. In the case of the finite difference method, the differential equations are converted into difference equations using Taylor's series expansion. Sometimes the conservation equation itself can be directly derived using the difference equation principle over the discrete domains by satisfying the required conservation principles. Generally, the given differential equations derived based on certain conservation principles are solved using the finite difference method. The principle of conservation of flux due to the variables being computed is employed in a finite volume method. In both finite difference and finite volume methods, the discrete domains are called *control volumes*, a differential volume, without using a differential. These methods have been widely used in thermal and fluids engineering fields for decades and their methodologies are well established in the literature. There are many commercial software packages that have been developed to solve many heat transfer and flow problems. It is worth mentioning at this stage that numerical solution techniques using numerical methods are entirely different from identifying solution techniques for the set of given differential equations based on the uniqueness and existence of the solution of the variables. Unless this stage is affirmed, the simple implementation of a numerical technique for the solution of the differential equations will not serve the purpose of getting the required solutions. That means for the given equations, the existence of a solution or uniqueness of solution needs to be ascertained. This can be done only with the help of mathematics because the differential equations have been derived using mathematics principles for a given conservation principle. Most of the differential equations employed in heat transfer and fluid mechanics have unique solutions except those equations derived for an entirely new problem which has not been solved to date. The uniqueness of the solution comes into picture to ensure the compatibility of the selected type of differential formulation to imply the conservation principle that has to be satisfied in the given situation. Stability is another important principle that needs to be assured in the case of transient problems when the equations are solved using numerical techniques. Many transient problems in heat transfer and fluid mechanics involve a variation of the variables in both time and spatial domains. While computing the variables, the phenomena that cause changes in the variables should be consistent in satisfying the spatial and temporal variations. For example, in transient heat conduction problems, during the given time step computed, the heat should be able to diffuse through the selected spatial domain. If the time step and spatial domain size do not cope with the physical phenomena of heat diffusion, then the resulting solution of temperature will lead to stability problems. That means that the temperature will start oscillating with time, leading to some unrealistic temperature distributions with time and space, and sometimes violating the basic principle of heat conduction through the medium. When implementing a numerical technique for the solution of transient problems, the stability criteria as specified by the method needs to be taken into account in order to get realistic solutions.

1.6.3 Computing Techniques

The final objective of any mathematical modeling exercise is to obtain simulation results. By using the final form of the mathematical model in simultaneous equations after implementing either the analytical technique or numerical technique, results can be computed for the futuristic conditions of the system under analysis. Simulation results are achieved by converting the final form of the discretized form of the mathematical model into required numbers by means of programming. Numerical methods cannot take us beyond transforming the given governing differential equations into simple algebraic equations. The number of algebraic equations, linear or non-linear, are equal to the number of nodal points or grid points at which the numerical solutions are pursued. Initially a computational algorithm is developed to solve the final form of the set of equations produced from the application of numerical technique. When a set of differential equations are solved in a modeling exercise for the solution of number of variables, a final set of algebraic equations will be produced as a result of application of the numerical technique. If the original differential equations are coupled and non-linear, then they need to be solved in a particular sequence which will ensure satisfaction of the imposed or expected conservation principles. This particular sequence of steps of computations is called *computational algorithm*, also known as the logical sequence of computations. Only when this sequence of computations is followed, the resulting simulation results will be meaningful for the given mathematical model, otherwise entirely unexpected results could be obtained. Once the computational algorithm is made ready, then the strategy for computer programming or coding is decided. In the recent past couple of decades, commercial software packages have become very popular for simulation in educational institutions and scientific laboratories. Packages such as ANSYS, COMSOL, Fluent etc. are some of the commercial software available to solve mathematical model problems in structural, thermal and fluid mechanics fields of engineering. A computer program can be built using different types of routines in built-in MATLAB to solve any particular mathematical model. In this textbook, the author will discuss developing computational algorithms for many example problems in thermal and fluids engineering so that computer codes using FORTRAN and C++could be easily developed.

The computational algorithms are represented in the form of flow charts or in simple computational steps. In the solution of the final form of the equations, there will be a number of variables used in the computational procedure. Some of the variables may be constants, and others may change with space and time. All these constants and variables have to be properly documented by assigning names. Suitable book-keeping is required in order to avoid the use of the same name for different variables or constants resulting in run-time errors. Development of error free computer code requires sufficient care, patience, meticulous planning and book-keeping. A beginner can develop a computer program even with basic knowledge of the syntax of a particular programming language. The style of programming also varies with an individual, just like language skill and style. However, the computer code should be syntax error free so that it can be compiled and executed to obtain the final results. It is worth mentioning here, that depending on the command to the particular programming language, for the same problem, the correct computer code can be written in say, 1000 lines or 1500 lines based on the details of the program. It is always a good habit to use a number of comments statements before the start of every computation. Without any comments statements the program may be very short and written in few lines, but still producing the required results. With the help of comments statements the programmer can easily understand every step of computation and verify whether the computations are being carried out according to the designed computational algorithm. It is well known that comments statements are non-executable statements and hence adding them will not have any effect on execution, it only guides the programmer on the process of computation. Many users avoid developing their computer programs due to lack of confidence or the difficulties they face during debugging. A well written computer program will lead to the least amount of debugging issue because all the syntax for computations using different constants and variables are considered during programming. The advantage of developing their own computer codes is that the designer is the master of the software, the brain behind the simulation results is their own and their self-confidence gets boosted. Unless an individual tests his/her programming skills by developing own computer code, he/she will never come to know his/her capacity and skills in software. This simply comes to anyone by practice. Developing one's own computer code means the developer has complete knowledge of all the computations being performed in a simulation program, and hence, if the results are not as expected, the code can be easily re-checked for errors. It is not necessary that everyone has to develop his/her own computer program for whatever problem they solve. However, wherever possible, one should develop his/her own code. Building one's own code gives complete insight into the problem and the simulation results, as the physics of the problem and computation procedure are combined together in the simulation.

REFERENCES

1. *Mechanics of Fluids SI Version*, Merle C. Potter, David C. Wiggert, Bassem H. Ramadan and Tom IP Shih, 4th Edition, Cengage Learning, Stamford, CT, 2012.
2. He, Y., Himeno, R., Liu, H., Yokota, H. and Gang Sun, Z. Finite element numerical analysis of blood flow and temperature distribution in three-dimensional image-based finger model, *International Journal for Numerical Methods for Heat & Fluid Flow*, 2008; 18(7/8): 932–953. https://doi.org/10.1108/09615530810899033.

EXERCISE PROBLEMS

Qn: 1 Develop a mathematical model to calculate the power input to a centrifugal pump to extract water at a depth of D_h and supply the same to a height of H. Mention the conservation principles considered while modeling the above system. Water can be assumed as an incompressible fluid with constant density, and zero flow losses can be considered inside the pump.

Specify the method employed, and whether it is a lumped parameter or continuum model.

Qn: 2 A cantilever suspension bridge has to be designed for a length of L and other dimensions can be assumed. Consider the bridge consisting of a number of structural elements that can be represented using the lumped parameter model. Develop the mathematical model for different types of structural elements for the bridge using the knowledge of solid mechanics.

Qn: 3 The water in a household geyser of a 20-liter capacity is heated using a 2 kW rated heater coil. The initial temperature of water in the geyser is 23°C. The final temperature of water at the end of 15 minutes heating needs to be determined. Develop a mathematical model for energy balance with the assumptions made. The exit water valve remains closed during the heating process. Assume a specific heat capacity of water of 4.18 kJ/kg K. Identify whether the heating process in this case satisfies either steady state condition or unsteady state condition.

Qn: 4 Consider the same data given for the geyser in Question 3. At the end of ten minutes of heating, the exit valve for hot water in the geyser is opened such that the water flow rate is 2.5 liters/min. It is necessary to find out the temperature of water exiting from the geyser at this flow rate. Extend the mathematical model developed in Question 3 and identify the transients in the problem. Suggest a method to solve this transient problem.

Qn: 5 Landmines buried under ground can be modelled as a diffusion heat and mass transfer with heat source problem. Consider that a two-dimensional domain is parallel to the ground source and consider that a landmine made of non-metal is buried under a depth of H m from the surface of the ground. A landmine made of 8 cm × 4 cm can be assumed as a constant heat-generating source. Comment on the development of a mathematical model to determine the location of the landmine buried below the ground.

Qn: 6 A rotameter is used to measure volume flow rate through a pipeline during an experiment. A mathematical model has to be developed to estimate the volume flow rate of the liquid flowing in the pipeline. Explain how the principle of continuity and fluid momentum will be applied while modeling this system. With the help of fundamentals of fluid mechanics, develop the final equation for volume flow rate, highlighting the mass and momentum conservation principles. List the assumptions made.

Qn: 7 The radiation heat falling on the side walls of a room has to be modelled. Heat transmitted inside the room due to solar radiation and by conduction and convection also has to be considered in the model. Apply energy conservation principles to develop governing equations that can be used to compute the time variation of temperature of air inside the room. Comment on whether the continuum principle or lumped parameter principle will be employed for developing the model.

Qn: 8 An iron box of capacity of 1 kW is used to iron cloths at a maximum temperature level. When the iron is pressing the cloth, what type of heat transfer takes place: conduction, convection or radiation? Develop a mathematical model to determine the interfacial temperature between the cloth and the iron bottom surface. Explain what type of boundary conditions can be considered for this problem.

Qn: 9 Consider some heat transfer problems in real life situations. Identify these problems either as lumped parameter problems or continuum problems. State the assumptions made in both cases and explain also how one type of problem cannot be considered as the other type and vice versa.

Qn: 10 Steam expands in high pressure and low pressure turbines in a steam power plant. Under what conditions can expansion of steam in these turbines be treated as a steady state problem, unsteady state or transient problem. What are the assumptions to be made with scaling of transients to treat these problems as steady state problems? Also explain the conditions under which the transients can be neglected.

QUIZ QUESTIONS

Qn: 1 Mathematical modeling of any engineering problem consists of principles derived from ____________________ and ____________________.

Qn: 2 The accuracy of results obtained from modelling and simulation programs mainly depends on (a) numerical methods, (b) analytical techniques, (c) differential equations or (d) conservation principles employed. Tick the right one.

Qn: 3 Physical modeling is always a part of mathematical modeling – yes/no. Tick the right one.

Qn: 4 Similitude analysis is the same as mathematical modeling – yes/no. Tick the right one.

Qn: 5 Structural strength of suspension bridges are analyzed using (a) a continuum approach, (b) steady state analysis, (c) lumped parameter analysis or (d) unsteady state analysis. Tick the right one.

Qn: 6 Heat transfer and flow problems can be modelled as (a) continuum problems, (b) lumped parameter problems, (c) either continuum or lumped parameter problems or (d) none of the above. Tick the right one.

Qn: 7 An air freshener gel is sprayed into the atmosphere. In order to analyze this situation with respect to the presence of material of the spray in air, the following conservation principle has to be employed along with other conservation equations: (a) energy conservation, (b) species conservation, (c) momentum conservation or (d) constitutive equation. Tick the right one.

Qn: 8 Numerical methods produce accurate results like analytical methods for the solution of differential equations. True/False.

Qn: 9 The finite difference method works on the principle of ____________________.

Qn: 10 The weak form solution is employed by Galerkin's finite element method. True/False.

2 Conservation Equations

The basic conservation principles required to model thermal and fluid systems are discussed in this chapter. During mathematical modeling it is required to understand energy interactions between systems so that the variation of attributes can be modeled. Most of the conservation equations that are derived in the following sections are available in standard text books of fluid mechanics and heat transfer. The very purpose of this chapter is to discuss the conservation laws from the perspective of modeling. Energy interaction in this text book mainly refers to heat energy. In thermodynamics heat was defined as energy interaction at the boundary of a system, arising as a result of temperature difference between the system and the boundary. Problems involving fluid flow with heat transfer are encountered in many engineering applications. Thermodynamics deals with systems that are in thermodynamic equilibrium and hence it does not go beyond defining heat as a form of energy interaction due to temperature difference. Rate of heat transfer takes place in a non-equilibrium situation, as noticed in internal combustion engines, a cycle of operation is completed in a fraction of a second during which heat transfer due to combustion takes place. Such systems cannot be analyzed in thermodynamics without making assumptions on energy interactions. There is a distinction between energy interaction and energy transfer; interaction is not a time-bound phenomenon whereas energy transfer is. Engineers are interested in designing and analyzing systems which perform an action or process within a specified period of time. Thermodynamics explains the natural constraint on conversion of heat energy into mechanical energy and the related propositions on the direction of processes. Hence, heat transfer is never discussed in thermodynamics, only heat interaction is considered. In order to understand the process of heat transfer, a separate scientific approach is required. It has been identified by experimentation that heat transfer takes place by three modes, called conduction, convection and radiation. Temperature difference is the potential that drives heat transfer from one medium to another medium. However, this temperature difference becomes a characteristic for these three modes of heat transfer. Let us consider conduction heat transfer, defined by the Fourier law of heat conduction as

$$q_x = -k\frac{dT}{dx} \tag{2.1}$$

The heat transfer per unit area, conducted across the solid is directly proportional to the temperature gradient $\frac{dT}{dx}$ normal to the direction of heat flux q_x. Heat conduction is governed by the temperature difference across the distance through which conduction takes place. The distance through which conduction takes place may be in a solid, liquid or gaseous medium. In the case of convective heat transfer, Newton's law of cooling states that

$$q_x = h_c(T_w - T_\infty) \tag{2.2}$$

DOI: 10.1201/9781003248071-2

The heat flux q_x is defined as heat transfer per unit area that is convected from a wall at temperature T_w to a fluid medium at T_∞ . Only the temperature difference becomes the driving potential for convective heat transfer from the wall to a fluid medium. The wall at a higher temperature than the fluid loses heat to the fluid. Convective heat transfer takes place between a surface and a fluid medium. Radiation heat transfer as defined by Stefan-Boltzmann's law states that

$$q_x = \varepsilon\sigma(T_w^4 - T_\infty^4) \tag{2.3}$$

Radiative heat flux is directly proportional to the fourth power of temperature. It is interesting to observe that heat transfer by convection is directly proportional to only temperature difference; in conduction, it is proportional to the gradient of temperature, and in radiation it is proportional to the fourth power. Hence, it is not always mere temperature difference that causes heat transfer as explained by thermodynamics. Heat energy is different from heat transfer; the first is a form of energy and the other is energy in transport. Heat and work are called energy in transit in thermodynamics; this means they get their meaning only when they flow. A system cannot be said to contain heat or work; in a way they are dynamic energy. However, thermodynamics does not elaborate on heat beyond this, and hence heat transfer is studied as a separate subject. Heat transport processes expressed by Equations (2.1) to (2.3) are postulated based on certain laws which are verified both theoretically and by experimental observations. These modes of heat transport may have existed for centuries; it is only after the physics behind these mechanisms were understood that they were represented in the form of equations. That means whenever there exists a temperature gradient across a medium, heat transfer by conduction takes place naturally. It is quantified after estimating the thermal conductivity of the material. Similarly, whenever a fluid flow takes place with a temperature difference in relation to a surface over which it is flowing, there will be convective heat transfer. The heat transfer coefficient in Equation (2.2) is a characteristic parameter specified for the given flow situation. Any material that exists at a temperature greater than 0 K radiates thermal energy in the form of electromagnetic wave spectrum.

Conduction heat transfer takes place across a temperature gradient of any material: solid, liquid or gas. However, the amount of conduction heat transfer also depends on thermal conductivity of the material through which heat is transported from one end to the other end. Thermal conductivity of materials decreases in the order of metals, non-metals, liquids and gases. Hence, heat conduction through solids is considered to be more significant compared to liquids and gases. In some engineering applications, it is effective heat transfer that is important, whereas in other cases it may be the reduction of heat loss. Most of the heat transfer problems deal with conduction through metals or non-metals or insulating materials. It is required to understand the governing equations related to conduction heat transfer through solids.

2.1 SOLID MEDIUM

Conduction heat transfer through solids are analyzed more in detail in heat transfer problems compared to conduction through liquids and gases. However, there are applications where conduction through liquids and gases cannot be ignored. It is worth appreciating that conduction heat transfer is defined only in one-dimension, in terms of the temperature gradient in a particular direction. That means in a three-dimensional solid, conduction heat transfer is defined in each direction independently in terms of the respective temperature gradient and thermal conductivity of the material. When thermal conductivity of a given solid remains constant in all directions, then it is called a homogeneous solid, and it is assumed that thermal conductivity in all the three-coordinate directions remains the same. In real life problems the solid under consideration for heat transfer may be rectangular, square, cylindrical, spherical or any other three-dimensional geometry, which is difficult to define by simple definitions. The solid may be symmetrical around an axis or it may be non-symmetrical. As engineers have to handle any type of geometry in real life problems, it is important to understand different coordinates systems used to represent a given geometry of a solid. For this reason, heat conduction equations are derived in Cartesian, cylindrical and spherical coordinates. These relations will provide knowledge about how conduction heat transfer is affected by different geometrical coordinates.

Implementation of certain physical laws with the help of mathematics will give rise to the governing equations. With regard to heat energy, thermodynamic law will be the physical law that has to be satisfied in the system through which heat transfer by conduction is assumed to take place. Heat conduction is directly proportional to the temperature gradient across the material in the given direction. Now, one has to decide the variables on which the governing equations need to be obtained. In heat conduction analysis, two quantities are defined, one is the conduction heat transfer, q_x (W/m^2) and the other one temperature, T (K). The heat transfer q_x itself is a derived quantity, and it is calculated and not measured as it is. In addition to this, temperature is a thermodynamic property of a system, whose variation in a given medium determines the type of heat transfer. Temperature within solids can be measured in the laboratory, and temperature plays an important role in finding out other properties. For example, thermal conductivity of solids is a strong function of temperature at high temperature ranges. Many thermo-physical properties such as density, specific heat capacity, viscosity etc. of refrigerants are strong functions of temperature. Temperature is the driving potential in any heat transfer problem. Hence, computation of variation of temperature in mathematical modeling is the most important part of simulation. Thermal energy conservation is controlled by mainly the temperature variation. In thermal and fluid problems, generally all the properties such as temperature, density, velocity and pressure are called field variables – this means variables in the continuum. In the derivation of a heat conduction equation, it is the temperature whose variation is important and hence the governing equations are obtained in terms of temperature at the end.

2.1.1 Energy Transport in Unsteady State Conditions

Heat conduction through solids in Cartesian, cylindrical and spherical coordinates will be derived by the application of the conservation principle. The main aim of developing governing equations in temperature is to compute the temperature variation in a given domain as a function of time and spatial coordinates. Let us assume that the modeler is interested in computing the temperature variation over the surface of an industrial chimney of 50 m height, or the temperature distribution over the wings of an aircraft, in order to calculate the thermal stresses. Temperature is a field variable, and its value at many number of locations is required for the analysis. Due to the interaction with hot gases or with the atmosphere, both the chimney and the aircraft wings will experience variation in temperature, and these variations have to be followed accurately. Irrespective of the size of the structure or the system under consideration, the requirement is a computation of temperature at different locations distributed over a larger surface area. This demands that the conservation law has to satisfy the physical principle over an infinitesimally small volume, so that the variation of temperature over such small volumes can be computed. Hence it is essential to make sure that the energy conservation principle is satisfied over a small volume, called control volume. This derivation is demonstrated for an unsteady state condition during which there is a variation of temperature of the control volume with time, other than heat transport through the solid by conduction. Due to temperature gradient across the solid, conduction heat transfer takes place. However, before the start of conduction, that is, entry of heat into the material, the control volume is at a particular temperature. In the process of energy balance over the control volume, all the quantities have to be considered in the form of thermal energy, not in terms of temperature. Hence, it is assumed that to start with time 't', the control volume is at an internal energy level of U_t. After a time lapse of Δt, the internal energy of the solid will change to $U_{t+\Delta t}$. During this time lapse, the rate of change of internal energy of the system has to be accounted for in the energy balance. With regard to the conduction through the spatial direction, heat quantity, q_x enters the control volume at x and leaves the control volume at $x+dx$ in the x-direction, similarly in the y- and z-directions. Heat generation also may take place within the solid due to electric heating, chemical or nuclear reactions, and this is accounted for by considering this heat as heat generation per unit control volume. Simultaneous heat conduction through a three-dimensional solid is considered, hence the temperature gradient in a particular direction will be a partial derivative. The variation of heat conducted in a particular direction also gives rise to partial derivative of heat quantity with respect to that particular direction. Considering a control volume $\Delta x\, \Delta y\, \Delta z$ in a Cartesian coordinates system (Figure 2.1(a)), the application of energy balance gives the following statement.

$$
\begin{aligned}
&(\text{Heat energy conducted into control volume at } x \\
&\quad + \text{heat energy conducted into control volume at } y \\
&\quad + \text{heat energy conducted into control volume at } z \\
&\quad + \text{heat generation per unit control volume} \\
&\quad + \text{internal energy of control volume at } t) \qquad (2.4) \\
&= (\text{Heat energy conducted out of control volume at } x \\
&\quad + \Delta x + \text{heat energy conducted out of control volume at } y + \Delta x \\
&\quad + \text{heat energy conducted out of control volume at } z + \Delta z \\
&\quad + \text{internal energy of control volume at } t + \Delta x)
\end{aligned}
$$

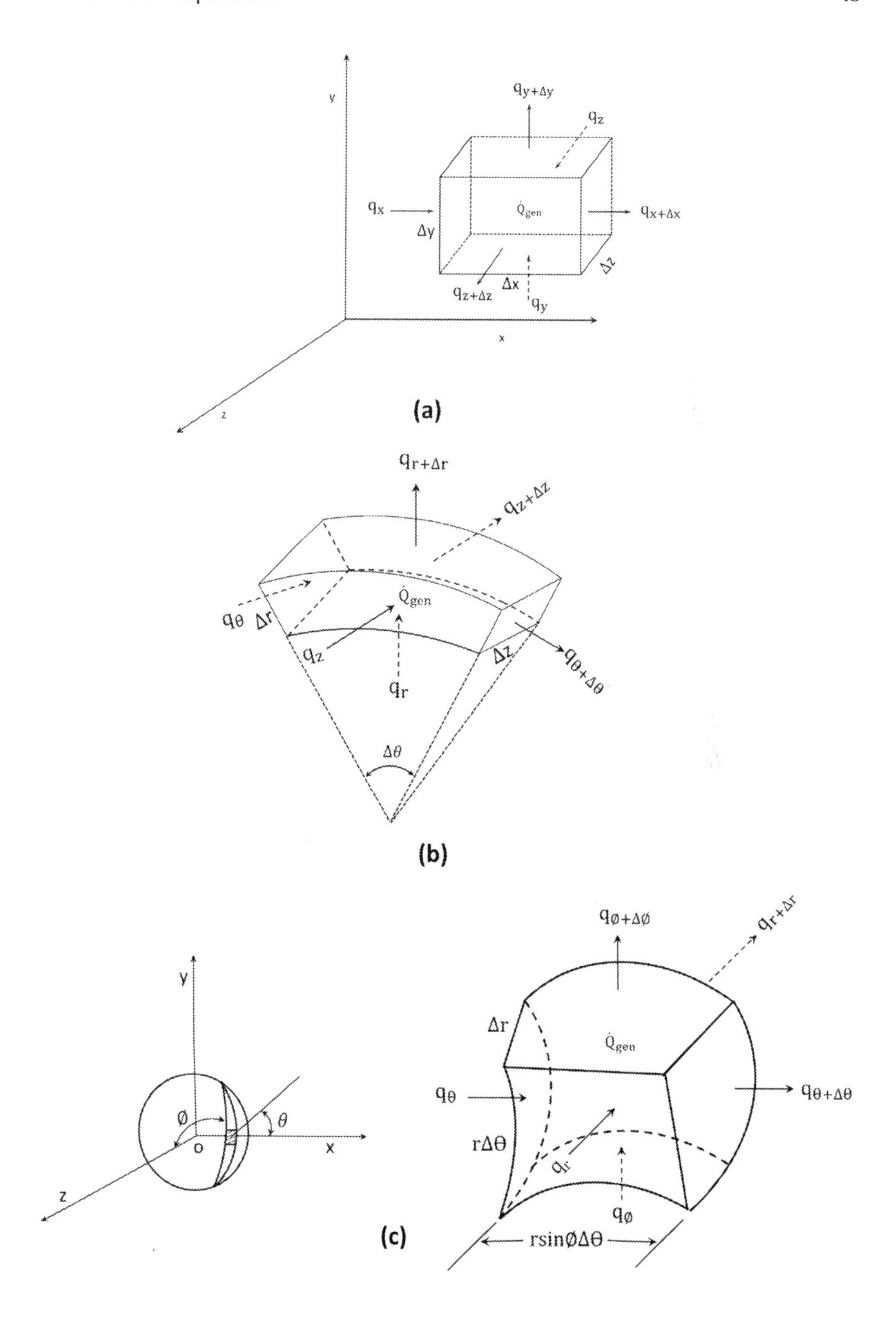

FIGURE 2.1 Conduction heat transfer in (a) Cartesian, (b) cylindrical and (c) spherical coordinates.

2.1.1.1 Generalized Conduction Equation in Cartesian Coordinates

Writing in mathematical form in Cartesian coordinates, we get

$$q_x + q_y + q_z + Q_{gen} + U_t = q_{x+\Delta x} + q_{y+\Delta y} + q_{z+\Delta z} + U_{t+\Delta t} \quad (2.5)$$

Re-arranging the heat quantities and internal energy terms, we get

$$(q_x - q_{x+\Delta x}) + (q_y - q_{y+\Delta y}) + (q_z - q_{z+\Delta z}) + Q_{gen} = U_{t+\Delta t} - U_t \quad (2.6)$$

Application of Taylor's series expansion to the heat quantities and simple differential for the time derivative during time interval Δt gives rise to

$$-\frac{\partial q_x}{\partial x}\Delta x - \frac{\partial q_y}{\partial y}\Delta y - \frac{\partial q_z}{\partial z}\Delta z + Q_{gen} = \frac{\partial U}{\partial t} \quad (2.7)$$

It is assumed that the time interval Δt considered satisfies the limiting value such that the variation of U with respect to time results in a derivative function. In the above expression Q_{gen} is assumed to be constant in W unit. This heat generation quantity can be treated as a function of spatial coordinates or time coordinate also. For example, in the case of nuclear waste disposal canisters, when conduction through the canisters is considered, the heat dissipation by the nuclear element will decay exponentially with time. In the present derivation, heat generation is assumed to be constant. Now the rate of heat conducted in each direction of the solid has to be represented using temperature, T, and this can be done using the Fourier law of heat conduction stated below.

$$q_x = -k_x A\frac{\partial T}{\partial x} = -k_x \Delta y \Delta z\frac{\partial T}{\partial x} \quad (2.8a)$$

$$q_y = -k_y A\frac{\partial T}{\partial y} = -k_y \Delta x \Delta z\frac{\partial T}{\partial y} \quad (2.8b)$$

$$q_z = -k_z A\frac{\partial T}{\partial z} = -k_z \Delta x \Delta y\frac{\partial T}{\partial z} \quad (2.8c)$$

where k_x, k_y *and* k_z are thermal conductivities in x, y and z-directions respectively and A is area of heat transfer.

Substituting Equation (2.8) in Equation (2.7), we get

$$\frac{\partial}{\partial x}(k_x\frac{\partial T}{\partial x})(\Delta x\Delta y\Delta z) + \frac{\partial}{\partial y}(k_y\frac{\partial T}{\partial y})(\Delta x\Delta y\Delta z) + \frac{\partial}{\partial z}(k_z\frac{\partial T}{\partial z})(\Delta x\Delta y\Delta z) + Q_{gen} = \frac{\partial U}{\partial t} \quad (2.9)$$

Now substituting $Q_{gen} = \dot{Q}_{gen}\Delta x\Delta y\Delta z$ and $\frac{\partial U}{\partial t} = \rho C\Delta x\Delta y\Delta z\frac{\partial T}{\partial t}$,

Equation (2.9) can be written as

$$\frac{\partial}{\partial x}(k_x \frac{\partial T}{\partial x})(\Delta x \Delta y \Delta z) + \frac{\partial}{\partial y}(k_y \frac{\partial T}{\partial y})(\Delta x \Delta y \Delta z) + \frac{\partial}{\partial z}(k_z \frac{\partial T}{\partial z})(\Delta x \Delta y \Delta z)$$

$$+ \dot{Q}_{gen}(\Delta x \Delta y \Delta z) = \rho C \frac{\partial T}{\partial t}(\Delta x \Delta y \Delta z)$$

$$\frac{\partial}{\partial x}(k_x \frac{\partial T}{\partial x})(\Delta x \Delta y \Delta z) + \frac{\partial}{\partial y}(k_y \frac{\partial T}{\partial y})(\Delta x \Delta y \Delta z) + \frac{\partial}{\partial z}(k_z \frac{\partial T}{\partial z})(\Delta x \Delta y \Delta z)$$

$$+ \dot{Q}_{gen}(\Delta x \Delta y \Delta z) - \rho C \frac{\partial T}{\partial t}(\Delta x \Delta y \Delta z) = 0$$

$$\left[\frac{\partial}{\partial x}(k_x \frac{\partial T}{\partial x}) + \frac{\partial}{\partial y}(k_y \frac{\partial T}{\partial y}) + \frac{\partial}{\partial z}(k_z \frac{\partial T}{\partial z}) + \dot{Q}_{gen} - \rho C \frac{\partial T}{\partial t}\right](\Delta x \Delta y \Delta z) = 0 \quad (2.10)$$

where ρ and C are density and specific heat capacity of solid.

In the above equation, the product of two terms is equated to zero, and this is arrived at after the application of energy balance over the control volume $\Delta x \Delta y \Delta z$ and differential mathematics. It has to be observed that the control volume $\Delta x \Delta y \Delta z$ cannot be equal to zero because this is the volume over which the energy balance has been applied. Hence, only the other term has to be equal to zero. That is

$$\left[\frac{\partial}{\partial x}(k_x \frac{\partial T}{\partial x}) + \frac{\partial}{\partial y}(k_y \frac{\partial T}{\partial y}) + \frac{\partial}{\partial z}(k_z \frac{\partial T}{\partial z}) + \dot{Q}_{gen} - \rho C \frac{\partial T}{\partial t}\right] = 0$$

$$\rho C \frac{\partial T}{\partial t} = \frac{\partial}{\partial x}(k_x \frac{\partial T}{\partial x}) + \frac{\partial}{\partial y}(k_y \frac{\partial T}{\partial y}) + \frac{\partial}{\partial z}(k_z \frac{\partial T}{\partial z}) + \dot{Q}_{gen} \quad (2.11)$$

Equation (2.11) shows the generalized unsteady heat conduction equation in Cartesian coordinates.

If $k_x = k_y = k_z = k$, called the anisotropic material, then Equation (2.11) is simplified as

$$\rho C \frac{\partial T}{\partial t} = k\left[\frac{\partial^2 T}{\partial x^2} + \frac{\partial^2 T}{\partial y^2} + \frac{\partial^2 T}{\partial z^2}\right] + \dot{Q}_{gen} \quad (2.12)$$

Let us review the energy conservation principle stated in Equation (2.4) by identifying different energy transports in the above equation. The left-hand side term indicates the rate of change of internal energy of the solid with time due to conduction heat transfer across the solid in the presence of heat generation. In the absence of heat generation, the net heat conducted expressed in the right-hand side of Equation (2.12) is equal to the rate of change of temperature of the solid. For example, in the case of cooling of a cubic solid in a convective atmosphere, the temperature of the solid keeps decreasing with time by losing heat to the convective fluid medium. For a constant value of fluid temperature and heat transfer coefficients, the rate of decrease of the solid temperature depends on the value of surface temperature of the cube with time. As heat is conducted from the core of the solid, the surface loses heat to the fluid by convection. Thus Equation (2.12) expresses the way heat energy is transported within the solid, and the energy balance is achieved by equating the heat lost by the solid to the heat conducted within the solid. However, if there is no medium at the boundary either to absorb or supply heat conducted through the solid, there will not be any heat flow across the solid.

By defining heat flux as the heat conducted per unit normal area to the direction of heat transfer, vector notation can be used for multi-dimensional heat flux. For material with constant thermal conductivity, heat flux in vector form is represented as

$$\underline{q} = -k\underline{\nabla} T = -k\left[\hat{i}\frac{\partial T}{\partial x} + \hat{j}\frac{\partial T}{\partial y} + \hat{k}\frac{\partial T}{\partial z}\right] \tag{2.13}$$

where $\hat{i}$, $\hat{j}$ and $\hat{k}$ are unit vectors in x, y and z-directions respectively.

However, for anisotropic material whose thermal conductivity depends on the direction of the coordinate, the heat flux vector can be written as

$$\underline{q} = -\underline{\underline{k}} \cdot \underline{\nabla} T \tag{2.13a}$$

where $\underline{\underline{k}}$ is the second order tensor indicating direction of heat flux and temperature gradient.

Equation (2.13a) can be expanded as

$$\begin{Bmatrix} q_x \\ q_y \\ q_z \end{Bmatrix} = \begin{bmatrix} k_{xx} & k_{xy} & k_{xz} \\ k_{yx} & k_{yy} & k_{yz} \\ k_{zx} & k_{zy} & k_{zz} \end{bmatrix} \begin{Bmatrix} \frac{\partial T}{\partial x} \\ \frac{\partial T}{\partial y} \\ \frac{\partial T}{\partial z} \end{Bmatrix} \tag{2.14}$$

2.1.1.2 Generalized Conduction Equation in Cylindrical Coordinates

The generalized heat conduction equation derived in Cartesian coordinates as expressed by Equation (2.12) can be derived for cylindrical coordinates. For the energy balance described in Equation (2.4), the following can be written for the control volume shown in Figure 2.1(b),

$$q_r + q_\theta + q_z + Q_{gen} + U_t = q_{r+\Delta r} + q_{\theta+\Delta\theta} + q_{z+\Delta z} + U_{t+\Delta t}$$

Re-arranging the above terms, we get

$$(q_r - q_{r+\Delta r}) + (q_\theta - q_{\theta+\Delta\theta}) + (q_z - q_{z+\Delta z}) + Q_{gen} = (U_{t+\Delta t} - U_t)$$

where r, θ and z indicate coordinate directions in cylindrical coordinates.

Application of Taylor's series expansion to the heat quantities and differential operator for the internal energy, U, during small interval of time Δt, we get

$$-\frac{\partial q_r}{\partial r}\Delta r-\frac{\partial q_\theta}{\partial \theta}\Delta\theta-\frac{\partial q_z}{\partial z}\Delta z+Q_{gen}=\frac{\partial U}{\partial t} \tag{2.15}$$

Rate of heat conduction in the coordinate direction can be represented in terms of temperature gradient using the Fourier law of heat conduction as

$$q_r=-kA_r\frac{\partial T}{\partial r}=-k(r\Delta\theta\Delta z)\frac{\partial T}{\partial r} \tag{2.16a}$$

$$q_\theta=-k\frac{A_\theta}{r}\frac{\partial T}{\partial \theta}=-k\frac{\Delta r\Delta z}{r}\frac{\partial T}{\partial \theta} \tag{2.16b}$$

$$q_z=-kA_z\frac{\partial T}{\partial z}=-k(r\Delta\theta\Delta r)\frac{\partial T}{\partial z} \tag{2.16c}$$

Substituting Equation (2.16) in Equation (2.15) we get

$$\frac{\partial}{\partial r}\left(kr\Delta\theta\Delta z\frac{\partial T}{\partial r}\right)\Delta r+\frac{\partial}{\partial \theta}\left(k\frac{\Delta r\Delta z}{r}\frac{\partial T}{\partial \theta}\right)\Delta\theta+\frac{\partial}{\partial z}\left(kr\Delta\theta\Delta r\frac{\partial T}{\partial z}\right)\Delta z+Q_{gen}=\frac{\partial U}{\partial t}$$

After substituting $Q_{gen}=\dot{Q}_{gen}r\Delta\theta\Delta r\Delta z$ and expanding U, we get

$$\frac{\partial}{\partial r}\left(kr\Delta\theta\Delta z\frac{\partial T}{\partial r}\right)dr+\frac{\partial}{\partial \theta}\left(k\frac{\Delta r\Delta z}{r}\frac{\partial T}{\partial \theta}\right)\Delta\theta+\frac{\partial}{\partial z}\left(kr\Delta\theta\Delta r\frac{\partial T}{\partial z}\right)\Delta z$$
$$+\dot{Q}_{gen}r\Delta\theta\Delta r\Delta z=\rho C\frac{\partial T}{\partial t}r\Delta\theta\Delta r\Delta z$$

Rewriting the above equation

$$\frac{\partial}{\partial r}\left(kr\Delta\theta\Delta z\frac{\partial T}{\partial r}\right)dr+\frac{\partial}{\partial \theta}\left(k\frac{\Delta r\Delta z}{r}\frac{\partial T}{\partial \theta}\right)\Delta\theta+\frac{\partial}{\partial z}\left(kr\Delta\theta\Delta r\frac{\partial T}{\partial z}\right)\Delta z$$
$$+\dot{Q}_{gen}r\Delta\theta\Delta r\Delta z-\rho C\frac{\partial T}{\partial t}r\Delta\theta\Delta r\Delta z=0$$

Taking out the control volume $r\Delta\theta\Delta r\Delta z$ as common, we get

$$\left(\frac{1}{r}\frac{\partial}{\partial r}\left(kr\frac{\partial T}{\partial r}\right)+\frac{\partial}{\partial \theta}\left(k\frac{1}{r^2}\frac{\partial T}{\partial \theta}\right)+\frac{\partial}{\partial z}\left(k\frac{\partial T}{\partial z}\right)+\dot{Q}_{gen}-\rho C\frac{\partial T}{\partial t}\right)r\Delta\theta\Delta r\Delta z=0$$

In the above equation the control volume $r\Delta\theta\Delta r\Delta z$ cannot be equal to zero and hence the other term must be equal to zero. Thus we get the final equation as

$$\rho C\frac{\partial T}{\partial t}=\left(\frac{1}{r}\frac{\partial}{\partial r}\left(kr\frac{\partial T}{\partial r}\right)+\frac{\partial}{\partial\theta}\left(k\frac{1}{r^2}\frac{\partial T}{\partial\theta}\right)+\frac{\partial}{\partial z}\left(k\frac{\partial T}{\partial z}\right)+\dot{Q}_{gen}\right) \tag{2.17}$$

Equation (2.17) represents the generalized unsteady state heat conduction equation in cylindrical coordinates. If the thermal conductivity of the solid remains constant in all the coordinate directions, then the above equation can be written as

$$\rho C\frac{\partial T}{\partial t}=k\left(\frac{1}{r}\frac{\partial}{\partial r}\left(r\frac{\partial T}{\partial r}\right)+\frac{\partial}{\partial\theta}\left(\frac{1}{r^2}\frac{\partial T}{\partial\theta}\right)+\frac{\partial}{\partial z}\left(\frac{\partial T}{\partial z}\right)+\dot{Q}_{gen}\right) \tag{2.18}$$

In many engineering applications, heat conduction through the radial direction finds more significance in the analysis compared to other directions. Hence the above equation can be written only in radial direction as

$$\rho C\frac{\partial T}{\partial t}=k\frac{1}{r}\frac{\partial}{\partial r}\left(r\frac{\partial T}{\partial r}\right)+\dot{Q}_{gen} \tag{2.19}$$

Modeling of problems related to chemical reactors, nuclear reactors etc. gives rise to equations as shown in the above equation.

2.1.1.3 Generalized Conduction Equation in Spherical Coordinates

Conduction heat transfer in spherical geometries becomes significant in storage of refrigerants and other liquids. Using the energy balance expressed by Equation (2.4) for the control volume $\Delta r r\Delta\phi r\sin\phi\Delta\theta$ shown in Figure 2.1(c), the final equation can be written as

$$\begin{aligned}\rho C\frac{\partial T}{\partial t}&=\frac{1}{r^2}\frac{\partial}{\partial r}\left(kr^2\frac{\partial T}{\partial r}\right)+\frac{1}{r^2\sin\theta}\frac{\partial}{\partial\theta}\left(k\sin\theta\frac{\partial T}{\partial\theta}\right)\\&+\frac{1}{r^2\sin^2\theta}\frac{\partial}{\partial\phi}\left(k\frac{\partial T}{\partial\phi}\right)+\dot{Q}_{gen}\end{aligned} \tag{2.20}$$

2.1.1.4 Initial and Boundary Conditions

Equation (2.12), (2.17) and (2.20) can be considered as mathematical models that represent the variation of temperature T in time and spatial domains in different coordinates. These equations are second order equations in space with first order derivatives in time, and hence the required initial and boundary conditions have to be specified. The time variation of temperature within the three-dimensional domain has to be obtained for a total period of time with some starting time, which is called the initial time. Spatial conditions have to be specified depending upon the order of the differential equation. For second order partial differential equations, two spatial boundary conditions with respect to the unknown variable $T(t,x,y,z)$ have to be enforced for each coordinate direction. In the x-direction, two boundary conditions, that is at $x=0$ and $x=L$ and similar values have to be provided for the other two directions, y and z. Boundary values of the variable may be in terms of its absolute value or its derivative. The solution of temperature is obtained for the constrained values of temperature or its derivatives as specified on the boundary of the domain. The type of solution for temperature will vary if these boundary values are changed. The following are the most common types of boundary conditions used in heat diffusion problems.

2.1.1.5 Initial Condition

At the start of the heat conduction phenomenon, the value of temperature throughout the computational domain must be specified and is expressed as

@ $t=0$, $T(0, x, y, z)$ within the computational domain $(L \times W \times H)$. Different types of boundary conditions can be explained using one-dimensional steady state heat conduction in a solid considering the boundary condition at $x=0$ (Figure 2.2).

2.1.1.6 Boundary Conditions

(i) ***Dirichlet boundary condition***

This type of boundary condition is also called boundary condition of *first kind* and this corresponds to a situation for which the temperature at the boundary is known and is assumed that boundary is maintained at a constant fixed temperature. For example, the surface over which boiling or condensation takes place can be treated as a boundary condition of first kind. For one-dimensional heat conduction along the x-direction, the boundary conditions at two points, $x=0$ and $x=L$, have to be specified. For steady heat conduction, the Dirichlet boundary condition is specified as

$$@x = 0, T(0) = T_w \tag{2.21a}$$

and this indicates the value of T_w will remain constant at this location throughout the computation of temperature during the solution process.

(ii) ***Neumann boundary condition***

Instead of the value of the unknown variable being computed, if its gradient is equated to a constant, then such a boundary condition is called Neumann boundary condition or boundary condition of *second type*. It is expressed mathematically as

$$-k\frac{\partial T}{\partial x}\bigg|_{x=0} = q''_w \tag{2.21b}$$

The heat flux at the boundary wall (q''_w) is known and remains constant and may be realized by the bonding of a thin electric heating element to the surface of the boundary wall. When the governing equation is solved for the solution of

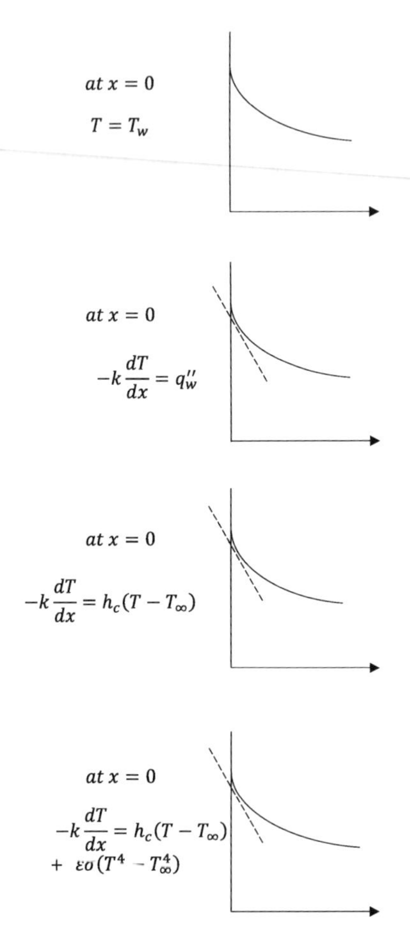

FIGURE 2.2 Different types of boundary conditions for conduction heat transfer.

temperature, the resulting temperature gradient from the temperature solution should satisfy the above condition. A special type of Neumann boundary condition is the insulation boundary condition where the temperature gradient becomes equal to zero as

$$-k\frac{\partial T}{\partial x}\Big|_{x=0} = 0 \tag{2.21c}$$

The inflexion point of temperature variation at maxima or minima will also result in such a situation. At the region of peak temperature variation, the temperature gradient is zero, hence heat conduction at that point is zero. For the purpose of simplicity, in the analysis of two- or three-dimensional regular solids, a line of symmetry is considered to minimize the computation. Along the line of symmetry, the above condition is satisfied.

(iii) ***Robin boundary condition***

This is also called a boundary condition of *third kind*, wherein the heat conducted at the boundary of the solid is dissipated in the form of convective heat transfer and can be expressed as

$$-k\frac{\partial T}{\partial x}\Big|_{x=0} = h_c\,(T_w - T_\infty) \tag{2.21d}$$

where h_c is convective heat transfer coefficient, T_w is wall temperature and T_∞ is free stream fluid temperature.

This is the most common type of situation observed in many applications involving cooling or heating of solid surfaces. Care must be taken regarding the direction of heat transfer from or to the solid surface for the negative sign that appears in the conduction term on the left side of the above equation. In this type of boundary condition, the temperature gradient on the left-hand side is equated to the unknown temperature on the right-hand side.

(iv) ***Combined heat transfer***

Convective boundary condition is commonly encountered in many real situations whenever fluid flow is involved. When the convecting surface is maintained at a high temperature above 300°C, the radiation heat transfer from the surface cannot be ignored. The conduction heat transfer emanating to or from the surface is equated to both convective and radiative heat transfer as

$$-k\frac{\partial T}{\partial x}\Big|_{x=0} = h_c\,(T_w - T_\infty) + \varepsilon\sigma(T_w^{\,4} - T_\infty^{\,4}) \tag{2.21e}$$

where ε is emissivity of surface and σ is Stefan-Boltzmann constant

If boundary conditions of the type discussed in cases (ii) to (iv) have to be satisfied, then the gradients of temperature obtained as a result of solution of the governing equation has to ensure that the solution satisfies the imposed boundary conditions. In other words, while forming the final solution matrix as part of solution of the governing equations, the constant terms appearing on the right-hand side of the above boundary conditions will contribute to the load vector of the matrix. Equation (2.21e) deliberates the condition of boundary to be satisfied with the fourth order temperature term at the solid boundary. The presence of the higher order term of the unknown temperature T results in a non-linear equation, and this poses a challenge while evolving the computational algorithm.

2.1.2 Energy Transport in Steady State Condition

It was shown that the generalized unsteady state heat conduction equation is expressed by Equation (2.12) with initial and boundary conditions given by Equation (2.21). There may be other types of boundary conditions in real problems during modeling that need to be taken into account to develop the solution procedure accordingly. Equation (2.12) can be reduced to many situations with certain modifications. For two-dimensional solids, conduction through x and y coordinates are considered to be significant. Let us consider one-dimensional conduction with heat generation in a plane wall, by simplifying Equation (2.12) to give

$$\rho C \frac{\partial T}{\partial t} = k \frac{\partial^2 T}{\partial x^2} + \dot{Q}_{gen} \tag{2.22}$$

The above equation can be solved with specified boundary conditions for temperature, $T\ (t,\ x)$. All real problems can be treated as transient because there is no quantity or property that can be maintained at a constant value without certain acceptable oscillation. Consider systems involving heating or cooling, evaporation or condensation: all of them produce temperature variation with time. It all depends on the specified temporal limits over which the system can be called steady state, otherwise it is very difficult, if not impossible, to define a system which is maintained fully at a steady state condition. However, in many heat transfer applications, transients are very important to understand, especially while controlling the temperature variation for certain processes. When a thermometer or thermocouple sensor is used to measure the temperature of a fluid at a constant temperature, the thermal response characteristic of the measuring sensor is explored to understand the nature of variation and time taken to reach the temperature of the fluid. Quenching of metal castings in manufacturing industries requires the knowledge of the temperature history of the solids being cooled. The nature of variation of temperature history with time plays a major role in determining the mechanical properties of the products. Electronic cooling systems need to be designed to dissipate heat from the devices within a fixed interval of time. The temperature of coolant used in a solar collector varies with time as the incident solar radiation varies with time in a day. Transient heat conduction equations can be solved either using analytical techniques or numerical methods.

2.1.2.1 Steady State Heat Conduction in Plane Wall

In the absence of a time derivative term, Equation (2.12) can be rewritten for steady state heat conduction as

$$k\left[\frac{\partial^2 T}{\partial x^2} + \frac{\partial^2 T}{\partial y^2} + \frac{\partial^2 T}{\partial z^2}\right] + \dot{Q}_{gen} = 0$$

$$\frac{\partial^2 T}{\partial x^2} + \frac{\partial^2 T}{\partial y^2} + \frac{\partial^2 T}{\partial z^2} = -\frac{\dot{Q}_{gen}}{k} \tag{2.23}$$

The above equation explains the energy balance that whatever heat generated within the solid is conducted across the solid. Without heat generation, Equation (2.23) reduces to

$$\frac{\partial^2 T}{\partial x^2} + \frac{\partial^2 T}{\partial y^2} + \frac{\partial^2 T}{\partial z^2} = 0 \qquad (2.24)$$

This equation merely indicates that whatever heat enters from all three directions into the solid, it comes out of the solid by conduction through the control volume. The solution of the above equation gives the temperature variation across the solid under steady state conditions. After switching on an air-conditioner in a room, the temperature is maintained as constant at the set temperature, then the room air is said to have reached a steady state condition. However, with a closer look at the heat transfer within the evaporator, condenser and compressor of the air-conditioner, one can notice the transient variation of temperature and pressure of the working fluid. This shows that many heat transfer systems are designed to operate at steady state conditions. Equation (2.24) is a Laplace equation for steady heat conduction, and similar equations can be derived for fluid flow systems. Steady state conduction equations can be solved using analytical techniques for the specified boundary conditions expressed by Equation (2.21). It is interesting to analyze the solution of the one-dimensional steady heat conduction equation. Let us consider the governing equation for the one-dimensional conduction

$$\frac{d^2 T}{dx^2} = 0 \qquad (2.25)$$

which implies that the rate of change of heat flux entering into the solid is zero because the derivative of a constant quantity of heat flux q_x is zero. Whatever heat enters the solid at x, it comes out of the solid at $x+dx$. Integrating twice Equation (2.25), the solution for T is obtained as

$$T = C_1 x + C_2 \qquad (2.26)$$

In order to determine the constants C_1 and C_2, two boundary conditions are required at $x=0$ and $x=L$. If we assume constant temperatures T_1 and T_2 at $x=0$ and $x=L$, respectively, then the solution for Equation (2.21) can be obtained as

$$T = \left(\frac{T_2 - T_1}{L} \right) x + T_1 \qquad (2.27)$$

which shows that the temperature variation along the length of the solid is linear and the temperature gradient is constant throughout the solid. From the Fourier law of heat conduction, the heat flux conducted into the solid under the steady state condition is directly proportional to the temperature gradient; this means the heat flux remains constant within the solid at any distance between $x=0$ and $x=L$ since the temperature gradient is constant. In modeling, it is important to understand the nature of variation of temperature; this is the solution of the governing equation. Now let us include the heat generation term in Equation (2.25) to get the governing equation as

$$\frac{d^2T}{dx^2}+\frac{\dot{Q}_{gen}}{k}=0 \tag{2.28}$$

Integrating twice the above equation gives the following solution

$$T=-\frac{\dot{Q}_{gen}x^2}{2k}+C_1x+C_2 \tag{2.29}$$

In the absence of the heat generation term, the solution becomes the same as that of Equation (2.26). The heat generation term, which is constant with respect to T and x, now has made the temperature variation non-linear. Here again, two boundary conditions are required to determine the constants C_1 and C_2. Let us assume the following boundary conditions:

@ $x=0$, $T=T_1$ and $x=L$, $T=T_2$, then the final solution of Equation (2.29) can be written as

$$T=-\frac{\dot{Q}_{gen}x^2}{2k}+\left\{\frac{\dot{Q}_{gen}L}{2k}+\frac{(T_2-T_1)}{L}\right\}x+T_1 \tag{2.30}$$

The above equation for the solution of T is of the form $T=a_1x^2+a_2x+a_3$, a second order polynomial in x, the spatial coordinate. Comparing the solution given by Equation (2.27) for steady state conduction without heat generation and the above one with heat generation, it is clear that the heat generation term is responsible for the non-linear temperature variation. Equation (2.30) is reduced to Equation (2.27) by making the heat generation equal to zero. The above exercise can be continued for the convective boundary condition, and symmetric and non-symmetric temperature boundary conditions to get different solutions. The temperature gradient in the presence of heat generation produces interesting situations in the case of nuclear reactors. As has been already shown in Equation (2.21(b)), by making the temperature gradient equal to zero, an insulating condition, the conduction heat transfer can be prevented. By suitably adjusting the value of heat generation within the solid, heat loss at one end of the solid can be achieved by making the temperature gradient at that end equal to zero.

2.1.2.2 Steady State Heat Conduction in Cylinder

The study of steady state heat conduction in the radial direction in pipes and tubes finds wide application in the design of heat exchangers. The following heat conduction equation can be reduced to steady state heat conduction in the radial direction without heat generation as

$$\rho C\frac{\partial T}{\partial t}=k\left(\frac{1}{r}\frac{\partial}{\partial r}\left(r\frac{\partial T}{\partial r}\right)+\frac{\partial}{\partial\theta}\left(\frac{1}{r^2}\frac{\partial T}{\partial\theta}\right)+\frac{\partial}{\partial z}\left(\frac{\partial T}{\partial z}\right)+\dot{Q}_{gen}\right)$$

$$\frac{1}{r}\frac{\partial}{\partial r}\left(r\frac{\partial T}{\partial r}\right)=0 \tag{2.31}$$

It is interesting to observe that unlike the case of conduction through a plane wall in one-dimension, the area of heat transfer varies at different radii in cylindrical geometry. For steady state heat conduction into a hollow cylinder of outer radius r_2 towards the inner radius r_1, the heat flux keeps increasing due to a decrease in the area of heat transfer. That means for the constant heat entering the solid at r_2, the temperature gradient does not remain constant as was observed in a plane wall case. Let us integrate Equation (2.31) to obtain the temperature gradient as

$$\frac{dT}{dr} = \frac{C_1}{r}$$

As the radius of the hollow cylinder from r_1, the inner radius, increases, the temperature gradient decreases, and vice versa, whereas the temperature gradient along a plane wall remains constant. Assuming Dirichlet boundary conditions, T_1 and T_2 at radii, r_1 and r_2, respectively, further integration of the above equation gives the following solution.

$$T = T_1 - (T_1 - T_2)\frac{\ln(r / r_1)}{\ln(r_1 / r_2)} \tag{2.32}$$

The temperature shows logarithmic variation between radius r_1 and r_2. The logarithmic variation of temperature in cylindrical solids is obtained due to the continuous variation of the normal area of heat conduction in the radial direction. Hence, it has to be realized that even with constant thermal conductivity of solids, temperature variation may be non-linear due to the geometric shape of the solid. The governing equation for cylindrical solids with heat generation can be obtained by simplifying Equation (2.18) for steady state radial heat conduction in a solid cylinder as

$$k\frac{1}{r}\frac{\partial}{\partial r}\left(r\frac{\partial T}{\partial r}\right) + \dot{Q}_{gen} = 0 \tag{2.33}$$

After integration, the solution for temperature variation is obtained as

$$T = -\frac{\dot{Q}_{gen} r^2}{4k} + C_1 \ln r + C_2$$

The heat generation is assumed to take place at the center of the cylinder at r=0. Hence the following boundary conditions are valid to determine the constants C_1 and C_2 in the above equation.

@ $r=0$, $\frac{dT}{dr} = 0$ and @ $r=R$, $T=T_s$

The solution for Equation (2.33) is obtained as

$$T = T_s + \frac{\dot{Q}_{gen}}{4k}(R^2 - r^2) \tag{2.34}$$

The above expression for T indicates the non-linear variation of temperature in the presence of heat generation. The presence of a heat generation term, which is constant, has resulted in second order functions in r, as was noticed for a plane wall.

2.1.2.3 Steady State Heat Conduction in Sphere

Heat conduction through spherical geometries, both hollow and solid, find applications in the storage of coolants and high pressure fluids. Considering only radial steady heat conduction through a spherical solid with constant thermal conductivity, Equation (2.20) can be modified as

$$\frac{1}{r^2}\frac{\partial}{\partial r}\left(r^2\frac{\partial T}{\partial r}\right)=0 \tag{2.35}$$

Integrating once with respect to r results in

$$\frac{dT}{dr}=\frac{C_1}{r^2}$$

which shows that the temperature gradient varies inversely with the square of r. As was noticed in the cylindrical geometry, the normal area of conduction heat transfer in the r direction varies in spherical geometry. Solving Equation (2.35) with the following boundary conditions, the expression for temperature can be obtained.

@ $r=R_1$, $T=T_1$ and @ $r=R_2$, $T=T_2$

$$T=T_1+\left(\frac{T_2-T_1}{R_2-R_1}\right)\frac{R_2}{r}(r-R_1) \tag{2.36}$$

The effect of geometrical variation of the area of heat transfer influences the temperature variation in the radial direction. When steady heat conduction with heat generation is considered, Equation (2.20) takes the following form.

$$\frac{1}{r^2}\frac{\partial}{\partial r}\left(kr^2\frac{\partial T}{\partial r}\right)+\dot{Q}_{gen}=0 \tag{2.37}$$

Integrating twice the above expression, the solution for temperature T is obtained as

$$T=-\frac{\dot{Q}_{gen}r^2}{6k}-\frac{C_1}{r}+C_2$$

Assuming the following boundary conditions, the solution for T can be obtained.

@ $r=0$, $\frac{dT}{dr}=0$ and @ $r=R$, $T=T_s$

$$T=T_s+\frac{\dot{Q}_{gen}}{6k}(R^2-r^2) \tag{2.38}$$

The solution resembles that of a cylinder except the constant with thermal conductivity. The solutions for temperature for a steady state conduction through plane wall, cylinder and sphere differ from each other due to the geometric variation of the solid. It is essential to understand how the temperature gradient within these solids varies as the heat conducted is directly proportional to the temperature gradient. In modeling heat transfer through solids under various boundary conditions, the reasons for the variation of temperature gradient has to be perceived so that interpretation of the results becomes meaningful.

2.2 FLUID MEDIUM

The study of heat and mass transfer through a fluid medium is commonly experienced in the design of heat exchangers, cooling towers, nuclear reactors etc. Convective heat and mass transfer are the most significant transport processes in these problems. As understood in basic heat transfer courses, convection is a phenomenon that takes place due to the *bulk movement of fluid*, transporting momentum, heat and species from one region to another in a given system. Convection involves contact of fluid with a solid surface, of course with certain exceptions. In flow problems, one has to distinguish between *fluid* and *flow field*. Fluid is well known from fluid mechanics; it may be a liquid or gas. Flow field is the resulting fluid domain obtained under the influence of various forces acting on the fluid in the system. These two terminologies should not be mixed up during mathematical modeling of flow problems. There are many classifications of convective transport of fluid momentum, heat and species, and these are organized based on experience in many fields of applications. The following are some of the divisions under which convective heat and mass transfer are studied.

(i) *Internal flow* – in this, fluid flow takes place in enclosures such as cavities, channels, ducts, pipes etc. In such flows, generally there is no free stream flow and the flow is well bounded with certain boundaries.
(ii) *External flow* – flow over flat plates, airplane wings, wind mill rotors etc. are studied as external flow where at least on one of the boundaries, there is a free stream of the fluid without having contact with any solid boundary.
(iii) *Free flow* – flow through spray cans or expanding nozzles in a fluid medium are studied under free flow – this is unbounded flow. The whole fluid domain is unconstrained by any solid boundary.
(iv) *Incompressible flow* – when density variation is not significant in the flow field, it is called incompressible flow. The fluid may be compressible, but still it may produce an incompressible flow field. For example, air can be treated as incompressible in applications such as air cooling devices, cooling towers, heat exchangers etc. where the change in density due to temperature and pressure variations is not significant.
(v) *Compressible* – variation of density due to pressure and temperature fields cannot be ignored in the analysis. In applications such as flow through

nozzles at a velocity equal to or higher than Mach number, the density variation is significant and it gives rise to shock waves in the flow field.

(vi) *One-dimensional flow* – the variations of field variables such as velocity, pressure and temperature are significant only in one direction. Flow through a pipe is the best example of one-dimensional flow. Though a pipe is a three-dimensional solid, the fluid flowing through a long pipe will contribute to a flow field with changes in the field variables only in the axial direction.

(vii) *Two-dimensional flow* – flow over aero foils, study of streamlined surfaces etc. are considered as two-dimensional. Some of the three-dimensional flow fields are analyzed as two-dimensional flow under certain assumptions.

(viii) *Three-dimensional flow* – analysis of space craft, turbomachine elements, wind load on building structures etc. involves three-dimensional modeling of flow field. Changes in the flow domain attributes in all three coordinate directions contribute to the performance of the system.

The different categories of fluid flow problems described above differ in some characteristics from others with respect to the flow pattern in the flow domain. Mathematical modeling of flow problems may focus on the calculation of fluid friction, pressure drop, heat transfer during heating or cooling and species transport, either all of them together or as a part thereof. Flow distribution within the computational domain is the most deciding factor on velocity gradients, temperature and species gradients at the boundary. It is essential to understand the factors and conditions that influence the formation of a particular flow pattern in a given geometry so that the required heat transfer or mass transfer is accomplished. Modeling and simulation become very handy tools to simulate the behavior of different types of fluids in a given geometry and compute the final parameters of interest. Convective heat transfer comprises the mechanism of transport of the fluid parcel due to inertial or other forces by carrying heat from the hot region towards the cold region. As convective heat transfer is a surface phenomenon, the relation between the solid surface and the fluid medium needs to be understood. In most of the flow analyses, a no-slip boundary condition is assumed at the wall of the surface over which the convection phenomenon is investigated. Of course, there are situations wherein there may be relative velocity between the solid surface and the fluid medium. In this textbook, only the problems with no-slip conditions will be considered for modeling purposes. When the no-slip boundary condition is satisfied at the solid wall, then it is expected that a thin layer of fluid is formed near the surface with zero velocity as that of the surface. By the very nature of fluid medium, which offers resistance for movement, a velocity gradient is set up between the free stream and the solid surface. The velocity gradient and viscosity of the fluid produce shear stress at the wall, also called fluid friction. When the wall is at a higher temperature compared to the temperature of free stream fluid, the existence of a velocity gradient of the fluid at the boundary results in a temperature gradient, which causes conduction heat transfer to take place within the thin fluid layer. Similar behavior by the fluid is exhibited for concentration difference also, thus setting up a concentration gradient at the wall, leading to species diffusion within the thin fluid layer. Hence, in convective flow analysis, computation of three gradients with respect to velocity, temperature and concentration is important to calculate fluid friction, heat transfer and mass transfer. The following definitions respectively highlight these quantities,

$$\tau_w = \mu \frac{du}{dy}\bigg|_{wall} \tag{2.39a}$$

$$q_w = -k_f \frac{dT}{dy}\bigg|_{wall} = h_c(T_w - T_\infty) \tag{2.39b}$$

$$\dot{m}_w = -D\frac{dC}{dy}\bigg|_{wall} = h_m(C_w - C_\infty) \tag{2.39c}$$

where τ_w is wall shear stress, μ is fluid viscosity, k_f is thermal conductivity, D is species diffusion coefficient, C is species concentration and h_m is convective mass transfer coefficient. Suffix w and ∞ indicate wall and free stream respectively.

Assuming the y-direction normal to the wall with x-direction along the flow direction and length of the wall, Equation (2.39a) is the definition for wall shear stress by the fluid flow over the surface, which causes fluid friction. As the conduction heat transfer and mass diffusion at the surface are responsible for the dissipation of heat and species from high temperature and high concentration at the wall to low temperature and low concentration free stream fluid, Equations (2.39b) and (2.39c) define convective heat transfer and mass transfer coefficients, h_c and h_m respectively. The complex mechanism of flow and heat transfer that involves the transport of heat and mass from the hot and high concentration wall to the low temperature and low concentration free stream fluid is accommodated in h_c and h_m such that a simple expression of the type shown on the right-hand side of Equations (2.39b) and (2.39c) is obtained. In order to take account of different sizes and geometries of flow systems involving heat and mass transfer, h_c and h_m are represented using non-dimensional numbers, Nusselt number, *Nu*, and Sherwood number, *Sh*, respectively. Accordingly, the temperature and concentration gradients in Equations (2.39b) and (2.39c) are also defined as non-dimensional quantities. It is well understood that for computing fluid friction over solid surfaces, convective heat and mass transfer, gradients of velocities, temperature and concentration have to be determined at the wall. These gradients are governed by the pattern of variation of velocity, pressure, temperature and concentration within the flow domain. Hence, computation of these variables becomes the first step in any problem involving fluid flow, heat and mass transfer. Fluid flow is governed by mass conservation and momentum conservation principles whereas heat and mass transfer has to satisfy energy and species conservation principles.

2.2.1 Mass Conservation

Mass conservation is the first principle that has to be satisfied in any flow problem followed by momentum conservation. As the distribution of velocity and pressure is required for analysis of flow problems, these conservation principles have to be stated in the form of differential equations, to obtain flow variables in the continuum.

Control volume is an infinitesimally small volume considered in the flow domain for the purpose of derivation of the differential equations from the conservation principles. When mass conservation is analyzed, it is said that only mass is conserved, not volume. At this juncture, it is worth relating the principles behind two types of coordinate systems, known as Lagrangian and Euler types of systems. The Lagrangian coordinate system focuses on a control mass and follows the same in the flow domain, whereas the Euler system of representation focuses on the flow of fluid passing through a fixed volume defined with respect to stationary spatial coordinates. Hence, the Lagrangian coordinate is a moving coordinate system tracing the same fluid parcel whereas the Euler coordinate is a fixed coordinate system observing the fluid passing over the fixed control volume. Referring to Figure 2.3, it is required to find out the status of change of mass from time 't' to time '$t+\Delta t$'.

Let us write the change of attribute ϕ per unit mass by treating it as contained within the control volume.

@ time t, the amount of ϕ is = amount of ϕ in control volume at time 't' + mass at inlet 'i'$\times\phi_i$

$$\phi\big|_t = (\phi)_{CV@t} + (\rho_i L_i A_i)\phi_i \tag{2.40a}$$

@ time $t+\Delta t$, the amount of ϕ is = amount of ϕ in control volume at time '$t+\Delta$' + mass at exit 'e'$\times\phi_e$

$$\phi\big|_{t+\Delta t} = (\phi)_{CV@t+\Delta t} + (\rho_e L_e A_e)\phi_e \tag{2.40b}$$

Writing Equations (2.40a) and (2.40b) in differential form during an interval of time 'Δt', we get

$$\left(\frac{d\phi}{dt}\right)_{\text{controlmass}} = \frac{\left[\phi_{CV}\big|_{t+\Delta t} - \phi_{CV}\big|_t\right]}{\Delta t} + \frac{(\rho_e L_e A_e \phi_e - \rho_i L_i A_i \phi_i)}{\Delta t} \tag{2.41}$$

Let L be the length of the duct travelled by the fluid with velocity V during time interval 'Δt', then $L_i = V_i \Delta t$ and $L_e = V_e \Delta t$. This assumption is valid because due to the interaction of the fluid within the control volume, the velocity at exit may change from the inlet velocity. After substituting these values, Equation (2.41) takes the form of the following:

$$\left(\frac{d\phi}{dt}\right)_{\text{controlmass}} = \frac{\left[\phi_{CV}\big|_{t+\Delta t} - \phi_{CV}\big|_t\right]}{\Delta t} + (\rho_e V_e A_e \phi_e - \rho_i V_i A_i \phi_i)_{CV} \tag{2.42}$$

$$\left(\frac{d\phi}{dt}\right)_{\text{controlmass}} = \left(\frac{d\phi}{dt}\right)_{CV} + (\text{efflux of fluid attribute through exit} - \text{influx of fluid attribute through inlet}) \tag{2.43}$$

The rate of change of attribute ϕ within the control mass is rewritten in terms of the rate of change of that variable within the control volume and the net efflux of the

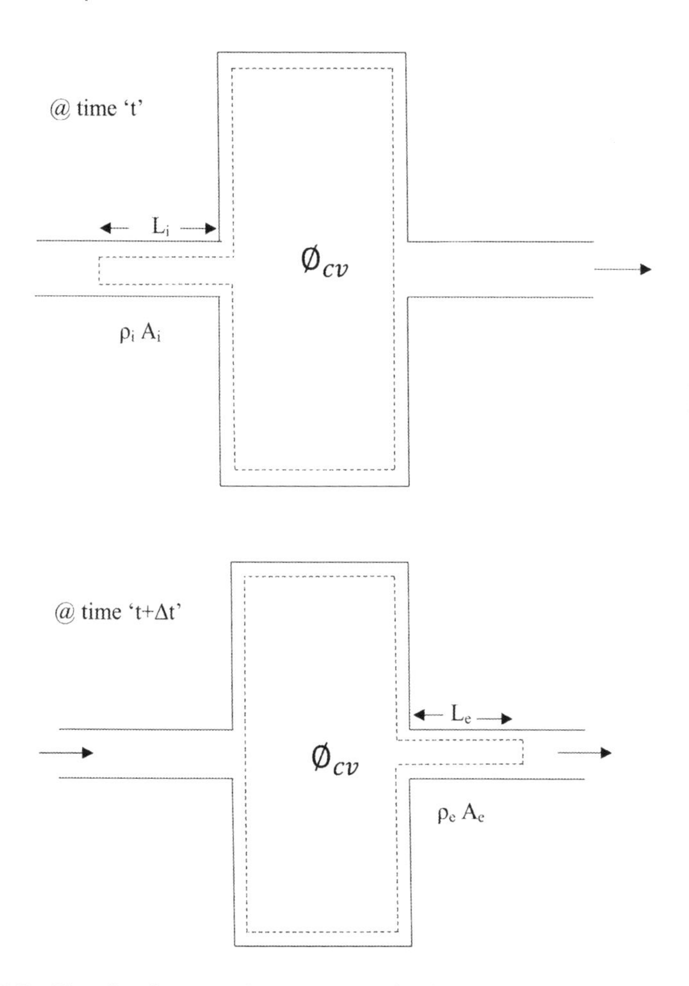

FIGURE 2.3 Transforming control mass to control volume.

variable. Now, this principle can be applied to the flow of fluid through a control volume (Figure 2.4) to understand mass conservation principles. Referring to the above figure, it can be written as

$$\begin{aligned} &\text{Rate of change of mass of control mass} = 0 \\ &\quad = \text{Rate of change of mass in CV} \\ &\quad\quad + \left(\text{efflux of mass from CV} - \text{influx of mass into CV}\right) \end{aligned} \tag{2.44}$$

The left-hand side of the above equation is zero because mass is conserved. Now, the two terms on the right-hand side have to be evaluated. The first term is as follows:

$$\text{Rate of change of mass in CV} = \frac{\partial\left(\rho\Delta x\Delta y\Delta z\right)}{\partial t} = \frac{\partial\rho}{\partial t}\left(\Delta x\Delta y\Delta z\right) \tag{2.45a}$$

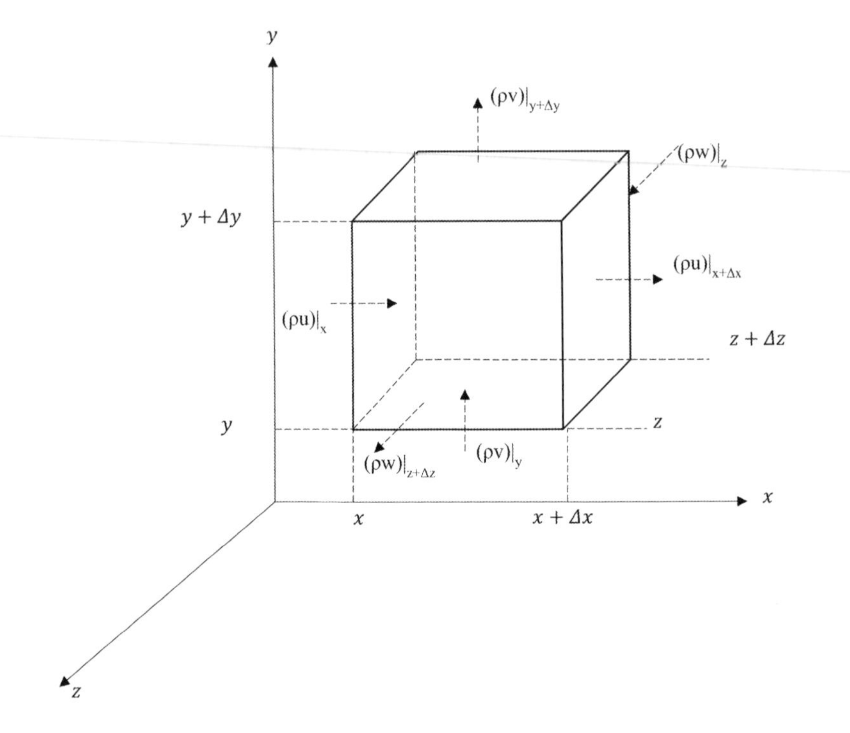

FIGURE 2.4 Mass conservation during fluid flow through a control volume.

The control volume does not change with time, only the density of the fluid changes with time. The second term of Equation (2.44) is as follows:

$$\left(\text{efflux of mass from CV} - \text{influx of mass into CV}\right)$$

The influx and efflux of mass in all the three directions, x, y and z, have to be considered.

$$\text{Efflux of mass in the } x\text{-direction} = \left(\rho u \Delta y \Delta z\right)\Big|_{x+\Delta x}$$

$$\text{Influx of mass in the } x\text{-direction} = \left(\rho u \Delta y \Delta z\right)\Big|_{x}$$

$$\text{Net efflux of mass in the } x\text{-direction} = \left(\rho u \Delta y \Delta z\right)\Big|_{x+\Delta x} - \left(\rho u \Delta y \Delta z\right)\Big|_{x}$$

$$= \left\{\left(\rho u\right)\Big|_{x+\Delta x} - \left(\rho u\right)\Big|_{x}\right\} \Delta y \Delta z$$

where u is velocity in x-direction.

From Taylor's series expansion, the above expression can be written as

$$= \frac{\partial\left(\rho u\right)}{\partial x} \Delta x \Delta y \Delta z \tag{2.45b}$$

$$\text{Similarly the net efflux of mass in the } y\text{-direction} = \frac{\partial(\rho v)}{\partial y}\Delta x \Delta y \Delta z \quad (2.45c)$$

$$\text{Net efflux of mass in the } z\text{-direction} = \frac{\partial(\rho w)}{\partial z}\Delta x \Delta y \Delta z \quad (2.45d)$$

where v and w are velocities in y- and z-directions respectively.

Now the mass conservation principle expressed by Equation (2.44) can be written in terms of Equations (2.45a) to (2.45d) as

$$\left\{\frac{\partial \rho}{\partial t} + \frac{\partial(\rho u)}{\partial x} + \frac{\partial(\rho v)}{\partial y} + \frac{\partial(\rho w)}{\partial z}\right\}\Delta x \Delta y \Delta z = 0$$

Finally, the mass conservation equation for compressible or incompressible fluid can be written as

$$\left\{\frac{\partial \rho}{\partial t} + \frac{\partial(\rho u)}{\partial x} + \frac{\partial(\rho v)}{\partial y} + \frac{\partial(\rho w)}{\partial z}\right\} = 0 \quad (2.46)$$

Equation (2.46) is called a continuity equation in a given fluid domain and is valid at every point which satisfies the continuum principle. It can be expressed in vector form as

$\frac{\partial \rho}{\partial t} + \nabla \cdot (\rho \underline{V}) = 0$ in which $(\rho \underline{V})$ can be expanded as

$$\rho \underline{V} = \hat{i}\rho u + \hat{j}\rho v + \hat{k}\rho w$$

The continuity equation given by Equation (2.46) can be rewritten in material derivative form and also can be simplified for incompressible fluid flow conditions.

2.2.1.1 Material Derivative Form

Equation (2.46) can be expanded in ρ and V as

$$\frac{\partial \rho}{\partial t} + \underbrace{\rho\frac{\partial u}{\partial x} + u\frac{\partial \rho}{\partial x}}_{\frac{\partial(\rho u)}{\partial x}} + \underbrace{\rho\frac{\partial v}{\partial y} + v\frac{\partial \rho}{\partial y}}_{\frac{\partial(\rho v)}{\partial y}} + \underbrace{\rho\frac{\partial w}{\partial z} + w\frac{\partial \rho}{\partial z}}_{\frac{\partial(\rho w)}{\partial z}} = 0$$

$$\frac{\partial \rho}{\partial t} + \left\{u\frac{\partial \rho}{\partial x} + v\frac{\partial \rho}{\partial y} + w\frac{\partial \rho}{\partial z}\right\} + \rho\left\{\frac{\partial u}{\partial x} + \frac{\partial v}{\partial y} + \frac{\partial w}{\partial z}\right\} = 0 \quad (2.47)$$

The first two terms on the left-hand side constitute the material derivative of ρ.

$\frac{\partial \rho}{\partial t}$ term corresponds to the rate of change of density of the fluid with time within the control volume, and it appears only for unsteady flow conditions. The density of fluid may change with respect to its spatial position.

$u\frac{\partial \rho}{\partial x}+v\frac{\partial \rho}{\partial y}+w\frac{\partial \rho}{\partial z}$ term corresponds to the convective derivative of density which represents the change in density due to the movement of fluid from one point to another spatial point by different velocity components. The summation of local derivative and the convective derivative is called the material derivative. Equation (2.47) can be written in vector form as

$$\frac{D\rho}{Dt}+\rho\underline{\nabla}\cdot\underline{V}=0 \tag{2.48}$$

where $\frac{D\rho}{Dt}=\frac{\partial \rho}{\partial t}+u\frac{\partial \rho}{\partial x}+v\frac{\partial \rho}{\partial y}+w\frac{\partial \rho}{\partial z}$

2.2.1.2 Incompressible Fluid Flow

For incompressible fluid flow, the variation in density is zero, that is

$\frac{D\rho}{Dt}=0$, making Equation (2.48) be in the form

$\underline{\nabla}\cdot\underline{V}=0$, which can be expanded as

$$\frac{\partial u}{\partial x}+\frac{\partial v}{\partial y}+\frac{\partial w}{\partial z}=0 \tag{2.49}$$

Liquid flows and gas flows at low speeds can be considered as incompressible. For fluid velocity with Mach number <0.3, incompressible flow condition is satisfied within 1% error in density variation. For steady flow conditions, Equation (2.46) can be expressed as

$$\underline{\nabla}\cdot\rho\underline{V}=0$$

During simulation of any type of flow problem, the continuity equation has to be satisfied. A definite velocity field is established in a flow domain only when all the forces acting on the fluid medium are balanced.

2.2.2 Momentum Conservation

Fluid flow takes place under the influence of various forces in a flow domain. Newton's second law of motion is employed to derive the momentum conservation equation [1]. It states that

$$\begin{aligned}&\big(\text{Rate of change of momentum of a fluid mass (control mass)}\big)\\&=\big(\text{Net forces acting on the fluid mass}\big)\end{aligned} \tag{2.50}$$

The net forces acting on a fluid mass may be classified as surface forces and body forces (Figure 2.5).

Surface forces are directly proportional to the area on which they act, such as viscous forces and pressure forces. Body forces are directly proportional to the volume of the fluid body, such as weight of the fluid, buoyancy force etc. Surface and body forces can be represented in vector form as

$$\iint \sigma \cdot \hat{n} ds = \underline{F}_s - \text{Surface force}$$

The normal surface force is indicated by the symbol σ, whereas the tangential surface force is represented by the symbol τ. These two surface forces have to be indicated by indices for their direction and surface area over which they act. In general

τ_{ij} - tangential surface force acts in '*i*' direction on a surface perpendicular to '*j*' direction.

σ_{ii} - normal surface force acts in '*i*' direction on the surface perpendicular to '*i*' direction.

$$\iiint \rho \underline{g} dV = \underline{F}_B \text{ - } \textit{Body force} \text{ where } dV = \Delta x \Delta y \Delta z$$

The body force generally considered is the one due to the gravitational force acting on the fluid volume, expressed as $\rho V \underline{g}$. Components of gravitational acceleration can be expressed as $\underline{g} \Rightarrow (g_x, g_y, g_z)$ for all three directions.

Then the right-hand side of momentum balance stated by Equation (2.50) becomes

$$\text{Rate of change of momentum of a fluid mass} = \iint \underline{\sigma} \cdot \hat{n} ds + \iiint \rho \underline{g} dx dy dz \quad (2.51)$$

In the case of mass conservation, the net mass flow through each direction, *x*, *y* and *z*, was considered, and a single mass conservation equation has been obtained. The

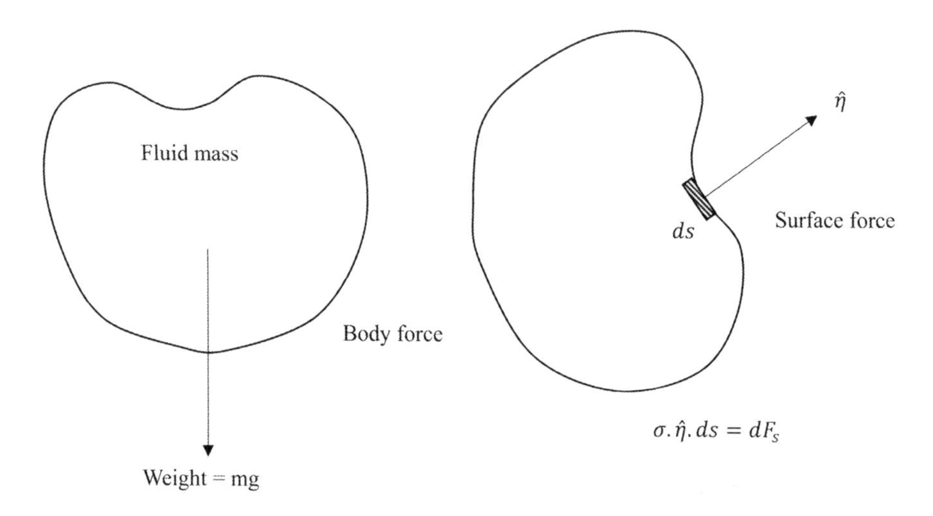

FIGURE 2.5 Surface and body forces acting on a fluid element.

entry and exit of fluid into a control volume may take place through more than one port, however, when mass conservation is considered, it is always the total mass put together in all three directions. However, in the case of momentum conservation, it is concerned with the rate of change of momentum, which results in force. Hence, the force balance of fluid mass has to be achieved in each direction separately. Some forces may be significant only in one direction and do not act in other directions. Buoyancy force always acts in the vertical direction. For this reason, the momentum conservation equation is derived for each direction separately in a flow field. The net velocity field in a flow domain is the result of force balance achieved on the fluid mass. The momentum conservation equation for *x*-, *y*- and *z*-directions can be derived as follows (Figure 2.6).

x-momentum balance equation

The left-hand side of Equation (2.51) can be expressed in terms of control volume as

$$\begin{aligned}&\left(\text{Rate of change of momentum of fluid mass}\right)\\&=(\text{Rate of change of momentum of fluid in CV}+\\&\text{Net efflux of fluid momentum from CV})\end{aligned} \tag{2.52}$$

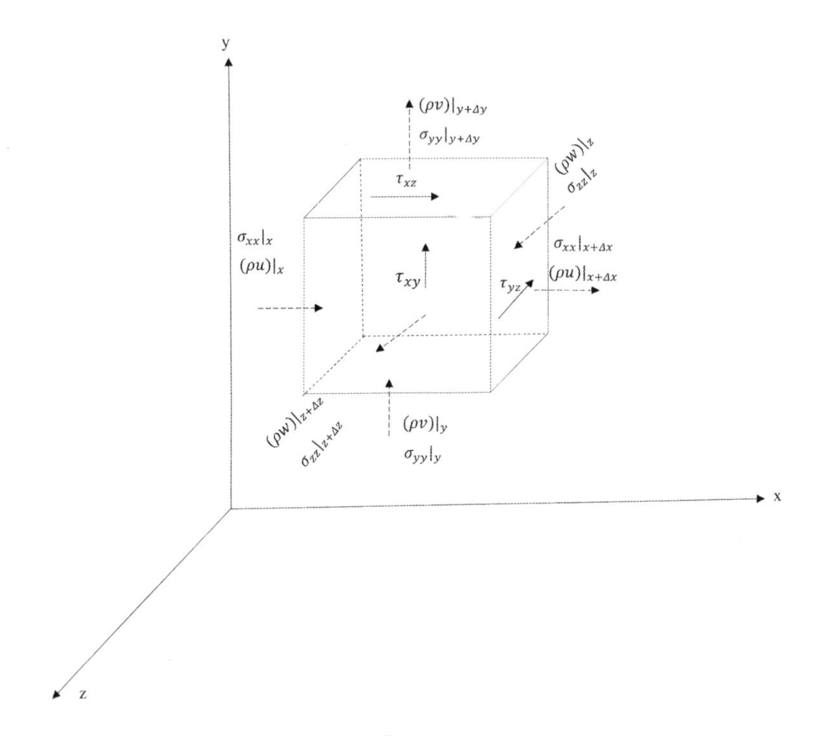

FIGURE 2.6 Momentum balance over a fluid control volume.

This means

$$\begin{aligned}&(\text{Rate of change of momentum of fluid in CV}\\&\quad + \text{Net efflux of fluid momentum from CV})\\&= \iint \underline{\sigma} \cdot \hat{n} ds + \iiint \rho \underline{g} dxdydz\end{aligned} \tag{2.53}$$

Surface forces and body forces in x-direction

Referring to Figure (2.6), the net surface forces acting in the x-direction can be written as

$$\left\{\sigma_{xx}\big|_{x+\Delta x}\Delta y\Delta z - \sigma_{xx}\big|_{x}\Delta y\Delta z\right\} + \left\{\tau_{xy}\big|_{y+\Delta y}\Delta x\Delta z - \tau_{xy}\big|_{y}\Delta x\Delta z\right\}$$
$$+\left\{\tau_{xz}\big|_{z+\Delta z}\Delta x\Delta y - \tau_{xz}\big|_{z}\Delta x\Delta y\right\}$$

Using Taylor's series expansion, we know that

$$F(x+\Delta x, y, z) = F(x, y, z) + \frac{\partial F}{\partial x}\Delta x + \frac{\partial^2 F}{\partial x^2}\frac{\Delta x^2}{2!} + ..$$

The net surface forces acting in the x-direction can be written in a differential equation form as

$$\frac{\partial \sigma_{xx}}{\partial x}\Delta x\Delta y\Delta z + \frac{\partial \tau_{xy}}{\partial y}\Delta x\Delta y\Delta z + \frac{\partial \tau_{xz}}{\partial z}\Delta x\Delta y\Delta z$$

The net body force acting in the x-direction is

$$\rho g_x \Delta x\Delta y\Delta z$$

Now the net surface and body forces acting in the x-direction can be expressed as

$$\frac{\partial \sigma_{xx}}{\partial x}\Delta x\Delta y\Delta z + \frac{\partial \tau_{xy}}{\partial y}\Delta x\Delta y\Delta z + \frac{\partial \tau_{xz}}{\partial z}\Delta x\Delta y\Delta z + \rho g_x \Delta x\Delta y\Delta z \tag{2.54}$$

Now the right-hand side of Equation (2.52) can be evaluated.

The rate of change of fluid momentum in the CV in the x-direction can be written as

$$\frac{\partial}{\partial t}(\rho u)\Delta x\Delta y\Delta z \tag{2.55a}$$

The rate of net efflux of fluid in the x-direction is produced by the mass of fluid entering in each direction and the change in u-velocity and this is given as

$$\left[(\rho u)u\big|_{x+\Delta x}\Delta y\Delta z-(\rho u)u\big|_{x}\Delta y\Delta z\right]+\left[(\rho v)u\big|_{y+\Delta y}\Delta x\Delta z-(\rho v)u\big|_{y}\Delta x\Delta z\right]$$
$$+\left[(\rho w)u\big|_{z+\Delta z}\Delta x\Delta y-(\rho w)u\big|_{z}\Delta x\Delta y\right] \tag{2.55b}$$

That is, the net rate of efflux of fluid momentum in the x-direction is written as

$$\left[\frac{\partial}{\partial x}(\rho u^2)+\frac{\partial}{\partial y}(\rho uv)+\frac{\partial}{\partial z}(\rho uw)\right]\Delta x\Delta y\Delta z$$

Now the total x-momentum balance of the fluid is expressed as

$$\underbrace{\frac{\partial}{\partial t}(\rho u)\Delta x\Delta y\Delta z}_{\text{Rate of change of momentum of fluid in CV in } x\text{-direction}}+\underbrace{\left[\frac{\partial}{\partial x}(\rho u^2)+\frac{\partial}{\partial y}(\rho uv)+\frac{\partial}{\partial z}(\rho uw)\right]\Delta x\Delta y\Delta z}_{\text{Net efflux of fluid momentum in CV in } x\text{-direction}}$$

$$=\underbrace{\frac{\partial\sigma_{xx}}{\partial x}\Delta x\Delta y\Delta z+\frac{\partial\tau_{xy}}{\partial y}\Delta x\Delta y\Delta z+\frac{\partial\tau_{xz}}{\partial z}\Delta x\Delta y\Delta z}_{\text{Net surface forces acting on the fluid in the } x\text{-direction}}+\underbrace{\rho g_x\Delta x\Delta y\Delta z}_{\text{Body force acting on the fluid in the } x\text{-direction}}$$

Thus the final form of the x-momentum balance equation takes the form

$$\frac{\partial}{\partial t}(\rho u)+\left[\frac{\partial}{\partial x}(\rho u^2)+\frac{\partial}{\partial y}(\rho uv)+\frac{\partial}{\partial z}(\rho uw)\right]=\frac{\partial\sigma_{xx}}{\partial x}+\frac{\partial\tau_{xy}}{\partial y}+\frac{\partial\tau_{xz}}{\partial z}+\rho g_x \tag{2.56}$$

The above equation can be simplified by expanding the terms on the left-hand side of the equation. The LHS of Equation (2.56) is expanded as

$$\frac{\partial}{\partial t}(\rho u)+\left[\frac{\partial}{\partial x}(\rho u^2)+\frac{\partial}{\partial y}(\rho uv)+\frac{\partial}{\partial z}(\rho uw)\right]$$

$$=\rho\frac{\partial u}{\partial t}+u\frac{\partial\rho}{\partial t}+\rho u\frac{\partial u}{\partial x}+u\frac{\partial(\rho u)}{\partial x}+\rho v\frac{\partial u}{\partial y}+u\frac{\partial(\rho v)}{\partial y}+\rho w\frac{\partial u}{\partial z}+u\frac{\partial(\rho w)}{\partial z}$$

$$=\underbrace{u\left\{\frac{\partial\rho}{\partial t}+\frac{\partial(\rho u)}{\partial x}+\frac{\partial(\rho v)}{\partial y}+\frac{\partial(\rho w)}{\partial z}\right\}}_{=0\text{ by continuity Equation (2.46)}}+\rho\left\{\frac{\partial u}{\partial t}+u\frac{\partial u}{\partial x}+v\frac{\partial u}{\partial y}+w\frac{\partial u}{\partial z}\right\}$$

Hence the left-hand side of Equation (2.56) is reduced to

$$=\rho\left\{\frac{\partial u}{\partial t}+u\frac{\partial u}{\partial x}+v\frac{\partial u}{\partial y}+w\frac{\partial u}{\partial z}\right\}$$

The above equation is of the material derivative form for any variable ϕ

$$\frac{D\phi}{Dt} = \frac{\partial \phi}{\partial t} + u\frac{\partial \phi}{\partial x} + v\frac{\partial \phi}{\partial y} + w\frac{\partial \phi}{\partial z}$$

Hence, the left-hand side of Equation (2.56) can be written as

$$\rho\frac{Du}{Dt} = \rho\left\{\frac{\partial u}{\partial t} + u\frac{\partial u}{\partial x} + v\frac{\partial u}{\partial y} + w\frac{\partial u}{\partial z}\right\}$$

A close look at the left-hand side of the above equation indicates that the term $\rho\frac{Du}{Dt}$ corresponds to mass of fluid particle $(\rho\Delta x\Delta y\Delta z)\times$ Acceleration $\left(\frac{Du}{Dt}\right) =$ Net forces.

This proves that whatever steps we have followed till this point are according to the underlying physics behind the derivation of the fluid momentum equation in the x-direction as stated in Equation (2.52). Now, the final form of the x-momentum equation is

$$\rho\frac{Du}{Dt} = \frac{\partial \sigma_{xx}}{\partial x} + \frac{\partial \tau_{xy}}{\partial y} + \frac{\partial \tau_{xz}}{\partial z} + \rho g_x \tag{2.57a}$$

Similarly, the fluid momentum equations in y- and z-directions can be derived as the

y-momentum balance equation

$$\rho\frac{Dv}{Dt} = \frac{\partial \tau_{yx}}{\partial x} + \frac{\partial \sigma_{yy}}{\partial y} + \frac{\partial \tau_{yz}}{\partial z} + \rho g_y \tag{2.57b}$$

z-momentum balance equation

$$\rho\frac{Dw}{Dt} = \frac{\partial \tau_{zx}}{\partial x} + \frac{\partial \tau_{zy}}{\partial y} + \frac{\partial \sigma_{zz}}{\partial z} + \rho g_z \tag{2.57c}$$

The right-hand side of Equations (2.57a), (2.57b) and (2.57c) can be expressed as

$$\nabla\cdot\underline{\underline{\sigma}} + \rho\underline{g}$$

The left-hand side of the above equations can be expressed in vector form as

$$\rho\frac{D\underline{V}}{Dt}$$

Now the momentum balance equations in x-, y- and z-directions can be written in vector form as

$$\rho\frac{D\underline{V}}{Dt} = \underline{\nabla}\cdot\underline{\underline{\sigma}} + \rho\underline{g} \tag{2.58}$$

The left-hand side of the above equation is written in terms of measurable properties, velocities in coordinate directions and their derivatives. However, the right-hand side of the equation consists of normal and shear stress components. The body force term will be considered later. Now the conversion of normal and shear stress components of the fluid element have to be dealt with such that they are represented in terms of measurable properties. Considering Newtonian fluids, which exhibit a linear relation between shear stress and rate of strain [2], the expressions for the stress terms can be evaluated.

2.2.2.1 Relation between Stress and Viscosity

From the definition of fluid, it is understood that fluids cannot withstand tensile stress. Pressure at a point of fluid in motion can be construed as the average of the normal stress acting in all three coordinate directions [2] and is treated as compressive in nature and indicated with a prefix of negative sign. Let us start with the relation between tangential shear stresses and normal stresses. From generalized Newton's law viscosity, we can express the shear stress components in terms of viscosity and velocity gradients as

$$\tau_{xy} = \tau_{yx} = \mu\left(\frac{\partial u}{\partial y} + \frac{\partial v}{\partial x}\right) \tag{2.59a}$$

$$\tau_{xz} = \tau_{zx} = \mu\left(\frac{\partial u}{\partial z} + \frac{\partial w}{\partial x}\right) \tag{2.59b}$$

$$\tau_{yz} = \tau_{zy} = \mu\left(\frac{\partial v}{\partial z} + \frac{\partial w}{\partial y}\right) \tag{2.59c}$$

Following similar definitions, the normal stresses are related to the pressure of fluid in motion using the following expression [2].

$$\sigma_{xx} = -p + 2\mu\frac{\partial u}{\partial x} + \lambda\left(\frac{\partial u}{\partial x} + \frac{\partial v}{\partial y} + \frac{\partial w}{\partial z}\right) \tag{2.60a}$$

$$\sigma_{yy} = -p + 2\mu\frac{\partial v}{\partial y} + \lambda\left(\frac{\partial u}{\partial x} + \frac{\partial v}{\partial y} + \frac{\partial w}{\partial z}\right) \tag{2.60b}$$

$$\sigma_{zz} = -p + 2\mu\frac{\partial w}{\partial z} + \lambda\left(\frac{\partial u}{\partial x} + \frac{\partial v}{\partial y} + \frac{\partial w}{\partial z}\right) \tag{2.60c}$$

in which λ is called liquid bulk viscosity analogous to Young's modulus and shear modulus defined for isotropic materials. For ideal gases, it can be proved that $\lambda = -\frac{2}{3}\mu$. For liquids, this parameter does not contribute significantly. Combining Equations (2.59) and (2.60) for tangential and normal stresses, a general equation can be obtained as

$$\sigma_{ij} = -p\delta_{ij} + \mu\left(\frac{\partial V_i}{\partial X_j} + \frac{\partial V_j}{\partial X_i}\right) + \lambda(V_{k,k})\delta_{ij} \tag{2.61}$$

where δ_{ij} is the Kronecker delta that takes values as

$$\delta_{ij} = 1 \text{ if } i = j$$

$$\delta_{ij} = 0 \text{ if } i \neq j$$

$V_{k,k}$ represents differentiation of V with respect to index k.

The generalized momentum balance Equation (2.58) can be written in tensor notation as

$$\rho\frac{DV_j}{Dt} = \frac{\partial \sigma_{ij}}{\partial X_i} + \rho g_j \tag{2.62}$$

Hence the differential of Equation (2.61) will produce the following expression

$$\frac{\partial \sigma_{ij}}{\partial X_i} = \frac{\partial}{\partial X_i}\left\{-p\delta_{ij} + \mu\left(\frac{\partial V_i}{\partial X_j} + \frac{\partial V_j}{\partial X_i}\right) + \lambda\left(\frac{\partial V_k}{\partial X_k}\right)\delta_{ij}\right\} \tag{2.63}$$

Now the generalized momentum equation can be expressed as

$$\rho\frac{DV_j}{Dt} = \frac{\partial}{\partial X_i}\left\{-p\delta_{ij} + \mu\left(\frac{\partial V_i}{\partial X_j} + \frac{\partial V_j}{\partial X_i}\right) + \lambda\left(\frac{\partial V_k}{\partial X_k}\right)\delta_{ij}\right\} \tag{2.64}$$

Equation (2.64) is the final form of momentum balance equation for all three directions in Cartesian coordinates. Further, this equation can be rewritten for incompressible and compressible fluids because the value of λ is different for these fluids.

2.2.2.2 Momentum Balance Equations for Incompressible Flow (μ=constant)

For incompressible fluids, $\nabla \cdot \underline{V} = 0$, hence there is no contribution of bulk viscosity to the normal stress and this assumption will result in the following momentum balance equations.

Let us consider the x-momentum equation

$$\rho\frac{Du}{Dt} = \frac{\partial}{\partial x}\left\{-p + 2\mu\frac{\partial u}{\partial x}\right\} + \frac{\partial}{\partial y}\left\{\mu\left(\frac{\partial u}{\partial y} + \underbrace{\frac{\partial v}{\partial x}}_{\text{By interchanging the order of differentiation}}\right)\right\}$$

$$+ \frac{\partial}{\partial z}\left\{\mu\left(\frac{\partial u}{\partial z} + \underbrace{\frac{\partial w}{\partial x}}_{\text{By interchanging the order of differentiation}}\right)\right\} + \rho g_x$$

$$= -\frac{\partial p}{\partial x} + \mu \frac{\partial^2 u}{\partial x^2} + \mu \frac{\partial}{\partial x}\left(\frac{\partial u}{\partial x}\right) + \mu \left\{ \frac{\partial}{\partial x}\left(\frac{\partial v}{\partial y}\right) + \left(\frac{\partial^2 u}{\partial y^2}\right) \right\} + \mu \left\{ \frac{\partial}{\partial x}\left(\frac{\partial w}{\partial z}\right) + \left(\frac{\partial^2 u}{\partial z^2}\right) \right\}$$

$$= -\frac{\partial p}{\partial x} + \frac{\partial}{\partial x}\left\{ \mu \underbrace{\left(\frac{\partial u}{\partial x} + \frac{\partial v}{\partial y} + \frac{\partial w}{\partial z}\right)}_{=0} \right\} + \mu \left(\frac{\partial^2 u}{\partial x^2} + \frac{\partial^2 v}{\partial y^2} + \frac{\partial^2 w}{\partial z^2}\right) + \rho g_x$$

$$= -\frac{\partial p}{\partial x} + \mu \left(\frac{\partial^2 u}{\partial x^2} + \frac{\partial^2 v}{\partial y^2} + \frac{\partial^2 w}{\partial z^2}\right) + \rho g_x$$

Finally, the x-, y- and z-momentum balance equations can be written in expanded form along with continuity equation as

Continuity equation

$$\frac{\partial u}{\partial x} + \frac{\partial v}{\partial y} + \frac{\partial w}{\partial z} = 0 \tag{2.65}$$

x-momentum equation

$$\rho\left(\frac{\partial u}{\partial t} + u\frac{\partial u}{\partial x} + v\frac{\partial u}{\partial y} + w\frac{\partial u}{\partial z}\right) = -\frac{\partial p}{\partial x} + \mu\left(\frac{\partial^2 u}{\partial x^2} + \frac{\partial^2 u}{\partial y^2} + \frac{\partial^2 u}{\partial z^2}\right) + \rho g_x \tag{2.66a}$$

y-momentum equation

$$\rho\left(\frac{\partial v}{\partial t} + u\frac{\partial v}{\partial x} + v\frac{\partial v}{\partial y} + w\frac{\partial v}{\partial z}\right) = -\frac{\partial p}{\partial y} + \mu\left(\frac{\partial^2 v}{\partial x^2} + \frac{\partial^2 v}{\partial y^2} + \frac{\partial^2 v}{\partial z^2}\right) + \rho g_y \tag{2.66b}$$

z-momentum equation

$$\rho\left(\frac{\partial w}{\partial t} + u\frac{\partial w}{\partial x} + v\frac{\partial w}{\partial y} + w\frac{\partial w}{\partial z}\right) = -\frac{\partial p}{\partial z} + \mu\left(\frac{\partial^2 w}{\partial x^2} + \frac{\partial^2 w}{\partial y^2} + \frac{\partial^2 w}{\partial z^2}\right) + \rho g_z \tag{2.66c}$$

Equations (2.65) and (2.66) represent mass conservation and momentum conservation equations to be solved for any flow field in order to obtain the velocity and pressure fields. They are called Navier-Stokes equations in primary variables form. There are four equations for the four unknowns, u, v, w and p. In these variables, pressure is a scalar field. The most interesting characteristic of these equations is that all three velocities appear in all four equations, however, pressure appears only in the momentum equations, not in the continuity equation directly. In the continuity equation, the pressure field is implicitly satisfied because of the assumption of incompressible flow, where the density remains constant. After deriving these conservation equations as differential equations, we need to understand the physical meaning of different terms that appear in these equations so that their importance could be appreciated during mathematical modeling. Now, considering the continuity equation, the first order derivative of each velocity term with respect to the respective direction indicates the net mass flow in each direction, and thus the continuity equation shows, that the sum of net mass flow in all three directions is zero, indicating mass is conserved.

In the momentum equations, the left-hand side of the equation indicates the force due to the acceleration of the fluid flowing through the control volume. In that, the first term corresponds to the rate of change of momentum of the fluid with time within the control volume, and the second term refers to the next efflux of momentum of fluid in the control volume. The total change of momentum of the fluid on the left-hand side must be equal to the net forces acting on the fluid control volume. The net forces acting on the fluid are shown on the right-hand side of the equation. These forces are surface forces and body force. Surface forces consist of forces due to normal and tangential stresses. The normal stress is equated to the pressure of fluid in motion and the tangential forces to viscous forces developed by the viscosity of the fluid and its velocity gradient in the flow field. The body force is directly proportional to the volume of the fluid, and hence forces such as buoyancy due to thermal and concentration gradients or magnetic force contribute to this term. The solution of Equations (2.65) and (2.66) confirms velocity and pressure fields which are obtained after satisfying mass conservation and momentum balance of the fluid in the control volume in all three directions. It is worth considering the units of these equations. For the continuity equation, the density term is ignored in the equation because it is constant. Hence if Equation (2.65) is multiplied by density of fluid, then the unit comes to be $\frac{kg}{s(m^3)}$, the mass flow rate per unit volume of the control volume. In the momentum Equation (2.66), all the terms on the left-hand side of the equation correspond to $\frac{N}{m^3}$, that is force per unit volume. Similarly, all the terms on the right-hand side of the equation give rise to $\frac{N}{m^3}$, thus the momentum conservation equation simply indicates the force balance on the fluid element in a given flow domain. It has to be noted that only through force balance in the fluid domain is the velocity field obtained, however, all the force terms appearing in the momentum balance equations are written in terms of velocities and velocity gradients with knowledge of some of the relevant and measurable fluid characteristics.

2.2.3 Energy Conservation

In thermal science analysis, fluid flow with heat transfer problems are studied in detail. Convective heat transport finds wide applications in thermal engineering field. Study of heat exchangers in chemical industries, HVAC systems, boilers, condensers etc. requires thorough knowledge of convective heat transfer. In fluid flow problems with heat transfer, the final objective may be to determine the capability of the fluid either to deliver or absorb heat to or from a system or to determine the temperature of the system which is limited by its working environment. During any heat transport process in a system, the system variables such as temperature, pressure etc. change according to some conservation principles. In the case of pure conduction or radiation heat transfer problems, it is only energy conservation that has to be satisfied. Whenever fluid flow is involved with heat transfer, then energy conservation has to be achieved along with mass and momentum conservation because of the role of the velocity field in energy transport. The application of the energy conservation principle for flow problems will be discussed in the following section.

2.2.3.1 Energy Balance

The energy conservation principle is stated by the first law of thermodynamics for a flow system as

$$\dot{Q} - \dot{W} = \frac{dE}{dt}$$

The above equation can be rewritten for the system when work is done on the system as

$$\dot{Q} + \dot{W} = \frac{dE}{dt} \tag{2.67}$$

where $\dot{Q}$ and $\dot{W}$ refer to heat and work interactions with the flow system and $\frac{dE}{dt}$ indicates the rate of change of energy of fluid in the control volume. Equation (2.67) gets modified when there is a flow of fluid through the control volume with convective energy transport. Unlike the equation for conduction heat transfer, in the case of convective heat transfer, the energy associated with the fluid also has to be considered since its contribution is significant in the overall energy transfer. In the simple energy balance equation in thermodynamics, energy only associated with the variation of fluid properties is considered. However, in convective heat transport, the energy transfer due to the change of momentum of the fluid plays an important role. Only an expansion type of work is considered in thermodynamic systems. In the case of fluid flow with a certain velocity, the shear and body forces also will contribute to the work transfer. Hence, the energy conservation with fluid flow through a control volume (Figure 2.7(a)), has to take into account all kinds of energy transfer as described in the following energy conservation statement.

$$\begin{aligned}
&(\text{Rate of change of energy in CV}) + (\text{Net efflux of energy from CV}) \\
&= (\text{Net work done by surface forces}) + (\text{Net work done by body forces}) \\
&+ (\text{Net heat added to CV}) + (\text{Net heat generation in CV})
\end{aligned} \tag{2.68}$$

Let us evaluate each term in the above equation one by one [1].

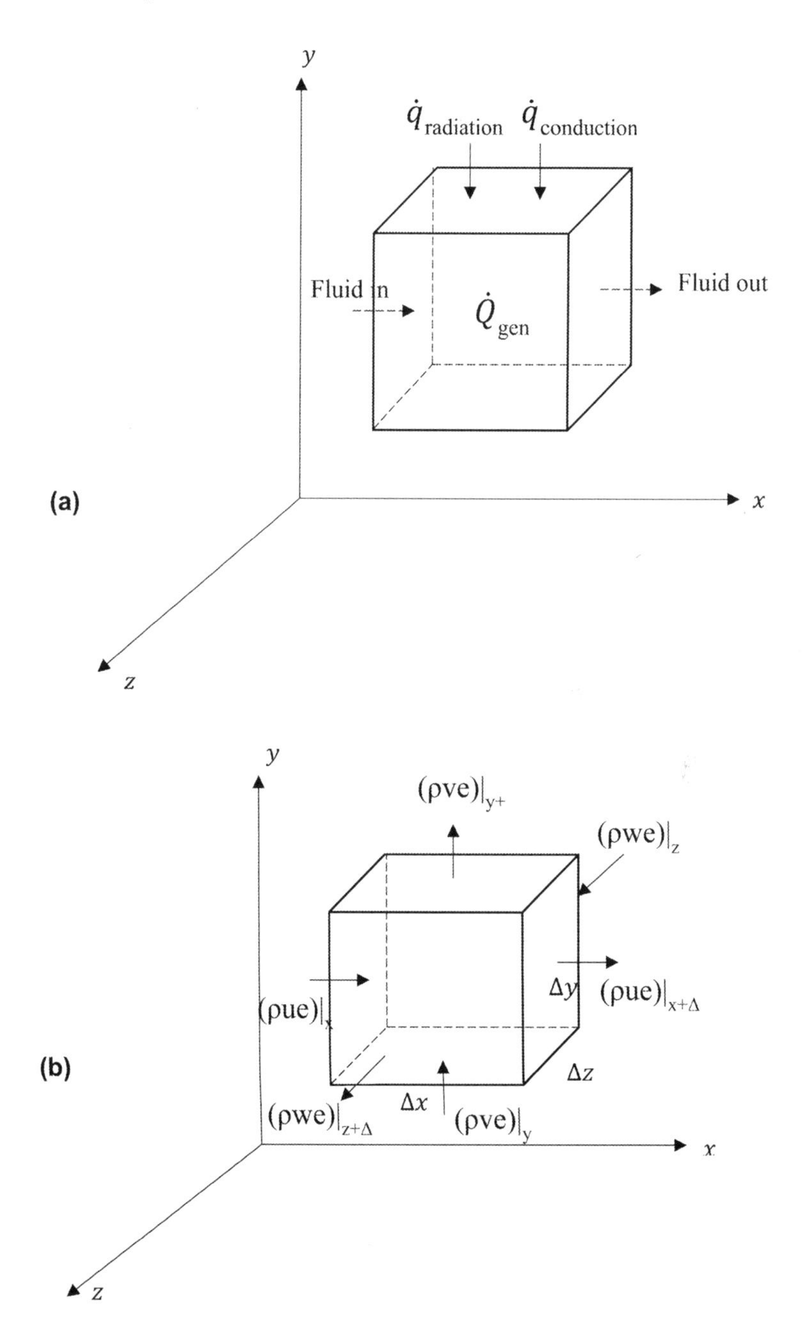

FIGURE 2.7 (a) Energy interactions in a fluid element and (b) energy balance in a fluid control volume.

2.2.3.2 Rate of Change of Energy in CV

In the total energy of fluid in CV, $e = i + \frac{V^2}{2} + gz$ in which 'i' refers to fluid internal energy, $\frac{V^2}{2}$ is kinetic energy due to velocity and gz is accounted for as work done by body forces.

The rate of change of total energy of fluid with time is shown as follows:

$$\mathrm{CV} = \frac{\partial}{\partial t}(\rho e)\Delta x \Delta y \Delta z \tag{2.69}$$

2.2.3.3 Net Efflux of Energy from CV

When fluid flow takes place through the control volume from inlet to outlet, there will be some change in this energy of fluid due to the distance it has travelled in x-, y- and z-directions) (Figure 2.7(b)), and this can be expressed as

$$\begin{aligned}&(\rho ue)\big|_{x+\Delta x}\Delta y\Delta z-(\rho ue)\big|_{x}\Delta y\Delta z+(\rho ve)\big|_{y+\Delta y}\Delta x\Delta z-(\rho ve)\big|_{y}\Delta x\Delta z\\&+(\rho we)\big|_{z+\Delta z}\Delta x\Delta y-(\rho we)\big|_{z}\Delta x\Delta y\end{aligned}$$

Using Taylor's series expansion, this can be written in differential form as

$$\left\{\frac{\partial}{\partial x}(\rho ue)+\frac{\partial}{\partial y}(\rho ve)+\frac{\partial}{\partial z}(\rho we)\right\}\Delta x\Delta y\Delta z \tag{2.70}$$

2.2.3.4 Rate of Work Done by Surface Forces

The surface forces acting in fluid element has already been discussed during the derivation of the fluid momentum conservation equation in the previous section. Due to the normal and tangential stresses acting on the surfaces of the control volume, work will be done by the fluid by the velocity components in the respective direction. Referring to Figure 2.8, let us find out the work done due to the normal and shear stress acting on the faces $\Delta y\Delta z$ at x and $x+\Delta x$, normal to the x-direction.

Looking at this figure, one can understand that the normal stress σ_{xx} which acts normal to the x-direction, will produce a force $\sigma_{xx}\Delta y\Delta z$, and due to the velocity u in the x-direction, there will be net work done in the x-direction. Similarly, the other two tangential stresses will produce forces, $\tau_{xy}\Delta y\Delta z$ and $\tau_{xz}\Delta y\Delta z$, which will in turn give rise to net work done due to the velocities v and w in the y- and z-directions respectively. Hence the work done due to surface forces for fluid flow in the x-direction can be written as

$$\begin{aligned}&(\sigma_{xx}\Delta y\Delta z)u\big|_{x+\Delta x}-(\sigma_{xx}\Delta y\Delta z)u\big|_{x}+(\tau_{xy}\Delta y\Delta z)v\big|_{x+\Delta x}-(\tau_{xy}\Delta y\Delta z)v\big|_{x}\\&+(\tau_{xz}\Delta y\Delta z)w\big|_{x+\Delta x}-(\tau_{xz}\Delta y\Delta z)w\big|_{x}\end{aligned}$$

Using Taylor's series expansion, the above equation can be written in differential form as

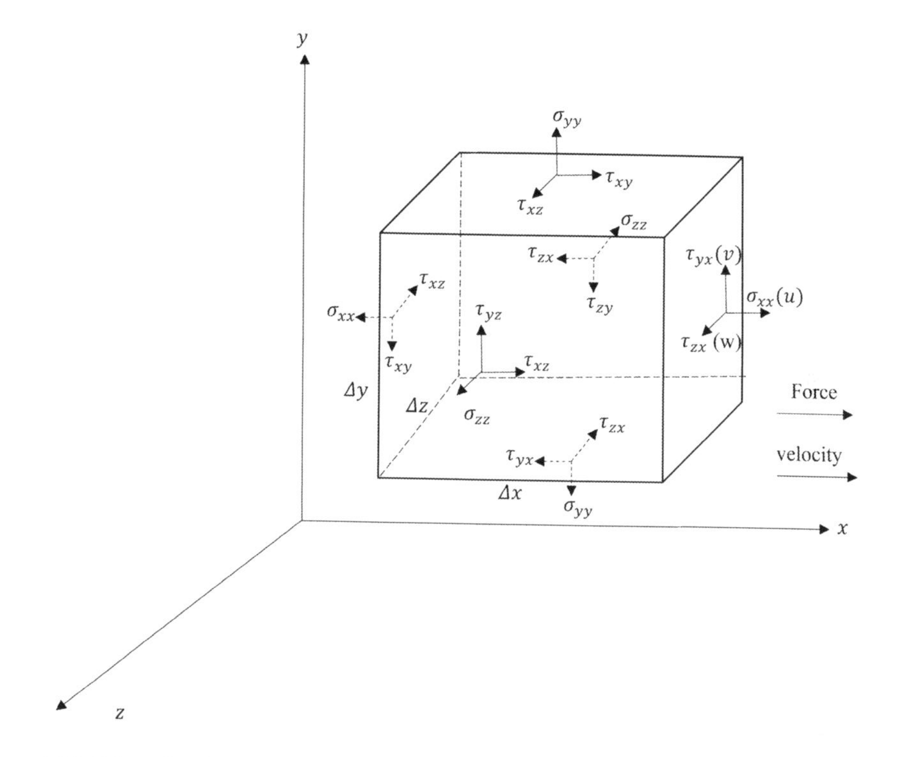

FIGURE 2.8 Work done due to normal and tangential stresses in a fluid control volume.

$$\left[\frac{\partial}{\partial x}(\sigma_{xx})u+\frac{\partial}{\partial x}(\tau_{xy})v+\frac{\partial}{\partial x}(\tau_{xz})w\right]\Delta x\Delta y\Delta z \tag{2.71a}$$

Similarly, the work done due to surface forces for flow of fluid in y-direction is written as

$$\left[\frac{\partial}{\partial y}(\tau_{yx})u+\frac{\partial}{\partial y}(\sigma_{yy})v+\frac{\partial}{\partial y}(\tau_{yz})w\right]\Delta x\Delta y\Delta z \tag{2.71b}$$

The expression for work done due to surface forces for flow fluid in z-direction is given as

$$\left[\frac{\partial}{\partial z}(\tau_{zx})u+\frac{\partial}{\partial z}(\tau_{zy})v+\frac{\partial}{\partial z}(\sigma_{zz})w\right]\Delta x\Delta y\Delta z \tag{2.71c}$$

2.2.3.5 Work Done by Body Forces

The total work done by the body forces acting in x-, y- and z-directions due to the respective velocities is written as

$$\begin{aligned}&(\rho g_x u+\rho g_y v+\rho g_z w)\Delta x\Delta y\Delta z\\&=(\rho \underline{g}\cdot \underline{V})\Delta x\Delta y\Delta z\end{aligned} \tag{2.72}$$

2.2.3.6 Net Addition of Heat due to Conduction and Radiation Heat Transfer

In the fluid control volume, other than energy transport due to the movement of fluid across the faces of the control volume, there may be conduction and radiation heat transfer within the fluid itself and this takes place due to the temperature difference between the surfaces. The net conduction and radiation heat transfer through the control volume is shown in Figure 2.9.

The respective equation for net heat transfer through the control volume can be evaluated as

$$-\left[\left(q_{c,x}\big|_{x+\Delta x}-q_{c,x}\big|_{x}\right)\Delta y\Delta z+\left(q_{c,y}\big|_{y+\Delta y}-q_{c,y}\big|_{y}\right)\Delta x\Delta z+\left(q_{c,z}\big|_{z+\Delta z}-q_{c,z}\big|_{z}\right)\Delta x\Delta y\right]$$

$$-\left[\left(q_{r,x}\big|_{x+\Delta x}-q_{r,x}\big|_{x}\right)\Delta y\Delta z+\left(q_{r,y}\big|_{y+\Delta y}-q_{r,y}\big|_{y}\right)\Delta x\Delta z+\left(q_{r,z}\big|_{z+\Delta z}-q_{r,z}\big|_{z}\right)\Delta x\Delta y\right]$$

Using Taylor's series expansion, the above equation can be expressed including radiation heat transfer as

$$-\left[\left(\frac{\partial q_{c,x}}{\partial x}+\frac{\partial q_{c,y}}{\partial y}+\frac{\partial q_{c,z}}{\partial z}\right)-\left(\frac{\partial q_{r,x}}{\partial x}+\frac{\partial q_{r,y}}{\partial y}+\frac{\partial q_{r,z}}{\partial z}\right)\right]\Delta x\Delta y\Delta z$$

$$=-(\nabla\cdot \underline{q}_c+\nabla\cdot \underline{q}_r)\Delta x\Delta y\Delta z \tag{2.73}$$

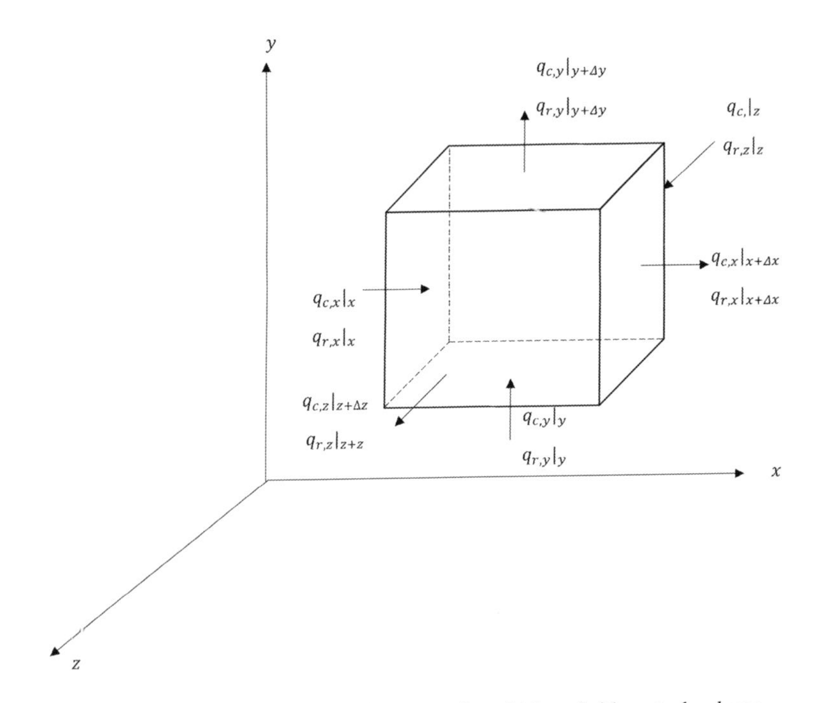

FIGURE 2.9 Conduction and radiation heat transfer within a fluid control volume.

2.2.3.7 Heat Generation within Control Volume

The heat generated within the control volume due to a chemical reaction, nuclear reaction or electrical heating can be expressed as

$$\dot{Q}_{gen}\Delta x\Delta y\Delta z \tag{2.74}$$

Now substituting Equations (2.69) to (2.74) in the energy balance Equation (2.68), we get

$$\left\{\frac{\partial(\rho e)}{\partial t}+\frac{\partial(\rho ue)}{\partial x}+\frac{\partial(\rho ve)}{\partial y}+\frac{\partial(\rho we)}{\partial z}\right\}\Delta x\Delta y\Delta z$$

$$=\left\{\frac{\partial(\sigma_{xx}u)}{\partial x}+\frac{\partial(\tau_{xy}v)}{\partial x}+\frac{\partial(\tau_{xz}w)}{\partial x}\right\}\Delta x\Delta y\Delta z$$

$$+\left\{\frac{\partial(\tau_{yx}u)}{\partial y}+\frac{\partial(\sigma_{yy}v)}{\partial y}+\frac{\partial(\tau_{yz}w)}{\partial y}\right\}\Delta x\Delta y\Delta z$$

$$+\left\{\frac{\partial(\tau_{zx}u)}{\partial z}+\frac{\partial(\tau_{zy}v)}{\partial z}+\frac{\partial(\sigma_{zz}w)}{\partial z}\right\}\Delta x\Delta y\Delta z$$

$$+(\rho g_x u+\rho g_y v+\rho g_z w)\Delta x\Delta y\Delta z$$

$$-(\underline{\nabla}\cdot\underline{q}_c)\Delta x\Delta y\Delta z-(\underline{\nabla}\cdot\underline{q}_r)\Delta x\Delta y\Delta z+\dot{Q}_{gen}\Delta x\Delta y\Delta z$$

Let us simplify the above equation. Now consider the left-hand side of the above equation,

$$\left\{\frac{\partial(\rho e)}{\partial t}+\frac{\partial(\rho ue)}{\partial x}+\frac{\partial(\rho ve)}{\partial y}+\frac{\partial(\rho we)}{\partial z}\right\}$$

$$=e\frac{\partial\rho}{\partial t}+\rho\frac{\partial e}{\partial t}+e\frac{\partial(\rho u)}{\partial x}+(\rho u)\frac{\partial e}{\partial x}+e\frac{\partial(\rho v)}{\partial y}+(\rho v)\frac{\partial e}{\partial y}+e\frac{\partial(\rho w)}{\partial z}+(\rho w)\frac{\partial e}{\partial z}$$

$$=\underbrace{e\left\{\frac{\partial\rho}{\partial x}+\frac{\partial(\rho u)}{\partial x}+\frac{\partial(\rho v)}{\partial y}+\frac{\partial(\rho w)}{\partial z}\right\}}_{=0\text{ by continuity equation}}+\rho\left\{\frac{\partial e}{\partial t}+u\frac{\partial e}{\partial x}+v\frac{\partial e}{\partial y}+w\frac{\partial e}{\partial z}\right\}$$

$$=\rho\left\{\frac{\partial e}{\partial t}+u\frac{\partial e}{\partial x}+v\frac{\partial e}{\partial y}+w\frac{\partial e}{\partial z}\right\}$$

$$=\rho\frac{De}{Dt}$$

Hence the energy equation can be written as

$$\rho\frac{De}{Dt}=\left\{\frac{\partial(\sigma_{xx}u)}{\partial x}+\frac{\partial(\tau_{xy}v)}{\partial x}+\frac{\partial(\tau_{xz}w)}{\partial x}\right\}+\left\{\frac{\partial(\tau_{yx}u)}{\partial y}+\frac{\partial(\sigma_{yy}v)}{\partial y}+\frac{\partial(\tau_{yz}w)}{\partial y}\right\}$$
$$+\left\{\frac{\partial(\tau_{zx}u)}{\partial z}+\frac{\partial(\tau_{zy}v)}{\partial z}+\frac{\partial(\sigma_{zz}w)}{\partial z}\right\}+(\rho g_x u+\rho g_y v+\rho g_z w) \tag{2.75}$$
$$-(\underline{\nabla}\bullet \underline{q}_c)-(\underline{\nabla}\bullet \underline{q}_r)+\dot{Q}_{gen}$$

Equation (2.75) is called *generalized energy equation* or *total energy equation*. The left-hand side represents thermal energy and the right-hand side also contains terms corresponding to thermal energy such as conduction and radiation heat transfer and heat generation and mechanical energy terms due to surface and body forces. The mechanical energy due to surface and body forces is converted into thermal energy in applications such as turning, drilling and milling processes and re-entry of space vehicles into the earth's atmosphere. That means Equation (2.75) can be employed for analyzing heat transfer in turning, drilling and milling processes. During metal cutting operations, work is spent overcoming the friction between two metal surfaces, wherein part of the work supplied is finally converted into heat. When space vehicles enter the earth's atmosphere, they encounter enormous fluid friction offered by the atmospheric layer surrounding the earth, and the impact of high-speed vehicles into the atmospheric layer gives rise to friction, which is finally converted into heat. In modeling thermal related problems, only thermal energy equation is required. Hence, in order to get the thermal energy equation, the mechanical energy equation has to be subtracted. This can be performed by rewriting Equation (2.75) in vector form so that the mathematics involved in subtracting the mechanical energy equation becomes simple.

The total energy balance expressed by Equation (2.75) can be expressed in vector form as

$$\rho\frac{De}{Dt}=\frac{\partial(\sigma_{ij}V_{\partial j})}{\partial X_i}+\rho g_j V_j-\frac{\partial q_{c,j}}{\partial X_j}-\frac{\partial q_{r,j}}{\partial X_j}+\dot{Q}_{gen} \tag{2.76}$$

Now the mechanical energy equation can be obtained by multiplying the velocity vector with the momentum equation. Consider the momentum equation expressed in vector form by Equation (2.62)

$$\rho\frac{DV_j}{Dt}=\frac{\partial\sigma_{ij}}{\partial X_i}+\rho g_j$$

Multiply the above equation with V_j on both sides to get the equation

$$\rho\frac{D\left(\frac{V_j^2}{2}\right)}{Dt}=V_j\frac{\partial\sigma_{ij}}{\partial X_i}+\rho g_j V_j \tag{2.77}$$

Equation (2.77) represents the mechanical energy equation. The thermal energy equation is obtained by subtracting this from the total energy equation. Now subtracting Equation (2.77) from Equation (2.76), we get

$$\rho \frac{De}{Dt} - \rho \frac{D\left(\frac{V_j^2}{2}\right)}{Dt} = \overbrace{\left(\rho g_j V_j - \rho g_j V_j\right)}^{=0} + \frac{\partial\left(\sigma_{ij} V_j\right)}{\partial X_i} - V_j \frac{\partial \sigma_{ij}}{\partial X_i} - \frac{\partial q_{c,j}}{\partial X_j} - \frac{\partial q_{c,j}}{\partial X_j} - \dot{Q}_{gen}$$

From the definition of $e = i + \frac{V^2}{2} + gz$ (gz is already accounted in body force term), $e - \frac{V_j^2}{2} = i$, and hence the above equation is reduced to

$$\rho \frac{Di}{Dt} = \frac{\partial\left(\sigma_{ij} V_j\right)}{\partial X_i} - V_j \frac{\partial \sigma_{ij}}{\partial X_i} - \frac{\partial q_{c,j}}{\partial X_j} - \frac{\partial q_{c,j}}{\partial X_j} - \dot{Q}_{gen}$$

Now the first term on the right-hand side of the above equation can be expanded and the equation can be further simplified as

$$\rho \frac{Di}{Dt} = V_j \frac{\partial \sigma_{ij}}{\partial X_i} + \sigma_{ij} \frac{\partial V_j}{\partial X_i} - V_j \frac{\partial \sigma_{ij}}{\partial X_i} - \frac{\partial q_{c,j}}{\partial X_j} - \frac{\partial q_{c,j}}{\partial X_j} - \dot{Q}_{gen}$$

After the first and third terms on the right-hand side of the above equation cancelled each other out, we get

$$\rho \frac{Di}{Dt} = \sigma_{ij} \frac{\partial V_j}{\partial X_i} - \frac{\partial q_{c,j}}{\partial X_j} - \frac{\partial q_{c,j}}{\partial X_j} - \dot{Q}_{gen} \tag{2.78}$$

The term $\sigma_{ij} \frac{\partial V_j}{\partial X_i}$ consists of normal and tangential stresses multiplied by a velocity gradient. The energy produced by the tangential stresses is called a viscous dissipation term, whereas the energy due to the normal stresses correspond to compressibility work due to the pressure of the fluid. One should understand that these terms become significant only when the normal and tangential stresses and the velocity gradients are significant and comparable to other energy terms.

Expansion of this term gives rise to the following expression

$$\sigma_{ij} \frac{\partial V_j}{\partial X_i} = \left(\sigma_{xx} \frac{\partial u}{\partial x} + \tau_{xy} \frac{\partial v}{\partial x} + \tau_{xz} \frac{\partial w}{\partial x}\right) + \left(\tau_{yx} \frac{\partial u}{\partial y} + \sigma_{yy} \frac{\partial v}{\partial y} + \tau_{yz} \frac{\partial w}{\partial y}\right)$$
$$+ \left(\tau_{zx} \frac{\partial u}{\partial z} + \tau_{zy} \frac{\partial v}{\partial z} + \sigma_{zz} \frac{\partial w}{\partial z}\right)$$

After summing up the normal and tangential stress terms separately we get

$$= \left(\sigma_{xx} \frac{\partial u}{\partial x} + \sigma_{yy} \frac{\partial v}{\partial y} + \sigma_{zz} \frac{\partial w}{\partial z} \right)$$

$$+ \left(\tau_{xy} \frac{\partial v}{\partial x} + \tau_{xz} \frac{\partial w}{\partial x} + \tau_{yx} \frac{\partial u}{\partial y} + \tau_{yz} \frac{\partial w}{\partial y} + \tau_{zx} \frac{\partial u}{\partial z} + \tau_{zy} \frac{\partial v}{\partial z} \right)$$

The first term in the above equation indicates the pressure work done due to the normal stress acting on the expansion of the fluid in the respective coordinate directions, whereas the second term relates to the viscous dissipation energy, which can be understood by expanding some of the tangential stress terms. Equations (2.59) and (2.60) can be employed respectively to relate the tangential stresses with velocity gradients and normal stresses with pressure and velocity gradients. As pressure is related to temperature and volume for a perfect gas, the term involving pressure is separated out, and all other terms involving viscosity are combined together to give rise to a viscous dissipation term. Now the thermal energy balance expressed by Equation (2.78) can be written as

$$\rho \frac{Di}{Dt} = -p\underline{\nabla} \bullet \underline{V} + \mu\Phi - \underline{\nabla} \bullet \underline{q}_c - \underline{\nabla} \bullet \underline{q}_r + \dot{Q}_{gen} \tag{2.79}$$

where $\mu\Phi = 2\mu \left[\left(\frac{\partial u}{\partial x} \right)^2 + \left(\frac{\partial v}{\partial y} \right)^2 + \left(\frac{\partial w}{\partial z} \right)^2 - \frac{1}{3} \left(\underline{\nabla} \cdot \underline{V} \right)^2 \right] +$

$$\mu \left[\frac{\partial v}{\partial x} + \frac{\partial u}{\partial y} \right]^2 + \mu \left[\frac{\partial v}{\partial z} + \frac{\partial u}{\partial y} \right]^2 + \mu \left[\frac{\partial w}{\partial x} + \frac{\partial u}{\partial z} \right]^2$$

Equation (2.79) can be further simplified in order to represent the energy balance equation in terms of temperature which is a measurable property. During mathematical modeling, it is important to obtain the final governing equations in terms of useful and measurable properties, then only the final simulation results can be used either for developing new products or improving the existing ones. Let us start with the left-hand side of Equation (2.79). The internal energy, 'i', can be rewritten as

$i = C_V T$ and

Enthalpy, $h = i + pv = i + \dfrac{p}{\rho}$

Then $i = h - \dfrac{p}{\rho}$

Hence $\dfrac{Di}{Dt} = \dfrac{D\left\{ h - \left(\dfrac{p}{\rho} \right) \right\}}{Dt} = \dfrac{Dh}{Dt} - \dfrac{D\left(\dfrac{p}{\rho} \right)}{Dt}$

Now $\dfrac{D\left(\dfrac{p}{\rho}\right)}{Dt} = \dfrac{1}{\rho}\dfrac{Dp}{Dt} - \dfrac{p}{\rho^2}\dfrac{D\rho}{Dt}$

Therefore, $\dfrac{Di}{Dt} = \dfrac{Dh}{Dt} - \dfrac{1}{\rho}\dfrac{Dp}{Dt} + \dfrac{p}{\rho^2}\dfrac{D\rho}{Dt}$

Now the energy balance expressed by Equation (2.79) becomes

$$\rho\left[\frac{Dh}{Dt} - \frac{1}{\rho}\frac{Dp}{Dt} + \frac{p}{\rho^2}\frac{D\rho}{Dt}\right] = -p\underline{\nabla}\bullet\underline{V} + \mu\Phi - \underline{\nabla}\bullet\underline{q}_c - \underline{\nabla}\bullet\underline{q}_r + \dot{Q}_{gen}$$

That is $\rho\dfrac{Dh}{Dt} - \dfrac{Dp}{Dt} + \left(\dfrac{p}{\rho}\dfrac{D\rho}{Dt} + p\underline{\nabla}\bullet\underline{V}\right) = \mu\Phi - \underline{\nabla}\bullet\underline{q}_c - \underline{\nabla}\bullet\underline{q}_r + \dot{Q}_{gen}$

$$\rho\frac{Dh}{Dt} - \frac{Dp}{Dt} + p\left(\frac{1}{\rho}\frac{D\rho}{Dt} + p\underline{\nabla}\bullet\underline{V}\right) = \mu\Phi - \underline{\nabla}\bullet\underline{q}_c - \underline{\nabla}\bullet\underline{q}_r + \dot{Q}_{gen} \qquad (2.80)$$

The continuity equation is written as

$$\frac{D\rho}{Dt} + \rho\underline{\nabla}\cdot\underline{V} = 0$$

Substituting the above continuity equation in Equation (2.80) we get

$$\rho\frac{Dh}{Dt} - \frac{Dp}{Dt} + \underbrace{p\left(\frac{1}{\rho}\frac{D\rho}{Dt} + p\underline{\nabla}\bullet\underline{V}\right)}_{=0 \text{ by continuity equation}} = \mu\Phi - \underline{\nabla}\bullet\underline{q}_c - \underline{\nabla}\bullet\underline{q}_r + \dot{Q}_{gen}$$

$$\rho\frac{Dh}{Dt} - \underbrace{\frac{Dp}{Dt}}_{\substack{\text{Conversion of mechanical work}\\ \text{due to pressure into thermal energy}}} = \underbrace{\mu\Phi}_{\substack{\text{Conversion of viscous dissipation}\\ \text{into thermal energy}}} - \underline{\nabla}\bullet\underline{q}_c - \underline{\nabla}\bullet\underline{q}_r + \dot{Q}_{gen} \qquad (2.81)$$

The above equation is the thermal energy equation including the effect of conversion of mechanical work due to the high-pressure variation and viscous dissipation effect, and these two terms make a significant contribution only at high-velocity flows.

For low-speed flows, Equation (2.81) can be modified as

$$\rho\frac{Dh}{Dt} = -\underline{\nabla}\bullet\underline{q}_c - \underline{\nabla}\bullet\underline{q}_r + \dot{Q}_{gen}$$

The contribution of radiation heat transfer becomes important only in high-temperature applications – for example, when modeling fire inside buildings, or when studying energy transport during combustion inside internal combustion engines or liquid or gaseous fuel combustors. In general, the radiation term is not included in the thermal energy equation. If it is assumed that enthalpy 'h' is only a function of temperature, which is true in most single-phase fluid flows, the above energy equation can be written as

$$\rho C_P\left(\frac{DT}{Dt}\right) = -\underline{\nabla}\bullet \underline{q_c} + \dot{Q}_{gen}$$

The heat conduction term in the above equation can be expanded and expressed in terms of temperature as

$$-\underline{\nabla}\cdot \underline{q_c} = -\left(\frac{\partial q_x}{\partial x} + \frac{\partial q_y}{\partial y} + \frac{\partial q_z}{\partial z}\right)$$

Now making use of the Fourier law of conduction equation in the respective coordinate directions, the final form of thermal energy equation for fluid with constant thermal conductivity is

$$\rho C_P\left(\frac{DT}{Dt}\right) = k\left(\frac{\partial^2 T}{\partial x^2} + \frac{\partial^2 T}{\partial y^2} + \frac{\partial^2 T}{\partial z^2}\right) + \dot{Q}_{gen}$$

Expanding the left-hand side of the energy equation for fluid at low speed and constant thermal conductivity and C_p values, the thermal energy equation is expressed as

$$\rho C_P\left(\frac{\partial T}{\partial t} + u\frac{\partial T}{\partial x} + v\frac{\partial T}{\partial y} + w\frac{\partial T}{\partial z}\right) = k\left(\frac{\partial^2 T}{\partial x^2} + \frac{\partial^2 T}{\partial y^2} + \frac{\partial^2 T}{\partial z^2}\right) + \dot{Q}_{gen} \quad (2.82)$$

Equation (2.82) is the final form of a thermal energy equation. In order to solve this equation, the velocity field should be known and hence the energy equation for fluid flow is solved along with continuity and momentum equations. In convective heat transfer models, there are five variables u, v, w, p and T to be solved in the flow field. There are five conservation equations consisting of continuity equation, three momentum balance equations and energy equation, thus the solution of these equations will provide unique solutions for the flow field and temperature field. It is worth verifying the units of all the terms in Equation (2.82), which comes to be $\frac{W}{m^3}$.

The left-hand side of Equation (2.82) shows the total thermal energy interaction within the control volume, in that the first term refers to the rate of change of thermal energy of fluid within the control volume with time, and the other three terms correspond to net efflux of thermal energy transport through the control volume. The net efflux energy term can be viewed as the net thermal energy carried away by the flow of fluid across the control volume due to mass flow in all three coordinate directions and the respective temperature difference across the control surfaces. This is also called the convective heat flow through the control volume. This change of thermal energy within the control volume takes place due to conduction heat transfer and heat generation within the control volume shown on the right-hand side of the equation. Though conduction in liquids and gases is not significant in comparison to conduction in solids, this mode of heat transfer cannot be ignored in flow problems. The conduction term has been included on the right-hand side of the equation using the Fourier law of heat conduction and is not derived like the work done due to surface and body forces. Similarly, the heat generation term is also included on the right-hand side of the equation for the purpose of energy balance. During mathematical modeling, care must be taken to include or exclude the relevant heat transport mechanisms as required for the given problem. In the case of systems involving high-speed flows or while dealing with fluids with high viscosity, the appropriate viscous dissipation term has to be included to make the model more realistic to the situation under consideration. The very purpose of showing the details of the above derivations is to demonstrate how the conservation principles take the form of mathematical equations.

2.2.4 Species Conservation

Species or mass transport problems find wide engineering applications in modeling atmospheric flows, drying of porous materials, storage of solar energy in solar ponds etc. Mass transfer operations is an important field of study in chemical engineering science. Species transport refers to the transport of one or two species in a fluid medium which is a mixture of stable vapors or gases giving rise to a homogeneous mixture. Such a mixture of gases will have its own average gas velocity. For example, atmospheric air is a mixture of gases such as oxygen, nitrogen, argon and water vapor. When we say air is flowing at a flow velocity, it means the whole homogeneous mixture is moving with a constant velocity. Transfer of moisture to atmospheric air is an example of species transport. In heat transfer, the temperature is the potential for heat flow; similarly, in species transport, species concentration is the potential for species or mass transfer. Convective mass transfer is directly proportional to the concentration difference between the concentration of the species at the wall and the concentration of the same species at the ambient. The constant of proportionality is called the convective mass transfer coefficient, similar to convective heat transfer coefficient. Mass diffusion is considered analogous to heat conduction, as both the transport processes are governed by the gradients of the respective potentials. Mass diffusion of a particular species in a medium is directly proportional to the concentration gradient across the thickness of the medium considered and the constant of proportionality is called mass diffusion coefficient. As discussed in basic heat transfer textbooks, under certain conditions or assumptions, both heat and mass transfer can be treated analogous to each other. Drying of porous materials involves simultaneous heat and mass transfer processes, and both the transport processes are coupled to each other. However, drying is generally a mass controlled process, wherein the concentration difference between the water vapor causes the vapor transport. In the example of an evaporative cooling process in which air is forced to flow over a wet porous medium, evaporation of water vapor causes the temperature of the air to drop a little, thus causing a cooling effect.

The conservation principle used to derive the species conservation equation is stated as follows:

$$\begin{aligned}&(\text{Rate of accumulation of species concentration within the CV}\\&\quad+\text{Net efflux of species concentration from the CV})\\&\quad=\left(\text{Diffusion of species concentration}+\text{mass of species generated in the CV}\right)\end{aligned} \tag{2.83}$$

It is understood from thermodynamics that all work can be converted into heat by friction; however, the reverse is not possible. Derivation of thermal energy conservation is required to consider work energy and its conversion into thermal energy. However, in the case of species concentration, there is only mass transport by diffusion, which is governed by Fick's law of mass diffusion and convective mass transfer. Hence, Equation (2.82) can be recast to provide the conservation equation for species concentration transport in a fluid medium, as follows.

$$\left(\frac{\partial C}{\partial t}+u\frac{\partial C}{\partial x}+v\frac{\partial C}{\partial y}+w\frac{\partial C}{\partial z}\right)=D\left(\frac{\partial^2 C}{\partial x^2}+\frac{\partial^2 C}{\partial y^2}+\frac{\partial^2 C}{\partial z^2}\right)+\dot{m}_{gen} \tag{2.84}$$

In the above equation, C represents species concentration in $\frac{\text{kmol}}{\text{sm}^3}$, D, mass diffusion coefficient in $\frac{\text{m}^2}{\text{s}}$ for the given set of species in the fluid medium and $\dot{m}_{gen}$ species generated in the CV in $\frac{\text{kmol}}{\text{sm}^3}$. The unit of each term in Equation (2.84) corresponds to $\frac{\text{kmol}}{\text{sm}^3}$. The first term on the left-hand side of Equation (2.84) refers to the rate of accumulation of species concentration within the control volume with time, and the second term corresponds to the convective transport of species concentration due to mass flow of fluid in all three coordinate directions. The first term on the right-hand side shows the mass diffusion of the species in the flow medium, and the second term is the species generated within the control volume.

REFERENCES

1. *Convective Heat and Mass Transfer*, S. Mostafa Ghiaasiaan, 2nd Edition, CRC Press, Boca Raton, FL, USA, 2018.
2. *An Introduction to Fluid Dynamics*, G. K. Batchelor, Cambridge University Press, Cambridge, UK, 2000.

EXERCISE PROBLEMS

Qn: 1 Water is pumped from a well and stored in an overhead tank in a household application. The house is provided with ten valves at different water outlets to regulate the water usage at different locations in the house.

Develop an imaginary situation wherein the flow of water from the pump and water usage can be treated as a steady state problem. Also derive expressions to represent the input and output water flow to achieve unsteady flow conditions. Make suitable assumptions if required.

Qn: 2 Consider a sidewalk which is heated in the daytime and cooled in the nighttime depending on the variation of the ambient temperature at the given location. Model the situation of heat transfer between the sidewalk and the ambient in the nighttime using basic heat transfer equations. Suggest a method to analyze this problem with an aim to determine the temperature at the top side of the sidewalk which is in direct contact with the ambient. Do your equations satisfy steady state condition or unsteady state conditions?

Qn: 3 Hot wire anemometer is employed to measure velocity of air in many applications. The principle involves the loss of heat by an electrically heated wire in the air medium. With the help of basic flow and heat transfer equations, develop a mathematical model using the conservation principles, to measure the velocity of air. Explain what type of conservation equations are employed highlighting the physics behind the principle of hot wire anemometer.

Qn: 4 A pedestal fan is used for air circulation in a living room. The power input required to run the fan needs to be computed for the given air swing of the fan blades. Develop a mathematical model to estimate the volume flow rate of air and power input to the fan. What are the basic conservation principles that must be considered for modeling this problem? The thermophysical properties of air can be considered constant during the analysis.

Qn: 5 Aerodynamic analyses of the outer body structure of a car have to be carried out using computational fluid dynamic analysis. Specify the type of approach, lumped parameter or differential formulation, that has to be implemented. Choose the required conservation principles and write down the governing equations to compute fluid friction due to flow of air during the running of the car. Suggest a suitable methodology to solve the governing equations.

Qn: 6 Water supply pipe network investigation for a household application has to be modelled using fluid mechanics principles. Is it required to employ differential formulation? If not, suggest a suitable methodology for the study. Draw a line diagram of a piping network for a house consisting of a ground floor and first floor considering suitable water requirements for the house.

Qn: 7 Wind load calculation of high-rise building structures have to be developed. Consider a hyperbolic shaped natural draft cooling tower as an example problem. Write down the conservation equations to be solved for the investigation. What are the assumptions to be made in the analysis?

Qn: 8 Research on fire hazards in high-rise buildings has received attention in the recent past. Consider a ten-storey building subjected to external fire due to some fire accident. It is essential to determine the time taken by the material of the building structure to collapse, so that the occupants can be evacuated without causing injuries. Hence, the building exposed to fire has to be modelled to determine the rate of heating of walls of the building structures. Develop a mathematical model to study the temperature variation across a plane wall of thickness L. Specify the type of conservation equations required to model this problem.

Qn: 9 A solar collector is used to convert solar radiation energy into useful thermal energy using a heat transfer fluid such as water. Develop a mathematical model for the thermal design of a flat plate solar collector. State the assumptions made while developing the governing equations. The main objective of the analysis is to determine the outlet temperature of heat transfer fluid. Identify the parameters that affect the heat absorption rate in the solar collector.

Qn: 10 Cooling of electronic devices can be considered as a number of heat sources stacked in a small channel. It is necessary to develop a mathematical model for the study of cooling of four electronic chips arranged in series in a channel using air as the working medium. Identify the governing equations to model this heat transfer problem with the main objective of computing the heat carried away by the cooling medium. Write down the governing equations with the boundary conditions assuming Dirichlet boundary conditions over the surfaces of the chips.

QUIZ QUESTIONS

Qn: 1 Heat transfer phenomenon can be described as a non-equilibrium/equilibrium process with regard to a thermodynamic definition. Tick the right one.

Qn: 2 A heat transfer mechanism that does not require any medium of transport is called ____________________.

Qn: 3 Heat flux/total heat transfer is a vector quantity. Tick the right one.

Qn: 4 In a steady state conduction heat transfer process, the presence of constant heat generation makes the temperature distribution linear/non-linear. Tick the right one.

Qn: 5 In a fluid momentum conservation equation, the summation of temporal and spatial variation of fluid momentum gives rise to ____________________.

Qn: 6 Pressure in a fluid medium is defined in terms of (a) body force, (b) tangential force, (c) normal force or (d) buoyancy force. Tick the correct one.

Qn: 7 Newton's law of viscosity in a fluid medium is defined over a (a) surface, (b) line segment, (c) volume of fluid or (d) at a point. Tick the correct one.

Qn: 8 Thermal energy equation is obtained from the total energy equation after subtracting the ____________________ and ____________________.

Qn: 9 Tick the correct statement in the following:
Thermal energy equation is derived from the mechanical energy equation. True/False.
The mechanical energy equation is derived from the thermal energy equation. True/False.

Qn: 10 Mass diffusion co-efficient is a characteristic property of (a) volume, (b) area, (c) material thickness or (d) none of the above. Tick the correct one.

3 Finite Difference and Finite Volume Methods

In mathematical modeling, the conservation principles that govern the changes in the field variables in a system are derived in the form of governing equations, represented as ordinary or partial differential equations. If not all the equations, at least some of these differential equations can be solved using analytical techniques, which will enable us to get the solution for the variables at an infinite number of points in the domain. However, due to complex and non-linear boundary conditions and irregular geometrical domain configurations, some of the governing equations in heat transfer and flow problems cannot be solved using analytical methods. Numerical methods are considered to be convenient and the easiest alternate methods for analytical solutions for dealing with differential equations. Now, before we get into the details of different numerical methods, it is essential to understand the type of solution required in the domain. Analytical solutions are expressed in terms of polynomials or a combination of different functions in terms of the spatial coordinates and other relevant properties or parameters. That means the value of the given variable can be obtained at any coordinate point of our choice in the domain of computation. In reality, information about the variables under consideration may not be required at any number of points but only at a certain definite number of points. Analytical methods and numerical methods differ in the selection of these points at which solution has to be obtained during simulation. While using analytical solutions, the solution points can be selected randomly to get the required solution, however, in the case of numerical methods, the number of points has to be decided based on certain propositions before proceeding to computations. Numerical methods provide solutions only at a definite number of points in the computational domain, pre-determined by some principles. That means the analytical solution realized for a continuum problem is getting reduced to solutions at certain discrete domains. When numerical methods are employed care must be taken that the continuity of variables accomplished by analytical technique in a continuum domain, is also satisfied, at least approximately. Numerical methods can provide solutions with certain approximations, deviating from the analytical solution, that is the exact solution. That is why numerical solutions are always validated with either an exact solution if available or with other accepted solutions in the literature.

In the implementation of a numerical method for solving differential equations, the given computational domain is discretized into a number of discrete domains, giving rise to a finite number of grid points or nodal points at which the solution will be secured. It has to be remembered that the very purpose of discretizing the domain is to get a finite number of grid points or solution points which can be achieved by following certain principles. The fundamentals on which the domain is discretized

DOI: 10.1201/9781003248071-3

to obtain the required number of grid points become the characteristic of the method. The most commonly used numerical methods, finite difference, finite volume and finite element methods are fundamentally based on two different approaches; the first is the differential and the second is the integral approach. Other than these popular numerical methods, there are many numerical techniques that can be used to solve differential equations for heat transfer and flow problems. In this textbook, attention will be focused only on the above three methods. Finite difference method is based on the differential approach, wherein the basic definition of the derivative of a continuous function itself is employed to approximate a differential equation. The rate of change of a dependent variable with respect to the independent variable is called a derivative, provided the dependent variable is a continuous function. Let us consider temperature gradient with respect to x-coordinate in one-dimensional heat conduction problems. The first order derivative of temperature represents how far the temperature changes with a small change in the distance traveled in the coordinate direction. In the finite difference method, the solution of the variables will be obtained only at a certain finite number of grid points which are designated using the coordinates. Now the derivative at a point is denoted by means of a difference equation; this means a difference of the nodal values of the variables across the distance considered. In a derivative, the difference of the function between two adjacent points is written as differential, assuming the distance between the adjacent points tends to zero or become very small. One can view the zero limit as the point of continuum and the definite value of the small distance as the discrete domain. Hence, in the finite difference method, the limit zero becomes a discrete distance, thus a differential equation is written as a difference equation. However, to represent the generalized expression for the finite difference method for first order and second order differential equations, Taylor's series expansion scheme is implemented with accepted truncation error by omitting higher order terms.

The domain discretization can be viewed from integral formulation as found in variational and weighted residual methods. These two approaches are the basis for finite element formulation. The variational formulation works on the principle of calculus of variation wherein the solution of the differential equation is equivalent to minimizing a quantity called *functional* derived for the given differential equations. This functional is minimized at the assumed grid points in the computational domain. The most challenging feature of this method is finding the functional for the given differential equation, and variational formulation in the finite element method is commonly used in stress analysis. The weighted residual method assumes a solution for the differential equation and the substitution of the assumed solution gives rise to a residual. As this residue is reduced to zero, then the assumed solution approaches the approximate numerical solution of the equation. Now the original differential equation but with the assumed solution becomes the residue, which needs to be minimized to zero over the discrete domain. However, this integration cannot be carried out because the residue is the result of our assumption of the solution. Hence, it needs to be normalized over the discrete domain using some weighting functions. When the weighting function is assumed to be one, then the resulting discrete domain is called either a subdomain or control volume. Now the residual will be minimized to zero over each control volume around a grid point. In the finite difference method, the differential is converted into difference equations involving the values of the variables at the neighboring grid points in the computational domain, whereas in the case of the finite volume method, a control volume surrounding a grid point is considered and the given equation is integrated over this control volume. In the finite volume method also, the values of the variables at the grid points are treated as the solution of the equation, just like the finite difference method. In both the finite difference and finite volume methods, a finite number of grid points are generated to get the solution at those grid points. Conversion of the differential equation into discrete equations at the grid points is called a discretization equation. It has to be noted that the differential equations completely get modified to difference equations in the finite difference method and control volume equations in the finite volume method, and they are called discretization equations. That means the differential equations developed in the continuum domain are now transformed into discrete equations at certain subdomains. In both these methods, the connectivity of the variables between two adjacent grid points is not well defined in the process of getting the discretization equations. Hence, the given computational domain is discretized such that a number of grid points are generated and the difference equations are written about each grid point using the values of the variables at a specified grid point and its neighboring grid points. Till now only the method of converting the differential equations with respect to the spatial coordinates has been discussed. However, some special care is taken while dealing with time derivatives. Heat transfer and flow problems involve generally first order derivative with time as the independent coordinate. Problems involving second order time derivatives can be treated as a special type of problem. The application of the finite difference method for first order and second order derivatives used in heat transfer and flow problems will be discussed in detail in the following section.

3.1 FINITE DIFFERENCE METHOD

Finite difference method is a well-established technique for solving heat transfer and flow problems [1]. One can easily understand the principle of the finite difference technique by the application of the method to solve different types of differential equations used in heat transfer and flow problems. Let us start with the steady state one-dimensional heat conduction equation shown below (Figure 3.1a).

3.1.1 One-Dimensional Conduction

$$\frac{d^2T}{dx^2} = 0 \tag{3.1}$$

Now, the governing equation expressed by Equation (3.1) has to be solved for temperature distribution within the solid for the specified boundary conditions. Implementation of finite difference method for the solution of temperature involves converting the differential equation into difference equations, for which the grid

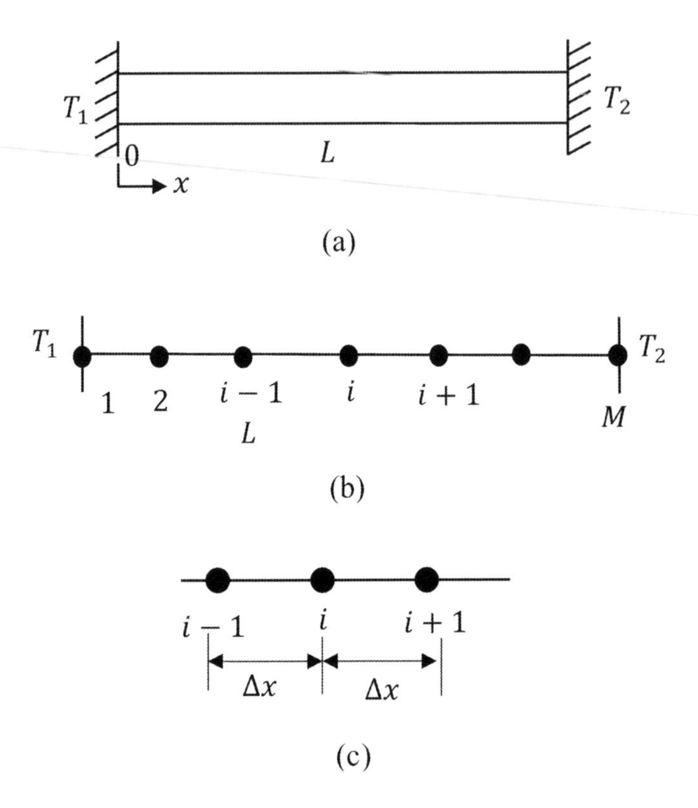

FIGURE 3.1 (a) Schematic diagram of one-dimensional heat conduction, (b) computational domain and (c) Interior node 'i'.

points have to be identified in the computational domain. Figure 3.1(b) represents the computational domain of the solid through which steady state conduction heat transfer is assumed to take place. The computational domain is a replica of the actual system, however, it is shown in a dimension or dimensions in which the variations of the variables are indicated by the governing equation. The solid through which one-dimensional heat conduction takes place is actually a three-dimensional solid, but the variation in temperature is significant only in one direction, hence the computational domain is drawn showing the temperature variation only in x-direction. Figure 3.1(b) shows the computational domain for one-dimensional heat conduction with grid points, starting from 1 to M in the x-direction. A generalized node 'i' is indicated separately highlighting its neighboring nodes, '$i-1$' and '$i+1$', separated by distance Δx from 'i' backward and forward directions respectively. As temperature gradient is significant only in the x-direction, the grid points for temperature variation are shown only in x-direction. At $x=0$, the node number is assigned value '1' and the coordinate is assumed to be zero. Now by adding Δx for the consecutive node, the coordinate of the second node can be obtained, and by repeating this process for all the other nodes, until M, the coordinates of all the nodes can be determined. In the finite difference method, the differential equation is converted into the difference equation. The Taylor series expansion scheme is employed as the generalized formulation for expressing the first order or second order ordinary or partial differential equations, as discussed below.

3.1.2 Taylor's Series Principle

Heat transfer and flow problems involve differential equations of first order or second order equations. The principle of finite difference method can be understood by implementing the Taylor's series expansion to represent first order differential equation. Let us denote T_i, T_{i+1} and T_{i-1} as temperatures at nodes '*i, i+1*' and '*i–1*' respectively along a computational domain for a one-dimensional problem (Figure 3.1(c)).

The first order derivative of temperature $\frac{dT}{dx}$ implies the rate of change of temperature with respect to the spatial coordinate, x as the spatial distance considered for the variation tends to zero or very small. That means across a small distance how the temperature change takes place has to be quantified, and this can be achieved using Taylor's series expansion for T_{i+1} as follows.

Expanding T_{i+1} around T_i, we get

$$T_{i+1} = T_i + \left(\frac{dT}{dx}\right)_i \Delta x + \left(\frac{d^2T}{dx^2}\right)_i \frac{\Delta x^2}{2!} + \left(\frac{d^3T}{dx^3}\right)_i \frac{\Delta x^3}{3!} + \left(\frac{d^4T}{dx^4}\right)_i \frac{\Delta x^4}{4!} + \ldots \tag{3.2}$$

Similarly T_{i-1} can be expanded about T_i to give the following expression.

$$T_{i-1} = T_i - \left(\frac{dT}{dx}\right)_i \Delta x + \left(\frac{d^2T}{dx^2}\right)_i \frac{\Delta x^2}{2!} - \left(\frac{d^3T}{dx^3}\right)_i \frac{\Delta x^3}{3!} + \left(\frac{d^4T}{dx^4}\right)_i \frac{\Delta x^4}{4!} + \ldots \tag{3.3}$$

Referring to Figure 3.1(c), one can understand that the temperature at the adjacent node of '*i*' can be expressed in terms of the temperature at node '*i*' and the derivatives at that node, using Taylor's series expansion. The expressions for first order and second order derivatives can be easily obtained from Equations (3.2) and (3.3) by suitable manipulations. For example, the first order derivative can be written using Equation (3.2) as

$$\left(\frac{dT}{dx}\right)_i = \frac{T_{i+1} - T_i}{\Delta x} - \underbrace{\left(\frac{d^2T}{dx^2}\right)_i \frac{\Delta x}{2}}_{\text{leading error term}} + \text{higher order terms} \tag{3.4a}$$

The same first order derivative using Equation (3.3) can be expressed as

$$\left(\frac{dT}{dx}\right)_i = \frac{T_i - T_{i-1}}{\Delta x} - \underbrace{\left(\frac{d^2T}{dx^2}\right)_i \frac{\Delta x}{2}}_{\text{leading error term}} + \text{higher order terms} \tag{3.4b}$$

Now subtracting Equation (3.3) from Equation (3.2) gives rise to

$$\left(\frac{dT}{dx}\right)_i = \frac{T_{i+1} - T_{i-1}}{2\Delta x} + \left(\frac{d^3T}{dx^3}\right)_i \frac{\Delta x^2}{3} + \text{higher order terms} \quad (3.4c)$$

Equation (3.4a) is called forward difference for the first order derivative as it is represented using temperatures at node 'i' and temperature after Δx in the direction of x-coordinate. Equation (3.4b) is known as backward difference as it involves temperature at node 'i' and temperature before Δx in the opposite direction of the x-coordinate. The above two equations are expressed in terms of temperature at node 'i'. However, Equation (3.4c) approximates the derivative at node 'i' using the temperatures at adjacent nodes of 'i', but with higher accuracy. The error associated with this difference equation is Δx^2 and this equation is called central difference. Now the difference equations can be expressed as

$$\left(\frac{dT}{dx}\right)_i = \frac{T_{i+1} - T_i}{\Delta x} + O(\Delta x) \quad \text{Forward difference equation} \quad (3.5a)$$

with error equivalent to Δx.

$$\left(\frac{dT}{dx}\right)_i = \frac{T_i - T_{i-1}}{\Delta x} + O(\Delta x) \quad \text{Backward difference equation} \quad (3.5b)$$

with error equivalent to Δx.

$$\left(\frac{dT}{dx}\right)_i = \frac{T_{i+1} - T_{i-1}}{2\Delta x} + O(\Delta x^2) \quad \text{Central difference equation} \quad (3.5c)$$

with error equivalent to Δx^2.

Equation (3.5c) shows that the differential can be expressed with higher order accuracy using only the temperatures at two nodes, forward and backward of the node 'i'. Taylor's series is a powerful mathematical series summation expression, which can be employed to represent the first order derivative in different combinations. Now, let us consider the Taylor series expansion of temperature after $2\Delta x$ from 'i' as

$$T_{i+2} = T_i + \left(\frac{dT}{dx}\right)_i 2\Delta x + \left(\frac{d^2T}{dx^2}\right)_i \frac{(2\Delta x)^2}{2!} + \left(\frac{d^3T}{dx^3}\right)_i \frac{(2\Delta x)^3}{3!} + \left(\frac{d^4T}{dx^4}\right)_i \frac{(2\Delta x)^4}{4!} + \ldots \quad (3.6)$$

Now. multiplying Equation (3.2) by 4, we get

$$4T_{i+1} = 4T_i + \left(\frac{dT}{dx}\right)_i 4\Delta x + \left(\frac{d^2T}{dx^2}\right)_i \frac{4\Delta x^2}{2!} + \left(\frac{d^3T}{dx^3}\right)_i \frac{4\Delta x^3}{3!} + \left(\frac{d^4T}{dx^4}\right)_i \frac{4\Delta x^4}{4!} + \ldots \quad (3.7)$$

Now, Equation (3.7) – Equation (3.6) results in

$$4T_{i+1} - T_{i+2} = 4T_i - T_i + \left(\frac{dT}{dx}\right)_i 4\Delta x - \left(\frac{dT}{dx}\right)_i 2\Delta x + \underbrace{\left(\frac{d^2T}{dx^2}\right)_i \frac{4\Delta x^2}{2!} - \left(\frac{d^2T}{dx^2}\right)_i \frac{(2\Delta x)^2}{2!}}_{=0} +$$

$$\left(\frac{d^3T}{dx^3}\right)_i \frac{4\Delta x^3}{3!} - \left(\frac{d^3T}{dx^3}\right)_i \frac{(2\Delta x)^3}{3!} + ...$$

$$4T_{i+1} - T_{i+2} = 4T_i - T_i + (4\Delta x - 2\Delta x)\left(\frac{dT}{dx}\right)_i - \left(\frac{d^3T}{dx^3}\right)_i \frac{4\Delta x^3}{3!} + ...$$

$$\left(\frac{dT}{dx}\right)_i = \frac{4T_{i+1} - 3T_i - T_{i+2}}{2\Delta x} + \left(\frac{d^3T}{dx^3}\right)_i \frac{2\Delta x^2}{3!} + ...$$

$$\left(\frac{dT}{dx}\right)_i = \frac{4T_{i+1} - 3T_i - T_{i+2}}{2\Delta x} + O(\Delta x^2) \tag{3.8}$$

Equation (3.8) expresses the approximation of first order derivative in terms of temperatures at node 'i' and other two consecutive nodes in the forward direction. It is found that the first order derivative can be expressed either using two nodal temperatures or three nodal temperatures adjacent to the node under consideration with maximum error in the order of Δx and minimum of Δx^2. Hence, the given differential equation is converted into a difference equation, here difference means the difference in temperatures of the adjacent nodes. In many numerical analyses, it is found that the approximate expressions derived for the first order differential equation provide acceptable solutions. Following the method of Taylor's series expansion, the approximate difference equations for second order differential equations can also be developed as follows. Consider the approximate equation for $\frac{d^2T}{dx^2}$ in the one-dimensional computational domain. Comparing the right-hand terms of Equations (3.2) and (3.3), one can understand that these two equations differ only by the negative sign of the odd numbered derivatives. By adding Equations (3.2) and (3.3), we get

$$T_{i+1} + T_{i-1} = 2T_i + 2\left(\frac{d^2T}{dx^2}\right)_i \frac{\Delta x^2}{2!} + 2\left(\frac{d^4T}{dx^4}\right)_i \frac{\Delta x^4}{4!}$$

$$\left(\frac{d^2T}{dx^2}\right)_i = \frac{T_{i+1} + T_{i-1} - 2T_i}{\Delta x^2} - \left(\frac{d^4T}{dx^4}\right)_i \frac{\Delta x^2}{12}$$

$$\left(\frac{d^2T}{dx^2}\right)_i = \frac{T_{i+1} + T_{i-1} - 2T_i}{\Delta x^2} + O(\Delta x^2) \tag{3.9}$$

Equation (3.9) is a difference equation for second order derivative with error equivalent to Δx^2. Similar to the approach followed to derive Equation (3.8) for first order derivative, a difference equation for second order derivative can also be obtained. The following manipulation of Equations (3.2) and (3.6) will give rise to the required expression.

$2 \times$ Equation (3.2) – Equation (3.6) produces

$$2T_{i+1} - T_{i+2} = 2T_i - T_i + 2\left(\frac{dT}{dx}\right)_i \Delta x + 2\left(\frac{d^2T}{dx^2}\right)_i \frac{\Delta x^2}{2!} + 2\left(\frac{d^3T}{dx^3}\right)_i \frac{\Delta x^3}{3!}$$

$$+2\left(\frac{d^4T}{dx^4}\right)_i \frac{\Delta x^4}{4!} + \ldots - \left(\frac{dT}{dx}\right)_i 2\Delta x - \left(\frac{d^2T}{dx^2}\right)_i \frac{(2\Delta x)^2}{2!}$$

$$-\left(\frac{d^3T}{dx^3}\right)_i \frac{(2\Delta x)^3}{3!} - \left(\frac{d^4T}{dx^4}\right)_i \frac{(2\Delta x)^4}{4!}$$

$$2T_{i+1} - T_{i+2} - T_i = +\left(2\frac{\Delta x^2}{2!} - 4\frac{\Delta x^2}{2!}\right)\left(\frac{d^2T}{dx^2}\right)_i + \left(2\frac{\Delta x^3}{3!} - 8\frac{\Delta x^3}{3!}\right)\left(\frac{d^3T}{dx^3}\right)_i$$

$$+\left(2\frac{\Delta x^4}{4!} - 16\right)\frac{\Delta x^4}{4!}\left(\frac{d^4T}{dx^4}\right)_i + \ldots \left(\frac{d^2T}{dx^2}\right)_i$$

$$= \frac{T_i - 2T_{i+1} + T_{i+2}}{\Delta x^2} - \Delta x\left(\frac{d^3T}{dx^3}\right)_i + \ldots$$

$$\left(\frac{d^2T}{dx^2}\right)_i = \frac{T_i - 2T_{i+1} + T_{i+2}}{\Delta x^2} + O(\Delta x) \tag{3.10}$$

Comparing Equations (3.9) and (3.10) one can notice that Equation (3.9) approximates the second order derivative with higher accuracy compared to the expression given by Equation (3.10).

3.1.3 Polynomial Method

The finite difference equations derived for the first order and second order derivatives in the previous section are based on the assumption that the spatial discretization Δx is uniform throughout the domain. However, there are many situations wherein non-uniform space discretization needs to be employed, like for the case of finding gradients of field variables accurately at the boundary. In convective heat transfer applications, the temperature gradient at the wall has to be computed with high accuracy in order to compute the Nusselt number which provides the convective heat transfer coefficient. Similar is the case with the determination of drag coefficient for flow over surfaces, in which determination of velocity gradients at the surface is crucial for computing wall shear stress. In another application in computational fluid dynamics, computation of vorticity at the solid boundaries becomes a big challenge because unless these boundary values of vorticity are computed accurately, a divergent free velocity field cannot be achieved while using the velocity-vorticity form of Navier-Stokes equations. In many applications these gradients are estimated during the post-processing stage of a simulation program. That means, after getting the solution of the field variables, certain derived quantities have to be computed as part of the simulation. These derived quantities are strong functions of derivatives of the field variables. In this section, derivation of first order and second order derivatives of a variable will be discussed using second order polynomial fitting, because in most of the flow and heat transfer problems, second order polynomial approximation gives results with acceptable error.

Consider a one-dimensional computational domain, which may be a part of the interior nodes up to the boundary node. Let us assume the temperature is approximated using second order polynomial in x as follows.

$$T(x) = ax^2 + bx + c \tag{3.11}$$

Now, let us consider a boundary node 'i' with the following notations.

$$@\,x = 0, T = T_i$$

$$@\,x = \Delta x, T = T_{i+1}$$

$$@\,x = 2\Delta x, T = T_{i+2}$$

Substitution of the above relations in Equation (3.11) will give the following expressions.

$$@\,x = 0, c = T_i \tag{3.12a}$$

$$@\,x = \Delta x, a\Delta x^2 + b\Delta x + c = T_{i+1} \tag{3.12b}$$

$$@\,x = 2\Delta x, a(2\Delta x)^2 + b(2\Delta x) + c = T_{i+2} \tag{3.12c}$$

Equation (3.12c)$-\,2\times$Equation (3.12b) produces

$$2a\Delta x^2 = T_{i+2} - 2T_{i+1} + c$$

After substituting for 'c' from Equation (3.12a) and simplification, we get

$$a = \frac{T_i - 2T_{i+1} + T_{i+2}}{2\Delta x^2} \tag{3.13a}$$

Substituting for 'a' and 'c' in Equation (3.12b), results in

$$\frac{T_i - 2T_{i+1} + T_{i+2}}{2\Delta x^2} \times \Delta x^2 + b\Delta x = T_{i+1} - T_i$$

After simplification, the expression for 'b' can be obtained as

$$b = \frac{4T_{i+1} - 3T_i - T_{i+2}}{2\Delta x} \tag{3.13b}$$

Now, differentiation of Equation (3.11) with respect to x results in

$$\frac{dT}{dx} = 2ax + b \tag{3.14a}$$

$$\frac{d^2T}{dx^2} = 2a \tag{3.14b}$$

As $x_i = 0$, Equation (3.14a) at point 'i' results in

$$\left.\frac{dT}{dx}\right|_i = b$$

Hence, $\left.\frac{dT}{dx}\right|_i = \frac{4T_{i+1} - 3T_i - T_{i+2}}{2\Delta x}$ which is the same equation obtained using Taylor's series expansion with Δx^2 accuracy with Equation (3.8).

Substitution of Equation (3.13a) in Equation (3.14b) gives rise to $\frac{d^2T}{dx^2} = \frac{T_i - 2T_{i+1} + T_{i+2}}{\Delta x^2}$, the expression obtained by Taylor's series expansion in Equation (3.10). Generally, in computer programs of computational fluid dynamics problems, separate subroutines will be developed to compute the gradients of different variables as a post-processing step, and depending on the accuracy required in the simulation, the polynomial fitting is assumed for the given variable.

3.1.4 Application to Ordinary Differential Equations

After deriving the difference equation for second order differential equation, it is time to implement the finite difference method for a specific problem. Let us consider heat transfer through a fin of rectangular cross-section, which is treated as one-dimensional heat conduction with convection heat transfer. Figure 3.2(a) depicts the fin problem with boundary conditions. The fin is subjected to temperature T_w at $x=0$ and if T_w is greater than T_∞, then the heat received by conduction at $x=0$ is conducted into the solid, which is in turn convected into the ambient. The governing equation for heat transfer through fin [2] can be expressed as

$$kA\frac{d^2T}{dx^2} - h_c P(T - T_\infty) = 0 \tag{3.15}$$

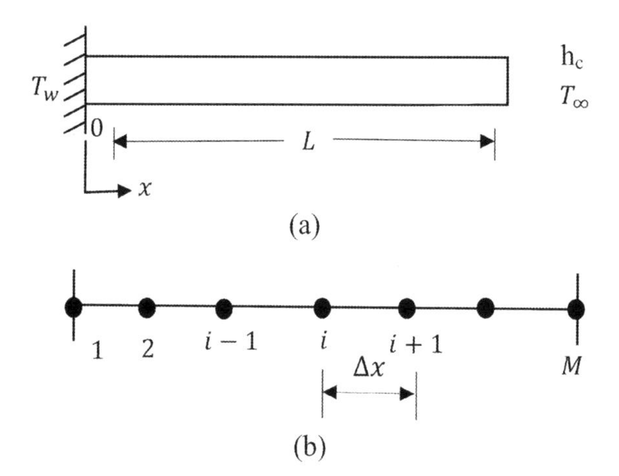

FIGURE 3.2 (a) Schematic diagram of a fin and (b) computational domain.

$$\frac{d^2T}{dx^2} - \frac{h_cP}{kA}(T - T_\infty) = 0 \tag{3.15a}$$

Boundary conditions:

$$@\, x = 0, T = T_w \text{ and } @\, x = L, -k\frac{dT}{dx} = h_c(T_M - T_\infty) \tag{3.15b}$$

Figure 3.2(b) represents the discretized computational domain of the fin.

It is assumed that the number of nodes varies from 1 @ $x=0$ and to M @ $x=L$. Initially the finite difference method will be applied to the governing equations over the discretized domain, that is at all the nodes from *1* to M and then the final set of equations will be modified to incorporate the boundary conditions. Now consider an interior node 'i' for which Equation (3.15a) is rewritten in finite difference form as

$$\frac{(T_{i-1} - 2T_i + T_{i+1})}{\Delta x^2} - \frac{h_cP}{kA}(T_i - T_\infty) = 0$$

$$(T_{i-1} - 2T_i + T_{i+1}) - \frac{h_cP\Delta x^2}{kA}(T_i - T_\infty) = 0$$

Now collecting the terms with different nodal temperatures, it can be written as

$$T_{i-1} - \left(2 + \frac{h_cP\Delta x^2}{kA}\right)T_i + T_{i+1} = -\frac{h_cP\Delta x^2 T_\infty}{kA} \tag{3.16}$$

The above equation is valid for nodes starting from *2* to $(M-1)$, except the boundary nodes at $x=1$ (node 1) and $x=L$ (node M). Equation (3.16) can be simplified by assuming constants $CC1 = \frac{h_c P \Delta x^2}{kA}$, $CC2 = 2 + CC1$ and $CC3 = CC1T_\infty$, Equation (3.16) can be written as

$$T_{i-1} - CC2T_i + T_{i+1} = -CC3 \tag{3.17}$$

The above equation can be written for nodes 2, 3 and 9 as

Node 2

$$T_1 - CC2T_2 + T_3 = -CC3 \tag{3.18a}$$

Node 3

$$T_2 - CC2T_3 + T_4 = -CC3 \tag{3.18b}$$

Node 9

$$T_8 - CC2T_9 + T_{10} = -CC3 \tag{3.18c}$$

It can be observed that writing the above nodal equations for all the nodes will result in a set of $(M-2)$ equations consisting only of three unknown temperatures, thus forming a tridiagonal matrix.

3.1.4.1 Equations for the Boundary Nodes 1 and M

The temperature at node 1 ($x=0$) is known, hence writing this known temperature in the form of Equation (3.18a) to maintain uniformity, we get the equation for

Node 1

$$T_1 + 0 + 0 = T_w \tag{3.18d}$$

As T_1 is known, the difference equation for node 2 will get modified by substituting for T_1 in Equation (3.18a) before solving the final $M \times M$ simultaneous equations. The boundary node at $x=L$ is subjected to the convective boundary condition, and this can be expressed as below.

$$-k\frac{dT}{dx} = h_c(T_M - T_\infty)$$

Let us write the above temperature derivative using the backward difference scheme to get the following equation.

$$-(T_M - T_{M-1}) = \frac{h_c \Delta x}{k}(T_M - T_\infty)$$

$$-T_M + T_{M-1} - \frac{h_c \Delta x}{k} T_M = -T_\infty \frac{h_c \Delta x}{k}$$

$$\left(1+\frac{h_c\Delta x}{k}\right)T_M - T_{M-1} = \frac{h_c\Delta x}{k}T_\infty$$

Collecting the coefficients of nodal temperatures, we get the following equation for *Node M.*

$$\left(1+\frac{h_c\Delta x}{k}\right)T_M - T_{M-1} = \frac{h_c\Delta x}{k}T_\infty$$

Introducing constant $CC4 = (1+\frac{h_c\Delta x}{k})$ and $CC5 = \frac{h_c\Delta x T_\infty}{k}$ the above equation can be written as

$$CC4T_M - T_{M-1} = CC5 \tag{3.18e}$$

Equations (3.18a) to (3.18e) will constitute a set of equations $M \times M$ and can be represented in matrix form as

$$\begin{bmatrix} 1 & 0 & 0 & .. & .. & 0 \\ 1 & -CC2 & 1 & 0 & .. & 0 \\ 0 & 1 & -CC2 & -1 & .. & 0 \\ 0 & 0 & 1 & -CC2 & 1 & 0 \\ . & . & . & . & . & . \\ . & . & . & . & -1 & CC4 \end{bmatrix} \begin{Bmatrix} T_1 \\ T_2 \\ T_3 \\ . \\ . \\ T_M \end{Bmatrix} = \begin{Bmatrix} T_w \\ -CC3 \\ -CC3 \\ . \\ . \\ CC5 \end{Bmatrix} \tag{3.19}$$

Solution of the above $M \times M$ simultaneous equations gives the values of temperatures at different nodes along the length of the fin. The best way of learning modeling and understanding simulation programs is to develop a computer code and get the results for input parameters and compare it with known solutions. For the purpose of developing the computer program, the following computational algorithm is explained:

(i) Read the input parameters, length of fin, cross section of the fin, thermal conductivity of the material of the fin, convective heat transfer coefficient, wall temperature and ambient temperature.

(ii) Assign variables for all the input parameters including number of nodes at which temperature has to be calculated. Evaluate Δx.

(iii) Compute all the constants used as coefficients in the matrix equations, both for the coefficient matrix and the load vector.

(iv) Initialize the coefficient matrix and load vector to zero.

(v) Form the coefficient matrix and load vector for rows starting from *2* to *(M–1)* if *M* is the total number of nodes.

(vi) Modify the coefficient matrix to include the effect of wall temperature at node 1. Similar modifications are carried out at nodes *(M–1)* and M to include the effect of convective boundary condition at *x=L*, that is at node *M*.

(vii) Modify the load vector to include the effect of wall temperature at node *1* and similar modification is required to take care of the convective boundary condition at node *M*.
(viii) Write a code for analytical solution for the convective boundary condition at the tip of the fin.
(ix) Call the simultaneous equations solver to solve the matrix equations to get temperatures at all the nodes along the length of the fin. Also compute the temperatures by analytical solution at the same nodes.
(x) Write the output of temperature in the formats: *x*-coordinate of nodes, finite difference solution of temperatures, analytical solution of temperatures.
(xi) Plot the temperature graphs for the purpose of comparison.
(xii) Repeat the simulation with a different number of nodes to identify the best number of nodes that give results closer to the analytical solution.
(xiii) Now, vary different parameters to obtain simulation results.

Example Problem on Heat Transfer through Fin (Finite Difference Method)

Modeling of a one-dimensional fin is demonstrated by developing a computer program in Fortran. A computer code *Fin_FDM.for* in *FORTRAN*, given in the computer program *CD_For* is used to get the results. The variables used in the above computer program for different input parameters are shown in brackets. The input parameters used are: length of fin (*fin_len*) – 10 mm, height of fin (*height*) – 1 mm, width of fin (*width*) – unit length, thermal conductivity (th_cond), k = 180 W/m K, convective heat transfer coefficient (*h*) = 100 W/m^2 K, wall temperature (*T_wall*) = 100°C and ambient temperature (*T_inf*) = 25°C. Figure 3.3 shows temperature distribution along the length of the fin for different grid points considered in the computational domain. Figure 3.3(a) shows the temperature variation with increase in number of nodes from *6* to *11* and *16*. All the three curves easily merge with each other in the region closer to the wall, that is *x=0* at which the temperature, that is, the unknown, is given. In other words, predictions near the boundary with Dirichlet boundary conditions converge much faster compared to the other end where boundary condition with temperature gradient is specified. An increase in number of nodes and the temperature predicted at *x=L* start merging with each other. As depicted in Figure 3.3(b), with further increase in number of nodes from *50* to *250* and *500*, the convergence at *x=L* looks much better. The plots corresponding to *250* and *500* nodes completely merge with each other even at *x=L*. Convergence of temperature predicted at *x=L* requires a finer mesh, because at that node the temperature gradient is equated to convective heat flux. As only the simple backward difference scheme is implemented to discretize the temperature gradient at *x=L*, accurate computation of gradient is required to satisfy the energy balance at this boundary point. That is why whenever boundary conditions involving the gradients of the variables being computed are specified, a finer mesh has to be used to get accurate results. Figure 3.3(c) shows the comparison between analytical results and numerical results obtained with *500* node points. Needless to say, both the curves coincide with each other.

The results discussed through Figures 3.3(a) and 3.3(b) are called mesh sensitivity analysis and Figure 3.3(c) illustrates validation of the computer program developed

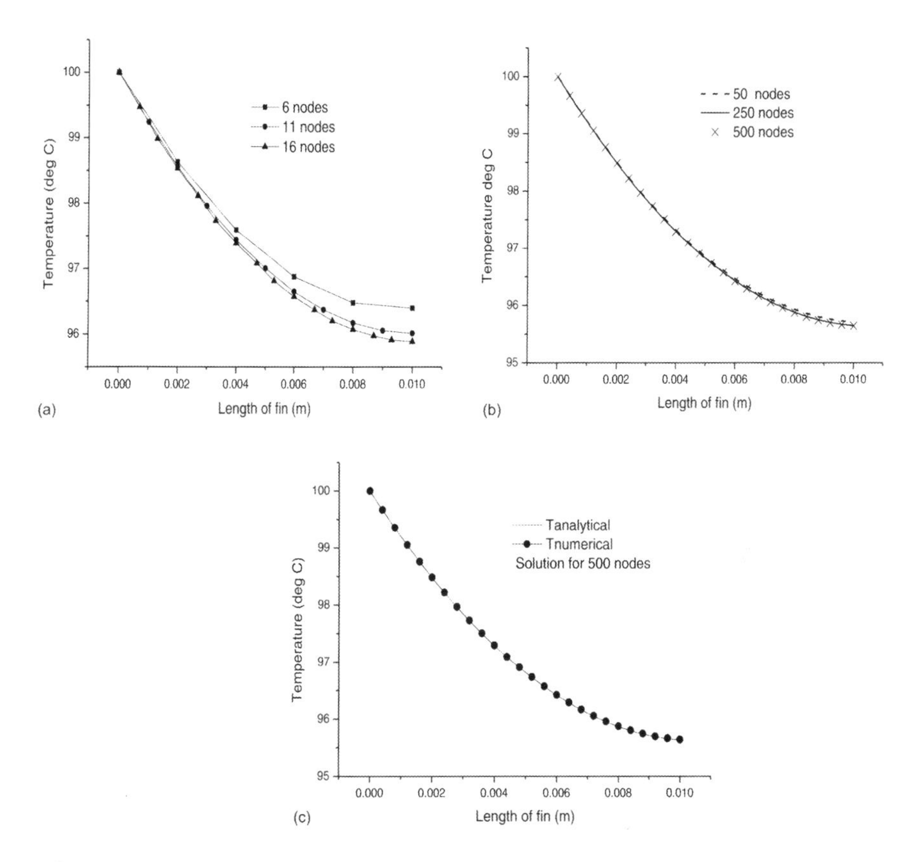

FIGURE 3.3 (a) Mesh sensitivity results for fin with 6 to 16 nodes, (b) results with 50 to 500 nodes and (c) validation results.

for the fin problem. Invariably in any simulation program, mesh sensitivity and validation results are important components that act as pre-conditions before performing any detailed simulations using the computer code developed. The Taylor series expansion discussed for the one-dimensional domain can be extended to the two-dimensional domain as discussed in the following section.

3.1.5 Application to Partial Differential Equations

3.1.5.1 Two-Dimensional Conduction Equation

Let us now implement Taylor's series expansion scheme for a two-dimensional conduction problem (Figure 3.4a). The governing equation for steady state heat conduction through a two-dimensional solid with constant thermal conductivity is expressed as

$$\frac{\partial^2 T}{\partial x^2} + \frac{\partial^2 T}{\partial y^2} = 0 \tag{3.20}$$

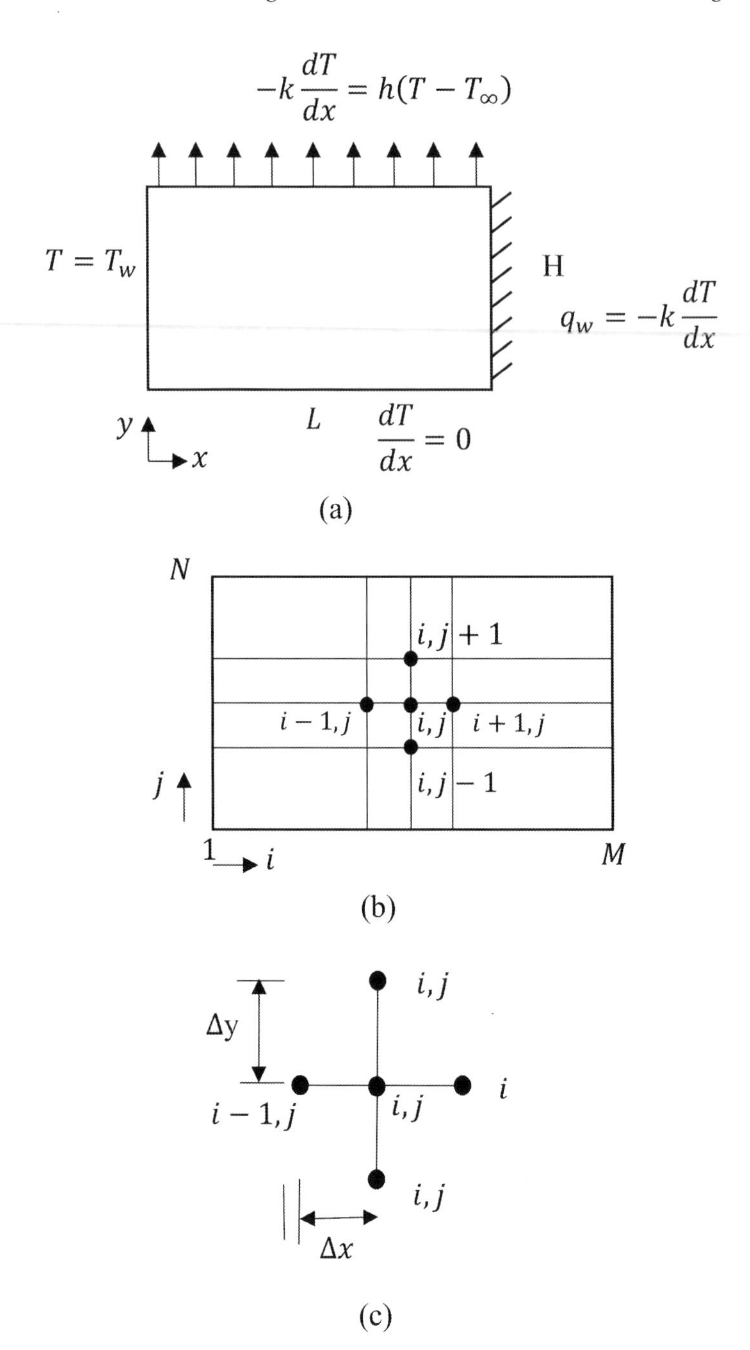

FIGURE 3.4 (a) Schematic diagram of two-dimensional heat conduction, (b) computational domain and (c) interior node 'i, j'.

Boundary conditions:

$$@\,y=0, \frac{\partial T}{\partial y}=0 \text{ for } 0<x<L, @\,y=H, -k\frac{\partial T}{\partial n}=h_c(T_w-T_\infty) \text{ for } 0<x<L$$

$$@\,x=0, T=T_w \text{ for } 0<y<H, @\,x=L, -k\frac{\partial T}{\partial x}=q_w \text{ for } 0<y<H \tag{3.20a}$$

Figure 3.4(b) depicts the computational domain for two-dimensional heat conduction case, in which the temperature gradients are significant in both the x- and y-directions. Any grid point is separated from its neighboring grid points by Δx in the x-direction and Δy in the y-direction and 'i' and 'j' are the indices used to indicate the position of the grid points with respect to x- and y-directions respectively. When a grid is moved through Δx in the x-direction, the index 'i' is increased by 1 and similarly for movement of Δy in the y-direction, the index 'j' increases by 1. In a two-dimensional computational domain, the reference node 'i' is generally considered at the center of the domain, hence in order to number all the grid points throughout the computational domain, one has to move in forward and backward directions while numbering the grid points. For example (Figure 3.4(c)), the four surrounding nodes for the reference node indicated by 'i, j' can be expressed as '$i+1, j$' for the right side node, '$i-1, j$' for the left side node, '$i, j+1$' for the top side node and '$i, j-1$' for the bottom side node. It has to be noted, for moving in the same coordinate direction either in x- or y-direction, '$+1$' is added and for the opposite direction '-1' is added.

Now, we can develop the finite difference equation for second order derivatives using Taylor's series expansion as follows. Consider the derivative $\frac{\partial^2 T}{\partial x^2}$; the difference equation for this derivative has to be obtained at every node point represented by the nodal indices 'i' and 'j'. Consider the expansion with respect to the variable in the x-direction,

$$T(x+\Delta x,y)=T(x,y)+\frac{\partial T(x,y)}{\partial x}\Delta x+\frac{\partial^2 T(x,y)}{\partial x^2}\frac{\Delta x^2}{2!}+\frac{\partial^3 T(x,y)}{\partial x^3}\frac{\Delta x^3}{3!}+\frac{\partial^4 T(x,y)}{\partial x^4}\frac{\Delta x^4}{4!}+\ldots \tag{3.21}$$

and

$$T(x-\Delta x,y)=T(x,y)-\frac{\partial T(x,y)}{\partial x}\Delta x+\frac{\partial^2 T(x,y)}{\partial x^2}\frac{\Delta x^2}{2!}-\frac{\partial^3 T(x,y)}{\partial x^3}\frac{\Delta x^3}{3!}+\frac{\partial^4 T(x,y)}{\partial x^4}\frac{\Delta x^4}{4!}+\ldots \tag{3.22}$$

Adding Equations (3.21) and (3.22), we obtain

$$T(x+\Delta x, y)+T(x-\Delta x, y)=2T(x, y)+\frac{\partial T(x, y)}{\partial x}\Delta x$$

$$-\frac{\partial T(x, y)}{\partial x}\Delta x+2\frac{\partial^2 T(x, y)}{\partial x^2}\frac{\Delta x^2}{2!}+$$

$$\frac{\partial^3 T(x, y)}{\partial x^3}\frac{\Delta x^3}{3!}-\frac{\partial^3 T(x, y)}{\partial x^3}\frac{\Delta x^3}{3!}+2\frac{\partial^4 T(x, y)}{\partial x^4}\frac{\Delta x^4}{4!}+\ldots$$

After simplification, it can be written as

$$\frac{\partial^2 T(x, y)}{\partial x^2}=\frac{T(x+\Delta x, y)-2T(x, y)+T(x-\Delta x, y)}{\Delta x^2}+\frac{\partial^4 T(x, y)}{\partial x^4}\frac{\Delta x^2}{12}+\ldots$$

That is, $\dfrac{\partial^2 T(x, y)}{\partial x^2}=\dfrac{T(x+\Delta x, y)-2T(x, y)+T(x-\Delta x, y)}{\Delta x^2}+O(\Delta x^2)+\ldots$

Referring to Figure (3.4(c)), the above equation can be written in nodal indices as

$$\left.\frac{\partial^2 T}{\partial x^2}\right|_{i,j}=\frac{\left(T_{i-1,j}-2T_{i,j}+T_{i+1,j}\right)}{\Delta x^2}+O(\Delta x^2) \tag{3.23}$$

Similarly, consider $\dfrac{\partial^2 T}{\partial y^2}$, the finite difference equation can be obtained with respect to the y variable by expanding the following terms.

$$T(x, y+\Delta y)=T(x, y)+\frac{\partial T(x, y)}{\partial y}\Delta y+\frac{\partial^2 T(x, y)}{\partial y^2}\frac{\Delta y^2}{2!}$$

$$+\frac{\partial^3 T(x, y)}{\partial y^3}\frac{\Delta y^3}{3!}+\frac{\partial^4 T(x, y)}{\partial y^4}\frac{\Delta y^4}{4!}+\ldots$$

and

$$T(x, y-\Delta y)=T(x, y)-\frac{\partial T(x, y)}{\partial y}\Delta y+\frac{\partial^2 T(x, y)}{\partial y^2}\frac{\Delta y^2}{2!}$$

$$-\frac{\partial^3 T(x, y)}{\partial y^3}\frac{\Delta y^3}{3!}+\frac{\partial^4 T(x, y)}{\partial y^4}\frac{\Delta y^4}{4!}+\ldots$$

Adding the above two equations and performing the simplifications, we can obtain

$$\left.\frac{\partial^2 T}{\partial y^2}\right|_{i,j}=\frac{\left(T_{i,j-1}-2T_{i,j}+T_{i,j+1}\right)}{\Delta y^2}+O(\Delta y^2) \tag{3.24}$$

The finite difference equation for Equation (3.20) can be expressed by adding Equations (3.23) and (3.24) as

$$\frac{\partial^2 T}{\partial x^2}+\frac{\partial^2 T}{\partial y^2}=\frac{\left(T_{i-1,j}-2T_{i,j}+T_{i+1,j}\right)}{\Delta x^2}+\frac{\left(T_{i,j-1}-2T_{i,j}+T_{i,j+1}\right)}{\Delta y^2}=0$$

Defining an aspect ratio, $\beta=\frac{\Delta x}{\Delta y}$, the above expression can be recast as

$$T_{i-1,j}+T_{i+1,j}+\beta^2\left(T_{i,j-1}+T_{i,j+1}\right)-2\left(1+\beta^2\right)T_{i,j}=0 \tag{3.25}$$

Generally, the value of β is closer to *1* and it cannot be much greater than one. This constant decides the gradients of temperature in x- and y-directions imposed by the choice of domain discretization in order to obtain the required number of grid points. A very high value of β may give rise to steep temperature gradients in the y-direction compared to the x-direction, which is an imposed one, not necessarily by the physics that govern the heat transfer process within the solid by the enforced boundary conditions. Hence, care must be taken in deciding this factor while discretizing the computational domain. If it is assumed that $\Delta x=\Delta y$, the above equation becomes

$$\left(T_{i-1,j}-2T_{i,j}+T_{i+1,j}\right)+\left(T_{i,j-1}-2T_{i,j}+T_{i,j+1}\right)=0$$

$$\left(T_{i-1,j}-4T_{i,j}+T_{i+1,j}+T_{i,j-1}+T_{i,j+1}\right)=0 \tag{3.26}$$

Equations (3.25) and (3.26) are valid for any interior node of the computational domain and they also represent that the difference equation for second order derivatives with respect to x- and y-directions is governed by the temperatures at the given node and the four surrounding nodes, two nodes separated by Δx in the positive and negative x-direction, and two nodes separated by Δy in the positive and negative y-direction. Irrespective of the numbering followed while discretizing the computational domain, the above principle can be applied for developing the difference equations. Using energy balance method also, the above finite difference equations can be developed [2].

3.1.5.2 Difference Equations for Boundary Conditions

The finite difference equations for all the boundary nodes have to be developed to complete the matrix equations for getting the solution for temperature. Figure 3.5a shows the computational domain with the main focus on the finite difference equations for the boundary conditions. The rectangular domain shown in Figure 3.5a consists of four corner nodes. The sides of the domain will be identified as given in the boundary conditions in Equation (3.20a). Any corner node is common for both the adjacent perpendicular surfaces, and hence it is up to the modeler to include a corner node in any one of the boundary surface. In a simulation program, the number of nodes used to discretize the domain is normally high enough to give an approximate solution in comparison with an analytical solution with acceptable

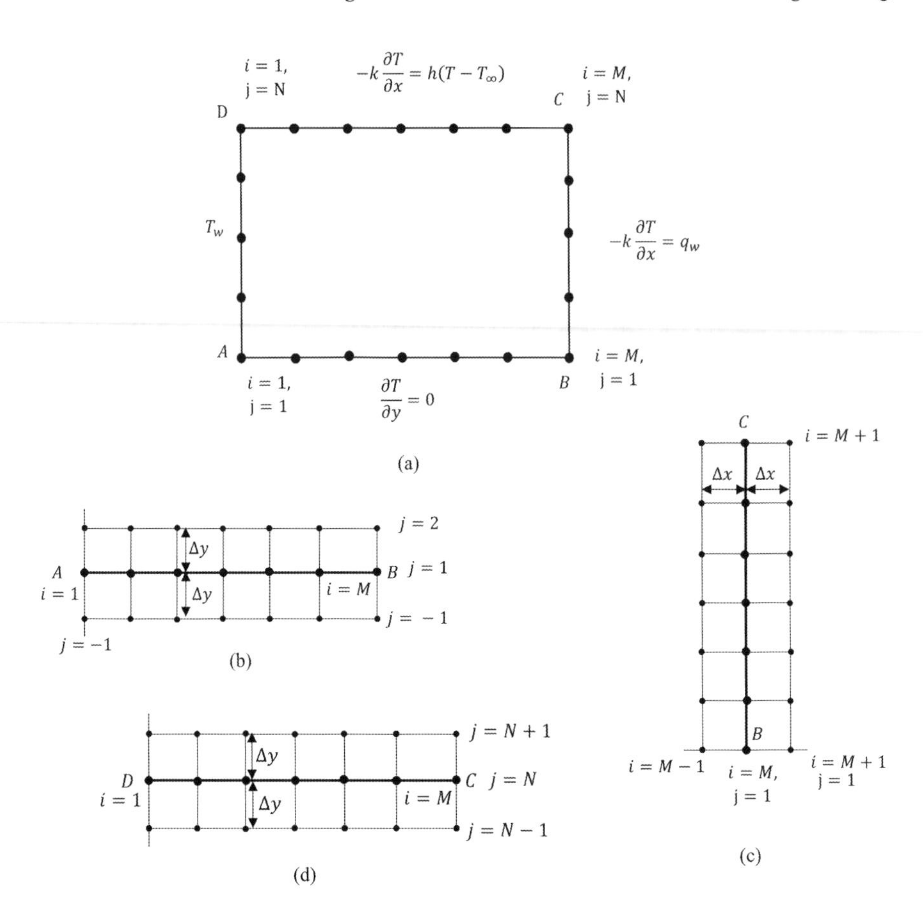

FIGURE 3.5 (a) Computational domain for two-dimensional heat conduction, (b) discretization for adiabatic boundary condition on side AB, (c) discretization for heat flux boundary condition on side BC and (d) discretization for convective boundary condition.

error. Hence, the inclusion of the corner node in any of the perpendicular surfaces will not affect the final solution. As it has been observed in one-dimensional fin simulation, whenever derivative boundary condition is specified, only finer mesh closer to that particular boundary surface will produce accurate results with a reasonable number of grid points. Hence, in the following derivation of finite difference equations for boundary conditions, the central difference scheme will be employed for the derivative of temperature. Treatment of boundary conditions at the corner nodes will be discussed separately.

Left Surface DA (Dirichlet Boundary Condition)

Boundary surface -

$$@\, x = 0, 0 < y < H$$

$$@\, i = 1, j = 2, \ldots\ldots \text{N-1}$$

$$T_{1,j} = T_w \tag{3.27a}$$

Bottom Surface AB (Adiabatic Boundary Condition)

Boundary surface - $@\, y = 0, 0 < x < L$
$@\, j = 1, i = 2, \ldots$ M-1

The bottom boundary surface, *AB* is subjected to the adiabatic boundary condition and hence the temperature gradient has to be determined accurately using the central difference scheme. This can be achieved by considering image nodes at a distance of Δy below the *i=M* and *j=1* boundary surface along the line, *j–1*, as shown in Figure 3.5b. Now writing the difference equation for adiabatic boundary condition, we get

$\frac{\partial T}{\partial y} = 0 = \frac{T_{i,j+1} - T_{i,j-1}}{2\Delta y}$ and this gives

$T_{i,j-1} = T_{i,j+1}$, that is at *j=1*, the bottom boundary surface, this becomes

$$T_{i,0} = T_{i,2}$$

The nodes on the boundary surface at *j=1* can be treated as interior nodes when considered along with nodes at a distance of Δy above and below these boundary nodes. After substituting the above expression in Equation (3.25), the adiabatic boundary condition for *j=1* and *i=2 to M–1* boundary surface can be expressed as

$$T_{i-1,1} + T_{i+1,1} + 2\beta^2 T_{i,2} - 2(1+\beta^2)T_{i,1} = 0 \tag{3.27b}$$

Right Surface BC (Heat Flux Boundary Condition)

Boundary surface - $@\, x = L, 0 < y < H$
$@\, i = M, j = 2, \ldots$ N-1

This surface is subjected to the heat flux boundary condition, wherein the temperature gradient has to be computed using the central difference scheme. By assuming image nodes on the right side of the surface, *i=M* and *j=2, N–2*, at a distance of Δx along the line *M+1* as shown in Figure 3.5c, the difference equation for the heat flux boundary condition can be expressed as

$-k\frac{(T_{M+1,j} - T_{M-1,j})}{2\Delta x} = q_w$ from which the expression for T_{M+1} can be found out as

$T_{M+1,j} = T_{M-1,j} - \frac{2\Delta x q_w}{k}$. After substituting this expression in Equation (3.25), the finite difference equation for the heat flux boundary condition along the right side of the domain can be written as

$$2T_{M-1,j} + \beta^2\left(T_{M,j-1} + T_{M,j+1}\right) - 2\left(1+\beta^2\right)T_{M,j} = \frac{2\Delta x q_w}{k} \tag{3.27c}$$

Top Surface CD (Convective Boundary Condition)

Boundary surface - $@\, y = H, 0 < x < L$
$@\, j = N, i = 2, \ldots$ M-1

The top surface is subjected to the convective boundary condition in which the temperature gradient needs to be evaluated using the central difference scheme with an image node at a distance of Δy above y=H, $0<x<L$ boundary surface (Figure 3.5d).

Using the central difference scheme, the convective boundary condition can be expressed as

$$-k\frac{(T_{i,N+1}-T_{i,N-1})}{2\Delta y}=h(T_{i,N}-T_{\infty})$$

From the above equation, the expression for the image node $T_{i,N+1}$ can be evaluated as

$$T_{i,N+1}=T_{i,N-1}-\frac{2\Delta yh}{k}(T_{i,N}-T_{\infty})$$

Now, after substituting the above expression the finite difference equation for all the boundary nodes for the top surface with the convective boundary condition can be written as

$$T_{i-1,N}+T_{i+1,N}+2\beta^2T_{i,N-1}-2\left(1+\beta^2+\frac{\beta^2\Delta yh}{k}\right)T_{i,N}=-\frac{2\beta^2\Delta yh}{k}T_{\infty} \quad (3.27d)$$

Now, Equations (3.27a) to (3.27d) represent the difference equations for the specified boundary conditions on the four sides of the computational domain. However, these equations are valid only from 2 to *M–1* in the *x*-direction and 2 to *N–1* in the *y*-direction, except the four corner nodes. Derivation of difference equations for the corner nodes are discussed below.

3.1.5.3 Corner Nodes

Bottom Left Corner Node (A)

The bottom boundary surface subjected to adiabatic boundary condition and left side with Dirichlet boundary condition makes this corner *A* (Figure 3.6a). As is already well established for an adiabatic boundary, the following equation can be written using the central difference scheme.

$$T_{1,0}=T_{1,2} \quad (3.28)$$

Along the side DA, the Dirichlet boundary condition is specified, hence there is no need to consider image node. In case, a derivative boundary condition is specified on DA, then an image node has to be assumed in the *x*-direction to the left of corner A. Hence, the temperature $T_{0,1}$ does not exist and is not considered in the final nodal equation.

Now, the finite difference expression for the interior node given by Equation (3.25) can be written at node *(1, 1)* as

$$T_{0,1}+T_{2,1}+\beta^2(T_{1,0}+T_{2,1})-2(1+\beta^2)T_{1,1}=0$$

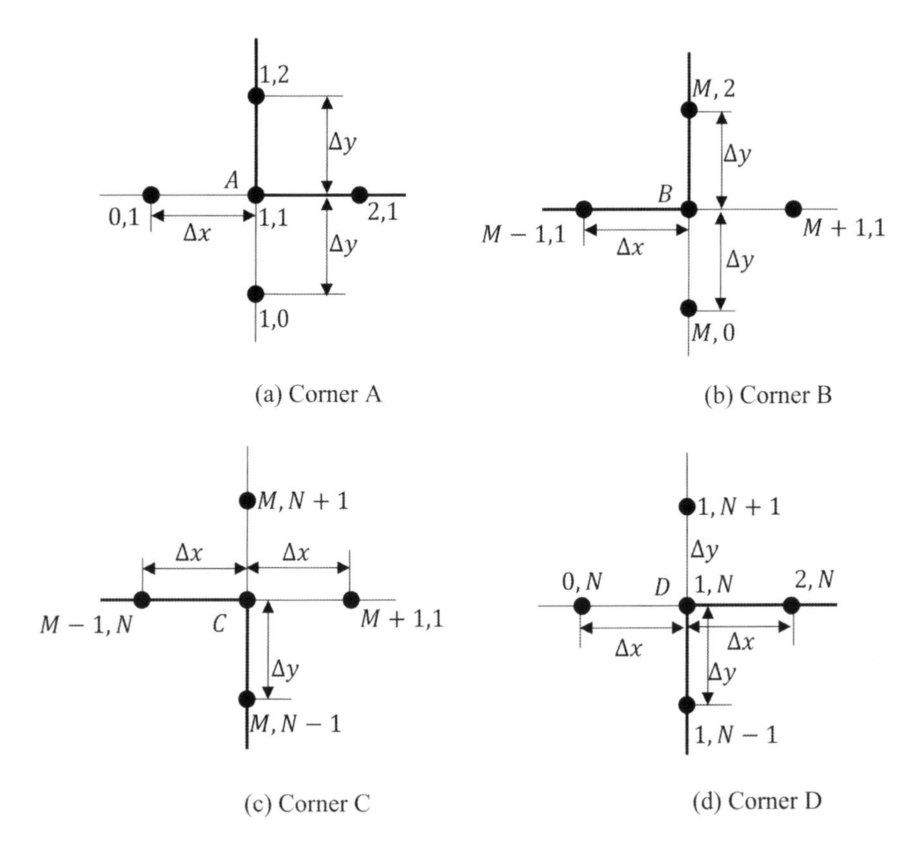

FIGURE 3.6 Finite difference equation for corners nodes (a) corner A, (b) corner B, (c) corner C and (d) corner D.

The substitution of expression given by Equation (3.28) in Equation (3.25), produces the following equation.

$$2\beta^2 T_{1,2} + T_{2,1} - 2(1+\beta^2)T_{1,1} = 0 \tag{3.29}$$

Bottom Right Corner Node (B)

This corner node shares the bottom boundary surface which is subjected to the adiabatic boundary condition and vertical boundary surface with the heat flux boundary condition. Consider the nodes *M–1, 1* on the bottom side, a node preceding the corner node. The temperature at this boundary node depends on the adiabatic boundary condition. Similarly, consider the node *M, 2* on the right vertical boundary surface, and its temperature is influenced by the temperature gradient established due to the constant heat flux entering this surface. The temperatures at both the nodes (*M–1, 1*) and (*M, 2*) have a direct influence on the temperature to be attained at this corner node (Figure 3.6b). However, the application of the image node principle for both the bottom surface and the right surface, while determining the derivatives using central difference, has given the following relations.

$T_{i,0} = T_{i,2}$ for the bottom side can be expressed at the corner node as

$$T_{M,0} = T_{M,2} \tag{3.30a}$$

Similarly, the relation

$T_{M+1,j} = T_{M-1,j} - \frac{2\Delta x q_w}{k}$ for the right side can be expressed at the corner node as

$$T_{M+1,1} = T_{M-1,1} - \frac{2\Delta x q_w}{k} \tag{3.30b}$$

Now, the generalized finite difference equation written for interior node as given by Equation (3.25) can be recast for the corner node, *(M, 1)* as

$$T_{M-1,1} + T_{M+1,1} + \beta^2\left(T_{M,0} + T_{M,2}\right) - 2\left(1+\beta^2\right)T_{M,1} = 0$$

The temperatures $T_{M,0}$ and $T_{M+1,1}$ in the above equation can be substituted from Equations (3.30a) and (3.30b) to include the effect of the respective boundary conditions on the bottom and right side boundary surfaces, on the corner node. After performing these substitutions, the finite difference equation for the corner node *B* is given as

$$2T_{M-1,1} + 2\beta^2 T_{M,2} - 2(1+\beta^2)T_{M,1} = \frac{2\Delta x q_w}{k} \tag{3.31}$$

Top Right Corner Node (C)

The boundary side *BC* exposed to heat flux boundary condition and the side *CD* subjected to convective boundary condition is shared by this corner node, *C*. Referring to Figure 3.6c, one can notice that the temperature gradients on the right side *BC* and top side *CD* can be determined using central difference scheme by assuming image nodes *(M, N+1)* and *(M+1, N)* near the corner at a distance of Δy and Δx respectively. Writing the boundary condition for side *BC* on corner *C*, we get

$-k\frac{(T_{M,N+1} - T_{M,N-1})}{2\Delta y} = q_w$, from which one can obtain the expression for $T_{M,N+1}$ as

$$T_{M,N+1} = T_{M,N-1} - \frac{2\Delta y q_w}{k} \tag{3.32a}$$

Similarly, the convective boundary condition on side *CD* using central difference scheme can be expressed as

$-k\frac{(T_{M+1,N} - T_{M-1,N})}{2\Delta x} = h(T_{M,N} - T_\infty)$, from which the expression for $T_{M+1,N}$ can be written as

$$T_{M+1,N} = T_{M-1,N} - \frac{2\Delta x h}{k}T_{M,N} + \frac{2\Delta x h}{k}T_\infty \tag{3.32b}$$

Now writing the finite difference Equation (3.25) at node *(M, N)*, we get

$$T_{M-1,N} + T_{M+1,N} + \beta^2\left(T_{M,N-1} + T_{M,N+1}\right) - 2(1+\beta^2)T_{M,N} = 0$$

After substituting the expressions for $T_{M,N+1}$ and $T_{M+1,N}$ from Equations (3.32a) and (3.32b) respectively, the above equation can be rewritten for the corner node C as

$$2T_{M-1,N} + 2\beta^2 T_{M,N-1} - 2\left(1+\beta^2+\frac{\Delta xh}{k}\right)T_{M,N} = \frac{2\beta^2\Delta y q_w}{k} - \frac{2\Delta xh}{k}T_\infty \quad (3.33)$$

Top Left Corner Node (D)

This corner node is formed by the side DC subjected to convective boundary condition and side DA with Dirichlet boundary condition (Figure 3.6d).

The convective boundary condition at node *(1, N)* can be expressed using the central difference scheme by assuming an image node *(1, N+1)* at a distance of Δy above the corner D as

$-k\dfrac{(T_{1,N+1} - T_{1,N-1})}{2\Delta y} = h(T_{1,N} - T_\infty)$ from which the expression for $T_{1,N+1}$ can be expressed as

$$T_{1,N+1} = T_{1,N-1} - \frac{2\Delta yh}{k}T_{1,N} + \frac{2\Delta yh}{k}T_\infty \quad (3.34)$$

Now, consider the side DA, subjected to Dirichlet boundary condition, where the temperatures at all the node points are known. As temperature derivative is not required to be obtained along this side, no image node is considered.

The generalized finite difference equation for the interior node given by Equation (3.25) can be expressed at $i=1$ and $j=N$ as

$$T_{0,N} + T_{2,N} + \beta^2 T_{1,N-1} + \beta^2 T_{1,N+1} - 2(1+\beta^2)T_{1,N} = 0$$

In the above equation $T_{0,N}$ does not exist and hence it is not considered in the nodal equation. After substituting the expressions for $T_{1,N+1}$ from Equation (3.34), the final expression for the corner node D can be written as

$$2\beta^2 T_{1,N-1} + T_{2,N} - 2\left(1+\beta^2+\frac{\Delta yh}{k}\right)T_{1,N} = -\frac{2\Delta yh}{k}T_\infty \quad (3.35)$$

Now, finally the finite difference equations for the two-dimensional heat conduction equation expressed by Equation (3.20) with boundary conditions given by Equation (3.20a) have been developed. These difference equations are derived separately for the interior nodes, boundary nodes and the corner nodes. Equation (3.25) represents the finite difference equation for all the interior nodes except the nodes on the boundary surfaces. Equations (3.27a) to (3.27d) express the boundary conditions over the four sides of the computational domain in terms of difference equations except the corner nodes. Finally, the difference equations for the four nodes are given by Equations (3.29), (3.31), (3.33) and (3.35). The finite difference principles discussed in the above section can be implemented to an example problem as discussed below.

Example Problem on Two-Dimensional Conduction Heat Transfer (Finite Difference Method)

The two-dimensional heat conduction equation expressed by Equation (3.20) with the specified boundary conditions as given by Equation (3.20a) can be solved using the finite difference method. Figure 3.7 shows the discretized computational domain of a square section solid with $\beta = 1$.

There are four interior nodes, eight boundary nodes and four corner nodes, thus consisting of a total of 16 nodes. The application of the finite difference method will produce 16 simultaneous equations to be solved for temperature at these 16 nodes. The nodal equations can be written as follows.

Interior Nodes – 6, 7, 10 and 11 (Equation (3.25)

Node 6

$$T_5 + T_7 + T_2 + T_{10} - 4T_6 = 0 \tag{3.36a}$$

Node 7

$$T_6 + T_8 + T_3 + T_{11} - 4T_7 = 0 \tag{3.36b}$$

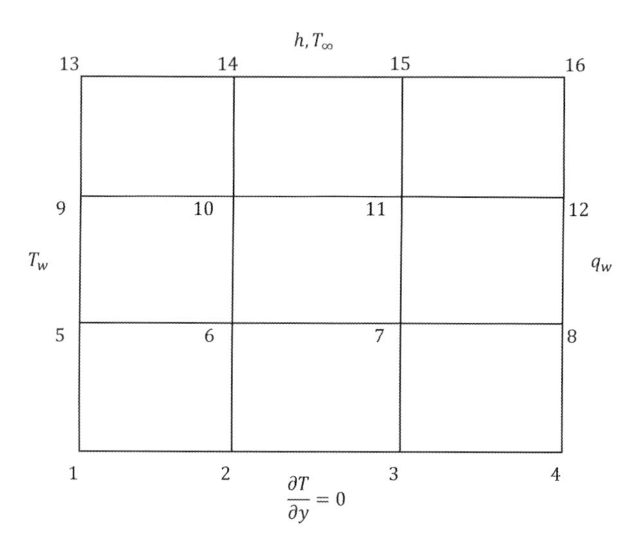

FIGURE 3.7 Two-dimensional conduction problem with 16 nodes.

Node 10

$$T_9 + T_{11} + T_6 + T_{14} - 4T_{10} = 0 \tag{3.36c}$$

Node 11

$$T_{10} + T_{12} + T_7 + T_{15} - 4T_{11} = 0 \tag{3.36d}$$

3.1.5.4 Boundary Nodes

Left Side DA (Equation (3.27a))

$$T_5 = T_w \tag{3.37a}$$

$$T_9 = T_w \tag{3.37b}$$

Bottom Side AB (Equation (3.27b))

Node 2

$$T_1 + T_3 + 2T_6 - 4T_2 = 0 \tag{3.38a}$$

Node 3

$$T_2 + T_4 + 2T_7 - 4T_3 = 0 \tag{3.38b}$$

Right Side BC (Equation (3.27c))

Node 8

$$2T_7 + T_4 + 2T_{12} - 4T_8 = \frac{2\Delta x q_w}{k} \tag{3.39a}$$

Node 12

$$2T_{11} + T_8 + T_{16} - 4T_{12} = \frac{2\Delta x q_w}{k} \tag{3.39b}$$

Top Side CD (Equation (3.27d))

Node 14

$$T_{13} + T_{15} + T_{10} - 2\left(2 + \frac{\Delta y h_c}{k}\right)T_{14} = -\frac{2\Delta y h_c}{k}T_\infty \tag{3.40a}$$

Node 15

$$T_{14} + T_{16} + T_{11} - 2\left(2 + \frac{\Delta y h_c}{k}\right)T_{15} = -\frac{2\Delta y h_c}{k}T_\infty \tag{3.40b}$$

Corner Nodes

Corner A (Node 1)

$$T_2 + 2T_5 - 4T_1 = 0 \tag{3.41a}$$

Corner B (Node 4)

$$2T_3 + 2T_8 - 4T_4 = \frac{2\Delta x q_w}{k} \tag{3.41b}$$

Corner C (Node 12)

$$T_{15} + T_{12} - 2\left(2 + \frac{\Delta x h}{k}\right)T_{16} = \frac{2\Delta y q_w}{k} - \frac{2\Delta x h_c}{k}T_\infty \tag{3.41c}$$

Corner D (Node 13)

$$2T_9 + T_{14} - 2\left(1 + \frac{\Delta y h}{k}\right)T_{13} = -\frac{2\Delta y h_c}{k}T_\infty \tag{3.41d}$$

Equations (3.36) to (3.41) represent the 16 simultaneous equations for the computational domain to be solved for all the nodal temperatures.

3.1.5.5 Comparison of Two-Dimensional Conduction Results with Analytical Solution

The finite difference formulation for the two-dimensional domain given by Equation (3.25) has been implemented for a square geometry of *2 m × 2 m* assuming Dirichlet boundary conditions over all the four sides of the solid. A flow chart detailing the computational algorithm for the solution of temperature is shown in Figure 3.8.

A computer program, *Twod_FDM.for* in *FORTRAN* has been developed to run the simulation results. The computational domain was discretized using 20 equal divisions in both x- and y-directions resulting in a total of 441 node points. Dirichlet boundary conditions have been assumed on all the four sides of the domain with 50°C (T_1) on all three sides except the top side where 150°C (T_2) is assumed. The finite difference equations have been developed using Equation (3.25) for all the nodes, and the resulting 441 simultaneous equations were solved for temperature. Analytical solution for the temperature has been obtained using the following equation. The temperature

$$\frac{T(x,y) - T_1}{T_2 - T_1} = \frac{2}{\pi}\sum_{n=1}^{\infty}\left(\frac{(-1)^{n+1} + 1}{n}\sin\left(\frac{n\pi x}{L}\right)\frac{\sinh\left(n\pi y/L\right)}{\sinh\left(n\pi H/L\right)}\right) \tag{3.42}$$

predicted along the x-direction at $y=0.1\ m$, that is the middle of the domain, was compared with the analytical solutions obtained at the same locations. In

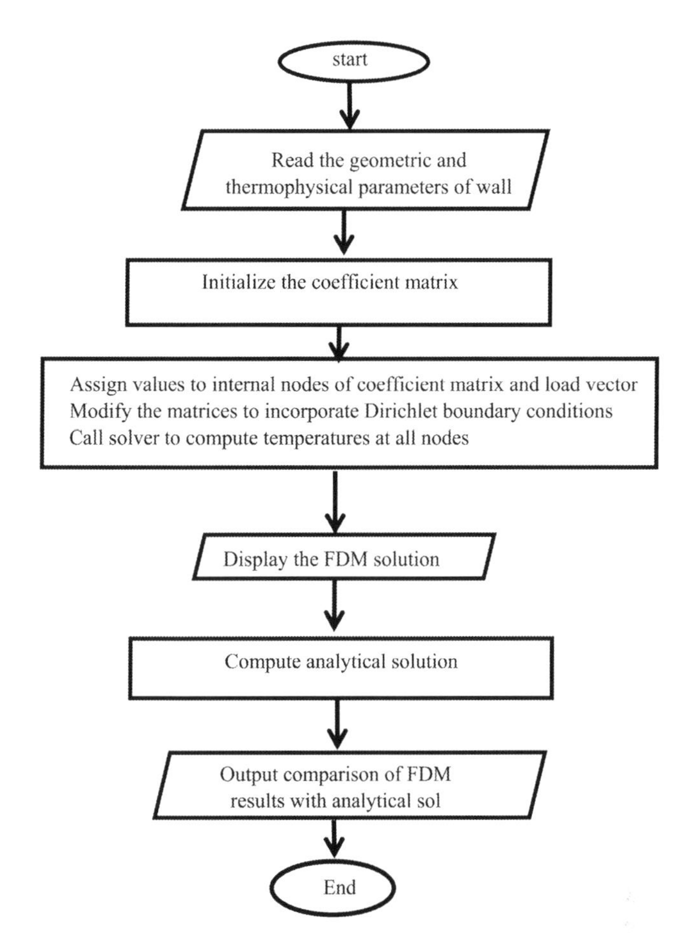

FIGURE 3.8 Flow chart for two-dimensional heat conduction problem using finite difference method (Dirichlet boundary conditions).

Equation (3.42), L and H represent the length and height of the computational domain. Figure 3.9a shows the comparison of temperatures simulated using the finite difference method and the analytical solutions at $y=1.0$ m. Numerical predictions using the finite difference method almost coincide with the analytical solution. The temperature distribution within the computational domain is depicted in Figure 3.9b.

The trends of temperature variations within the domain satisfy the underlying physics as imposed by the Dirichlet boundary conditions on the four sides of the domain. For the known temperatures by Dirichlet boundary conditions, the simultaneous nodal equations obtained have to be modified by including the values of these temperatures. The computational algorithm for including the Dirichlet boundary conditions will be discussed in Chapter 4. The final simultaneous equations obtained in the finite difference technique are diagonally dominant matrices and they are generally solved using iterative techniques.

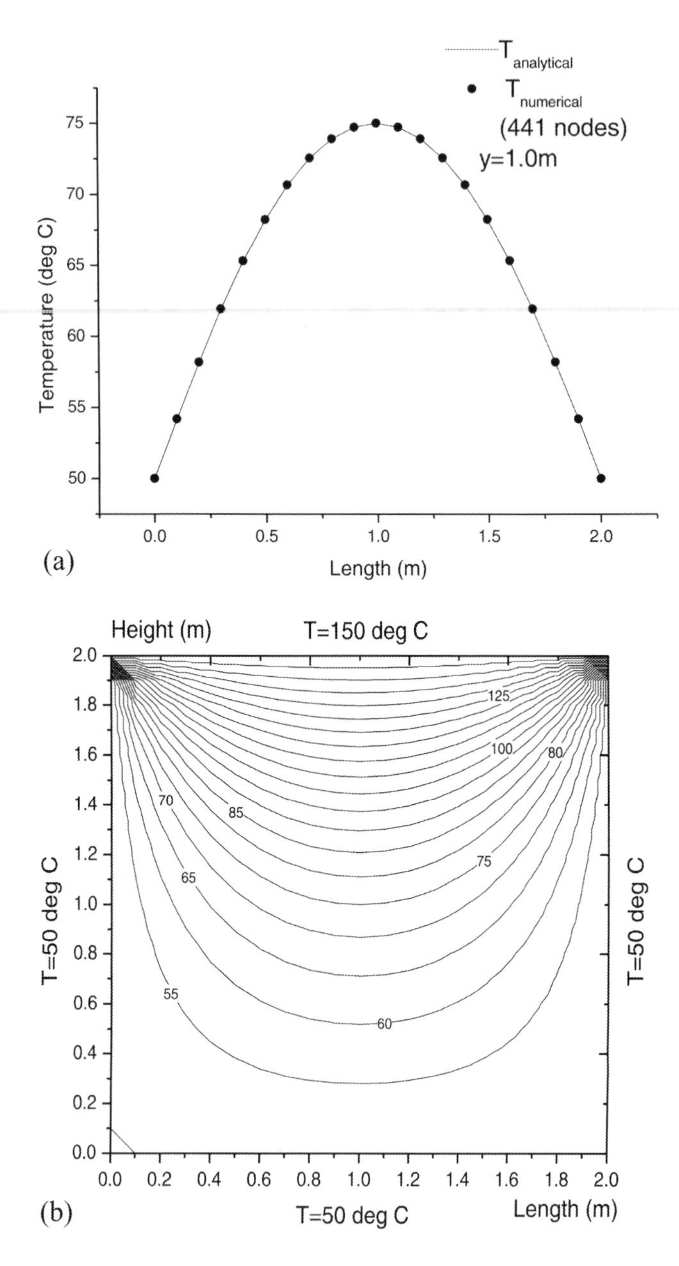

FIGURE 3.9 (a) Comparison of temperatures for finite difference solution and analytical solution along x-direction @ y=1.0 m and (b) temperature contours.

Once the computer program for two-dimensional heat conduction equation solved by finite difference method is validated using analytical solution as discussed in the previous section, the code is modified to handle mixed boundary conditions expressed by Equations (3.27a) to (3.27d). For the boundary conditions shown in Figure 3.5a, the required changes in the code have been included. The computer

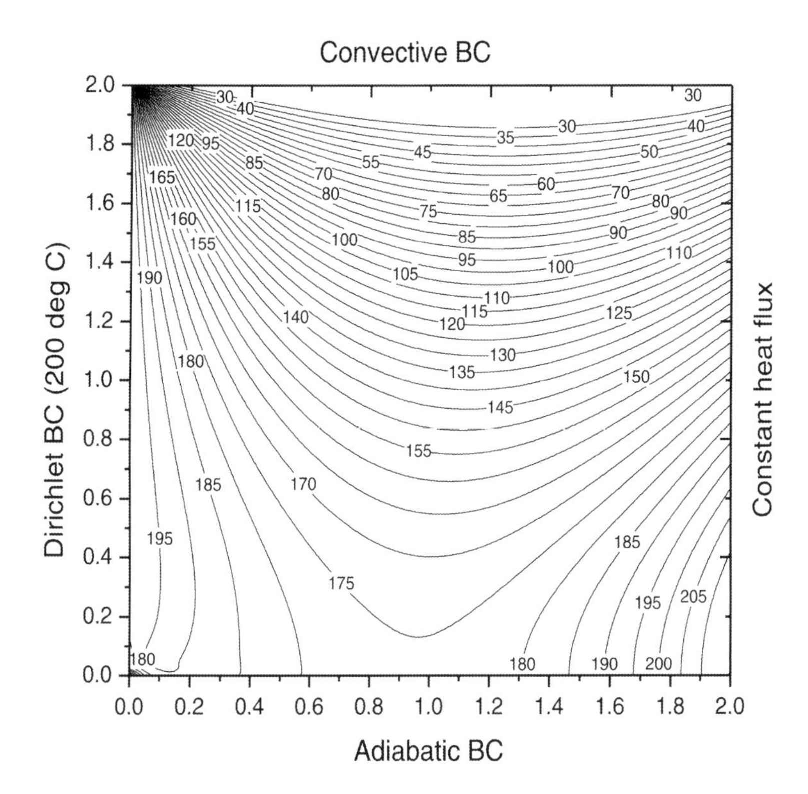

FIGURE 3.10 Temperature distribution for 2-D wall with mixed boundary conditions.

program, *Twod-mbc_FDM.for* in *FORTRAN* has been used to obtain the temperature contours within the computational domain using the input parameters: T_w=200°C on the side *DA*, adiabatic boundary condition on the side *AB*, heat flux, q_w=20 kW on the side *BC* and h_c=10 W/m^2 K and T_∞ =25°C on the boundary *CD*. Figure 3.10 shows the temperature contours within the 2 $m \times 2$ m computational domain discretized into 2601 nodes (51 × 51).

As the bottom side *AB* is subjected to the adiabatic boundary condition, the temperature contours become normal to the boundary as seen in Figure 3.10. With constant heat flux on the right side, *BC*, the temperature gradient becomes constant along the length of this side. Figure 3.10 depicts constant temperature lines with constant slopes along the boundary of *BC*. Thus the temperature contours simulated by the code satisfy the expected underlying physics of the problem in the example considered.

3.2 FINITE VOLUME METHOD

The finite volume method is an integral method, which works on the principle of the weighted residual method [3]. In the finite difference method, the given differential equations are converted into difference equations using Taylor's series expansion scheme. However, in the finite volume method, the governing equations are integrated over a number of discretized subdomains to get the solution of the field variables. The weighted residual method is based on the proposition of an assumed solution to the solution field, wherein the substitution of the assumed solution gives rise to a residue, as the assumed solution is not the exact solution, or at least the expected numerical solution. The minimization of the residue to zero will yield the required solution, however, the residue is multiplied by a weighting function, because the residue is the outcome of our own making due to the assumed solution. There are different types of weighted residual methods, such as collocation method, subdomain method and Galerkin's method, that are well known in the finite element method. The finite volume method can be treated as a subdomain method, in which the weighting function is one. In this method, the given governing equation is integrated over each discretized subdomain with the weighting function equal to *1* for that subdomain and zero for the rest of the domain. The main advantage of the finite volume method is there is no restriction that the subdomains have to be uniform throughout the computational domain; the only condition is that the flux through the interface between the control volumes has to be continuous. The finite volume method is very widely used for solving heat transfer and flow problems, and the method can be demonstrated with the following example.

Consider the fin problem which is described by the governing equation given by Equation (3.15) along with the boundary conditions as expressed by Equation (3.15b). Figure 3.11a shows the schematic diagram of a fin with boundary conditions and Figure 3.11b depicts a control volume of an interior node P considered in the computational domain of the fin. The governing equation is

$$kA\frac{d^2T}{dx^2} - h_cP(T - T_\infty) = 0$$

Let us assume $\tilde{T}(x)$ the solution of the above equation and hence after substituting the solution, we get

$$kA\frac{d^2\tilde{T}}{dx^2} - h_cP(\tilde{T} - T_\infty) = R_\Omega \tag{3.43}$$

The right-hand side of Equation (3.43) is not equated to zero because we are not sure about the assumed solution is the exact solution or the expected approximate numerical solution. However, as $R_\Omega \Rightarrow 0$, then the assumed solution $\tilde{T}(x)$ becomes the numerical solution of the equation. Now, Equation (3.43) can be integrated using the subdomain weighted residual principle as

$$\int_\Omega W_i R_\Omega dx = 0 \tag{3.44}$$

where W_i is the weighting function with unknowns equal to the constants to be determined for the assumed temperature profile for the solution $\tilde{T}(x)$ and its value is equal to *1* for the subdomain method. After substituting for R_Ω from Equation (3.43), the above equation is rewritten as

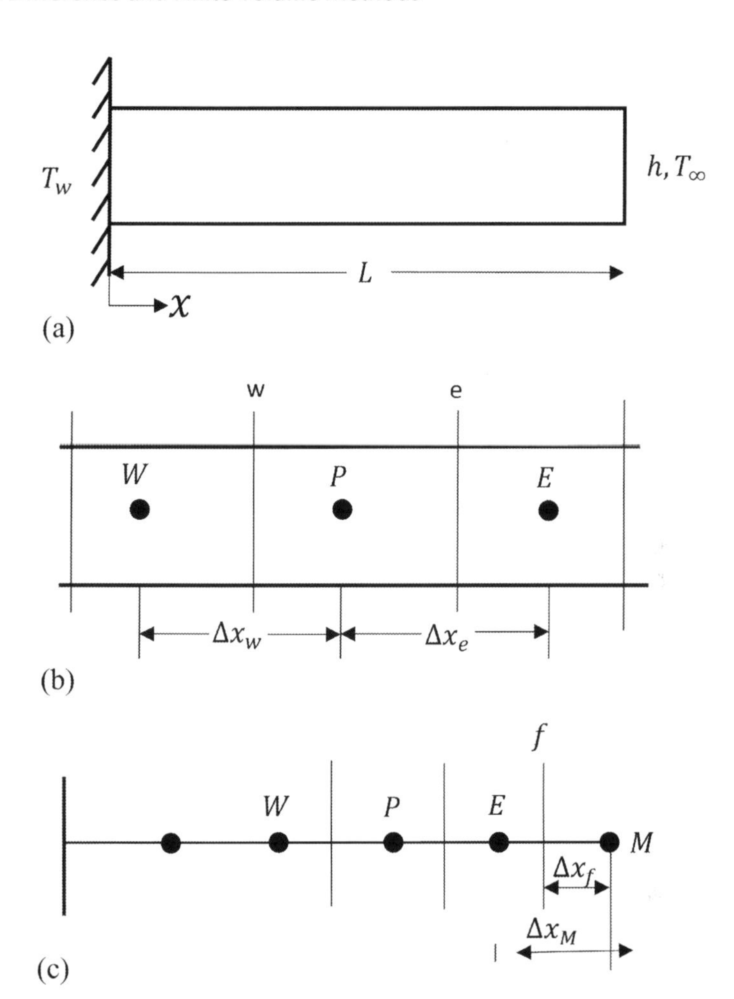

FIGURE 3.11 (a) Schematic diagram of fin, (b) control volume for grid point P and (c) control volume for boundary node M at x=L.

$$\int_{\Omega} 1.\left(kA\frac{d^2T}{dx^2} - h_c P(T - T_\infty) \right) dx = 0 \tag{3.45}$$

For simplicity, $\tilde{T}(x)$ is written as *T(x)* in the above expression. The above equation has to be integrated over the control volume shown in Figure 3.11b. As per the convention followed in finite volume method, a control volume around point P is considered with '*w*' (west) and '*e*' (east) as left and right interfaces with the respective neighboring control volumes around points *W* (west) and *E* (east) respectively. The temperature at point *P* is the representative temperature of the control volume obtained as a result of satisfaction of the conservation equation over the control volume. It is worth remembering that all the conservation equations in differential form are derived by considering a control volume of dimensions equivalent to the elemental distance considered as the limiting values for the differential operators. In the present method, the elemental distance is considered as the distance between the nodal points P and W and P and E. It has to be observed that Δx_w is measured across the interface along 'w', and Δx_e is measured along the interface along 'e', and this is the basic difference of the control volume approach from the finite difference method. In the finite difference method, the distance between nodal points is measured as the discretized domain distance, whereas in the finite volume method, it is the distance across the interface of the control volume around the node under consideration. In the present example of the fin problem, it is the interface across which heat is conducted from one control volume to the neighboring control volume. As the heat is entering through the interface along 'w' and leaving the control volume through the interface along 'e', Equation (3.45) can be integrated between these interfaces as

$\int_w^e kA\frac{d^2T}{dx^2}dx - \int_w^e h_cP(T-T_\infty)dx = 0$, which can be written as

$$kA\frac{dT}{dx}\Big|_e - kA\frac{dT}{dx}\Big|_w - \int_w^e h_cP(T-T_\infty)dx = 0 \tag{3.46}$$

In the above expression, it is the temperature difference between the nodal points E and P that drives the heat conducted across the interface along 'e', and similarly, the temperature between P and W causes the conduction heat transfer across the face along 'w'. Regarding the convective heat transfer term, that is the third term in the above equation, the surface area over which convective heat transfer takes place has to be considered. The surface area of the control volume around the point P contributes to this term. As Δx_w and Δx_e are measured in terms of distances between W and P, and P and E, the arithmetic average of Δx_w and Δx_e will give the surface area of the control volume. Now, Equation (3.46) can be expanded as

$$kA\frac{(T_E-T_P)}{\Delta x_e} - kA\frac{(T_P-T_W)}{\Delta x_w} - h_cP(T_P-T_\infty)\Delta x_{we} = 0$$

where $x_{we} = \dfrac{\Delta x_e + \Delta x_w}{2}$ and Δx_w and Δx_e need not be equal. This is another advantage of the finite volume method. Now assuming $\Delta x_w = \Delta x_e = \Delta x$ and after simplification, the above equation can be written as

$$T_E + T_W - 2T_P - \frac{h_cP}{kA}(T_P - T_\infty)\Delta x^2 = 0$$

After rearrangement of the temperature coefficients and the load term, the above equation is written in the standard finite volume format as

$$a_ET_E + a_WT_W + a_PT_P = b_P \tag{3.47}$$

where $a_E = a_W = 1$, $a_P = -\left(2 + \dfrac{h_cP}{kA}\Delta x^2\right)$ and $b_P = -\dfrac{h_cP}{kA}\Delta x^2 T_\infty$

Equation (3.47) represents the equation for any interior node P within a control volume. In the implementation of the finite volume method, the given computational domain is discretized into small subdomains, generally called control volumes, at the center of which a nodal temperature is assigned which is the representative of that control volume. All the control volumes located in the interior of the computational domain are called interior grid points or control volumes, whereas the control volumes discretized at the boundaries are called boundary grid points or control volumes. Hence, Equation (3.47) can be employed to develop the discretized equation for any interior control volume and this equation can be applied at all the interior control volumes. Now, the procedure to incorporate the boundary conditions has to be evolved. Consider Figure 3.11c; it depicts the discretized computational domain for the fin problem with regard to discussion on implementation of boundary condition at node M at $x=L$.

If temperature at this node is known, then it becomes straightforward to implement the boundary condition using the Dirichlet boundary condition concept. Let us consider heat flux and convective boundary conditions at this boundary node M.

3.2.1 Heat Flux Boundary Condition at M (x=L)

Consider a generalized node point P at the interior through which conduction heat transfer takes place from interface 'w' and leaves at the interface 'e'. The node M is separated by a distance of Δx_M from node E. If the boundary node M is considered as the center node of a control volume, like that of the interior node P, then an interface 'f' separates the nodes M and E. In other words, a half control volume is considered around the boundary node M. Now the energy balance across this half control volume formed by the face 'f' around M will give rise to an equation for T_M. Assuming q_M as positive, energy balance on this half control volume will give

$$(q_f - q_M)A - h_c P \Delta x_f (T_M - T_\infty) = 0 \tag{3.48}$$

$$kA\frac{(T_E - T_M)}{\Delta x_M} - q_M A - h_c P \Delta x_f (T_M - T_\infty) = 0$$

Collecting the coefficient terms for the nodal temperatures, we get

$$a_{Eq} T_{Eq} + a_{Mq} T_M = b_{Mq} \tag{3.49}$$

where $a_{Eq} = \dfrac{kA}{\Delta x_M}$, $a_{Mq} = -\left(\dfrac{kA}{\Delta x_M} + h_c P \Delta x_f\right)$ and $b_{Mq} = -q_M A + h_c P \Delta x_f T_\infty$ in which A is the cross-sectional area of the fin and P is the perimeter of the fin. In the finite difference method, the expression for heat conduction at a given boundary node is directly equated to the heat flux. However, in the case of the finite volume method, it is the energy balance over the control volume around the boundary node that produces the equation for the boundary node.

3.2.2 Convective Boundary Condition at Node M (x=L)

The energy balance over the half control volume around the boundary node M will give rise to Equation (3.48) as

$$(q_f - q_M)A - h_c P \Delta x_f (T_M - T_\infty) = 0$$

If the heat flux q_M is directly known, then Equation (3.49) is the final expression for the boundary node M for heat flux boundary condition. However, in the case of the convective boundary condition, this heat flux becomes equal to the convective heat transfer and hence, the above equation can be rewritten as

$$(q_f - q_M)A - h_c P \Delta x_f (T_M - T_\infty) = 0$$

$$kA \frac{(T_E - T_M)}{\Delta x_M} - h_c A (T_M - T_\infty) - h_c P \Delta x_f (T_M - T_\infty) = 0$$

The above equation indicates that the heat conducted through face 'f' is convected at the boundary node M. In the meantime, the convective heat transfer that takes place over the fin surface in this half control volume across the distance Δx_f cannot be neglected. After collecting the terms for the nodal temperatures, the final equation for convective boundary condition at boundary node M can be expressed as

$$a_{Ec} T_E + a_{Mc} T_M = b_{Mc} \tag{3.50}$$

where $a_{Ec} = \dfrac{kA}{\Delta x_M}$, $a_{Mc} = -\left(\dfrac{kA}{\Delta x_M} + h_c A + h_c P \Delta x_f\right)$ and $b_{Mc} = -(h_c A + h_c P \Delta x_f) T_\infty$

3.2.3 Example Problem for Finite Volume Method – Fin

The finite volume equations derived for the fin problem in the previous section are converted into a computer program in *FORTRAN* language. The same fin problem considered for the finite difference method is considered in the present case also. Program *Fin_FVM.for* is used for validation of the computer code developed and the finite volume formulation derived. Figure 3.12 shows the results obtained using the above computer program. Mesh sensitivity results are shown in Figure 3.12a, which indicate that with refinement of grid points, the consistency of results of temperature distribution along the length of the fin increases. Validation results depicted in Figure 3.12b also confirm that the temperature predicted by the finite volume method perfectly matches with the analytical solution, thus establishing the underlying physics of the problem.

3.2.4 One-Dimensional and Two-Dimensional Applications

3.2.4.1 One-Dimensional Application

The fundamentals of the finite volume method discussed in the above section can now be implemented for a specific problem, and all the steps involved in developing the equations for the interior control volume and boundary control volume can be

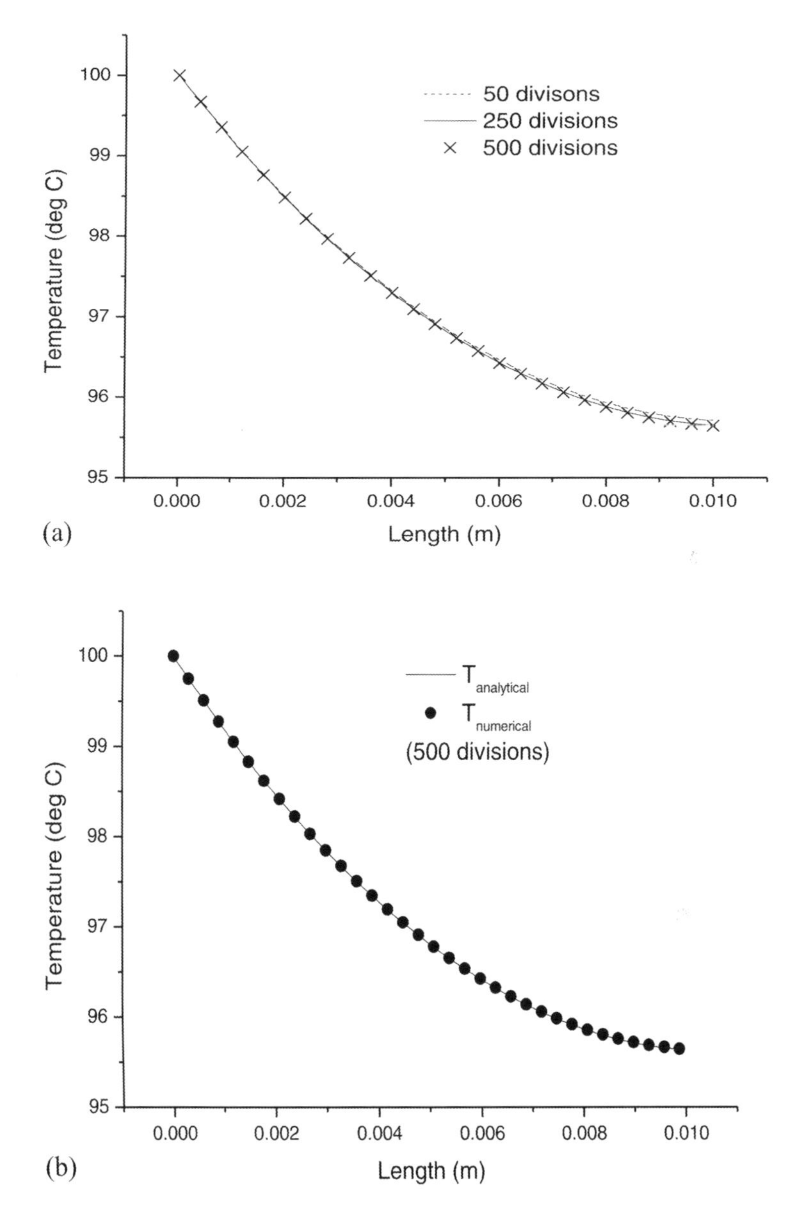

FIGURE 3.12 (a) Mesh sensitivity results for fin and (b) validation results.

explained in this section. One-dimensional steady state heat conduction with heat generation is considered as an example problem. Figure 3.13a shows the schematic diagram of the solid through which conduction is assumed to take place in x-direction with heat generation, $\dot{Q}_{gen}$. The governing equation and boundary conditions are

$$\frac{d}{dx}\left(k\frac{dT}{dx}\right)+\dot{Q}_{gen}=0 \tag{3.51}$$

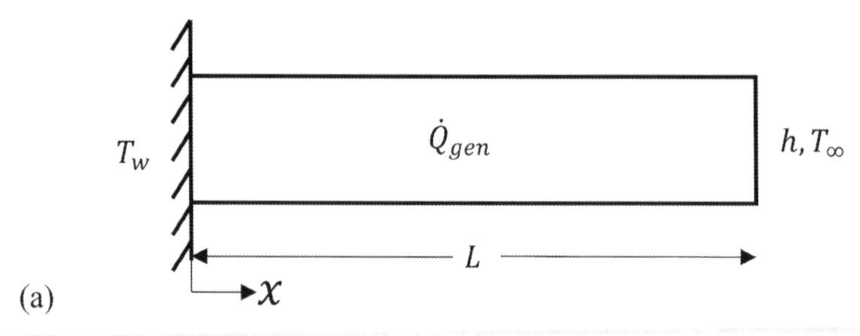

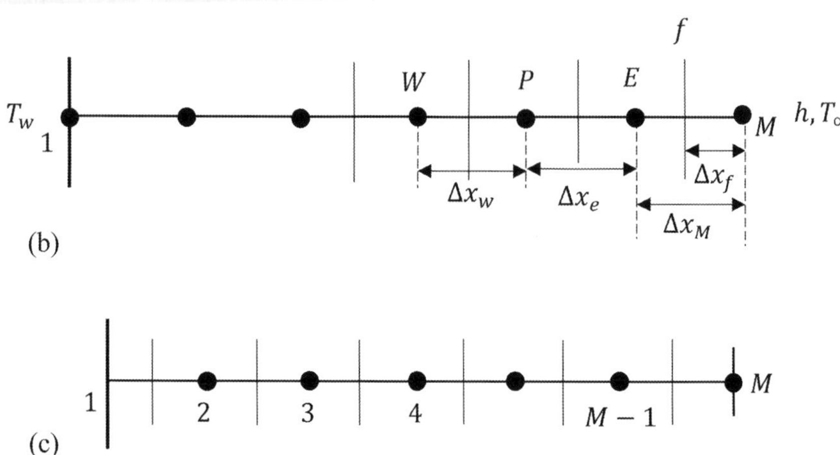

FIGURE 3.13 (a) Schematic diagram of one-dimensional heat conduction with heat generation, (b) control volume for interior node and boundary node M and (c) control volumes from 1 to M.

Boundary conditions

$$@\, x = 0, T = T_w \text{ and } @\, x = L, -k\frac{dT}{dx} = h_c(T - T_\infty) \tag{3.51a}$$

Referring to Figure 3.13b, implementation of finite volume method over the control volume around the interior node P yields

$$k\frac{(T_E - T_P)}{\Delta x_e} - k\frac{(T_P - T_W)}{\Delta x_w} + \dot{Q}_{gen}\Delta x_{we} = 0 \tag{3.52}$$

where $x_{we} = \dfrac{\Delta x_e + \Delta x_w}{2}$.

After collecting the terms of the nodal temperatures, Equation (3.52) can be expressed as

$$a_E T_E + a_W T_W + a_P T_P = b_P \tag{3.53}$$

where $a_E = \dfrac{k}{\Delta x_e}$, $a_W = \dfrac{k}{\Delta x_w}$, $a_P = \left(-\dfrac{k}{\Delta x_w} - \dfrac{k}{\Delta x_e}\right)$ and $b_P = \dot{Q}_{gen}\Delta x_{we}$

Equation for the Convective Boundary Condition at Node M (x=L)

Energy balance over the half control volume from the face f to the boundary node M gives the following expression.

$$q_f - q_M + \dot{Q}_{gen}\Delta x_f = 0$$

After substituting the expressions for the heat quantities in the above equation, we get

$$k\frac{(T_E - T_M)}{\Delta x_M} - h_c(T_M - T_\infty) + \dot{Q}_{gen}\Delta x_f = 0$$

After collecting the terms of the nodal temperatures, the above equation can be written as

$$a_E T_E + a_M T_M = b_M \quad (3.54)$$

where $a_E = \dfrac{k}{\Delta x_M}$, $a_M = -(\dfrac{k}{\Delta x_M} + h_c)$ and $b_M = -(h_c T_\infty + \dot{Q}_{gen}\Delta x_f)$

Now, with the help of Equations (3.53) and (3.54) the nodal equations for all the control volumes discretized in the computational domain can be obtained. For example, consider the computational domain shown in Figure 3.13c with a number of control volumes, resulting in interior grids from 2 to $M-1$ with two boundary nodes, 1 and M. Let us write the equations for the interior nodes 3 and 4.

Node 3

$$a_2 T_2 + a_4 T_4 + a_3 T_3 = b_3 \quad (3.55a)$$

Node 4

$$a_3 T_3 + a_5 T_5 + a_4 T_4 = b_4 \quad (3.55b)$$

Let $\Delta x_w = \Delta x_e$ and constant thermal conductivity, k for the entire solid, then the coefficients, $a_E = a_W$ for all the nodes from 2 to $M-1$ and then, $\Delta x_w = \Delta x_e = \Delta x$ and $\Delta x_{we} = \Delta x$. Application of Equations (3.53) and (3.54) to all the interior and boundary control volumes will result in a set of $M \times M$ simultaneous equations which have to be solved to obtain the temperature field.

3.2.4.2 Two-Dimensional Application

The finite volume method has been widely applied in computational fluid dynamics problems, and this method is highly suitable to solve two-dimensional flow and heat transfer problems. The control volume principle discussed for one-dimensional problems can be extended to two-dimensional problems also. Consider the two-dimensional computational domain shown in Figure 3.14a for steady state conduction problem. The gird point P is surrounded by four control surfaces, 'w'; 'e' in the x-direction; and 's' (south), 'n' (north) in the y-direction; and P lies at a distance of Δx_w and Δx_e from the grid points W and E respectively in the x-direction and Δy_s, Δy_n from the grid points S and N in the y-direction. It is clearly understood from this figure that the surrounding grid points W, E, S and N are all representative of their own control volumes. Application of the finite volume method for two-dimensional

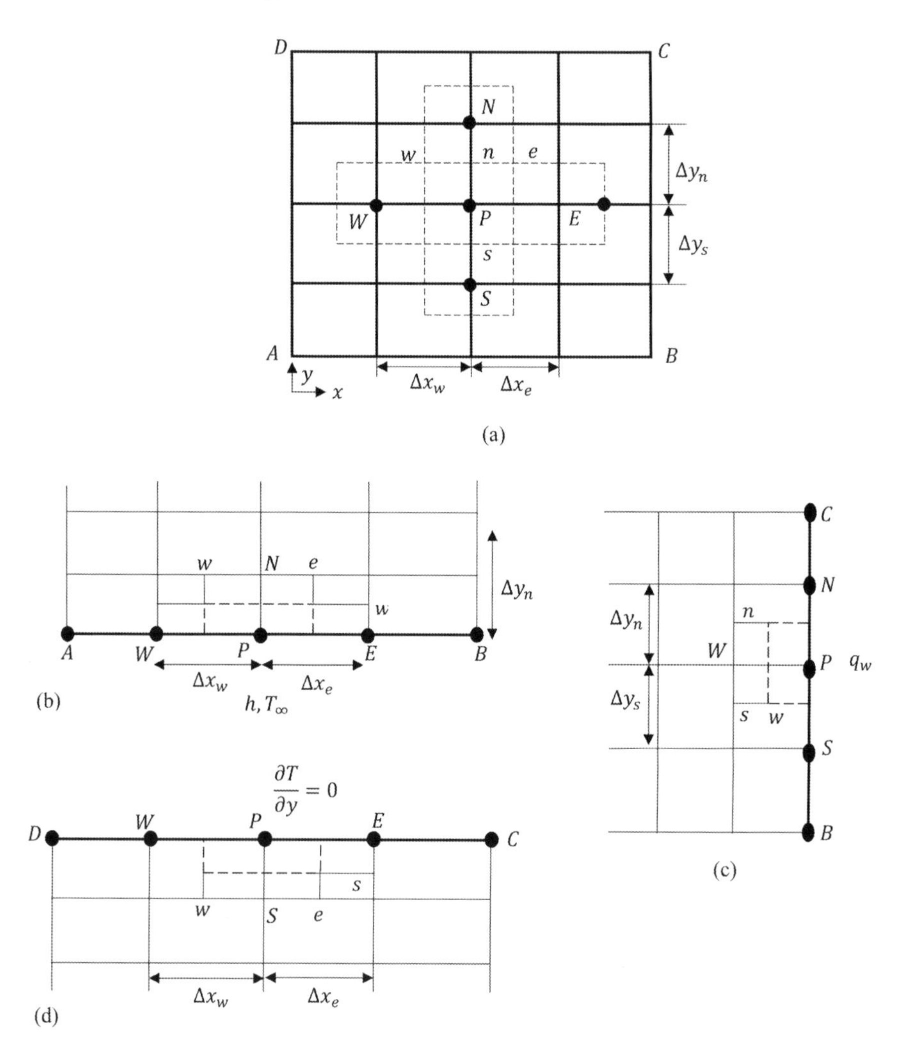

FIGURE 3.14 (a) Control volume for interior node in two-dimensional computational domain, (b) control volume for convective boundary condition, (c) control volume for flux boundary condition and (d) control volume for adiabatic boundary condition.

problems is explained with the help of the two-dimensional steady state conduction equation with heat generation. The governing equation is given as

$$\iint_{CV}\left(\frac{\partial}{\partial x}\left(k\frac{\partial T}{\partial x}\right)+\frac{\partial}{\partial y}\left(k\frac{\partial T}{\partial y}\right)\right)dxdy+\iint_{CV}\dot{Q}_{gen}dxdy=0 \tag{3.56}$$

Initially the above equation will be solved using the finite volume method, and then the implementation of different types of boundary conditions will be discussed. Application of the finite volume method to the above equation will give rise to the following expressions.

$$\iint_{CV}\frac{\partial}{\partial x}\left(k\frac{\partial T}{\partial x}\right)dxdy+\iint_{CV}\frac{\partial}{\partial y}\left(k\frac{\partial T}{\partial y}\right)dxdy+\iint_{CV}\dot{Q}_{gen}dxdy=0$$

$$k\frac{\partial T}{\partial x}\Big|_e\Delta y.1-k\frac{\partial T}{\partial x}\Big|_w\Delta y.1+k\frac{\partial T}{\partial y}\Big|_n\Delta x.1-k\frac{\partial T}{\partial y}\Big|_s\Delta x.1+\dot{Q}_{gen}\Delta x_P\Delta y_P=0$$

where $\Delta x_P=\dfrac{(\Delta x_w+\Delta x_e)}{2}$ and $\Delta y_P=\dfrac{(\Delta y_s+\Delta y_n)}{2}$

Now expanding the heat flux terms, we get

$$k\frac{(T_E-T_P)}{\Delta x_e}-k\frac{(T_P-T_W)}{\Delta x_w}+k\frac{(T_N-T_P)}{\Delta y_n}-k\frac{(T_P-T_s)}{\Delta y_s}+\dot{Q}_{gen}\Delta x_P\Delta y_P=0$$

After collecting the terms corresponding to the nodal temperatures, we get

$$a_ET_E+a_WT_W+a_NT_N+a_ST_S+a_PT_P=b_P \tag{3.57}$$

where $a_E=\dfrac{k}{\Delta x_e}$, $a_W=\dfrac{k}{\Delta x_w}$, $a_N=\dfrac{k}{\Delta y_n}$, $a_S=\dfrac{k}{\Delta y_s}$, $a_P=\left(-\dfrac{k}{\Delta x_e}-\dfrac{k}{\Delta x_w}-\dfrac{k}{\Delta y_n}-\dfrac{k}{\Delta y_s}\right)$
and $b_P=\dot{Q}_{gen}\Delta x_P\Delta y_P$

Equation (3.57) can be used to obtain nodal equations for the finite volume method for interior control volume nodes from 2 to $M-1$ in the x-direction and 2 to $N-1$ in the y-direction. The boundary conditions can be derived as follows.

3.2.4.3 Boundary Conditions

Convective Boundary Condition on Side AB (@j=1, i=2 to $M-1$)

Due to the convective boundary condition on the boundary node P as shown in Figure 3.14b, it is required to determine an equation for temperature T_P on the boundary AB after satisfying an energy balance around the control volume of P. Writing the energy balance we get

$$\left(\text{Heat conducted from grid point } W \text{ to } P\right)$$
$$+\left(\text{Heat conducted from grid point } E \text{ to } P\right)$$
$$+\left(\text{Heat conducted from grid point } N \text{ to } P\right)$$
$$+\left(\text{Heat generated within the control volume surrounding point } P\right)$$
$$=\left(\text{Heat convected to the ambient}\right)$$

$$kA_w\frac{(T_W-T_P)}{\Delta x_w}+kA_e\frac{(T_E-T_P)}{\Delta x_e}+kA_n\frac{(T_N-T_P)}{\Delta y_n}+\dot{Q}_{gen}A_P=h_cA_c(T_P-T_\infty)$$

$$k\left(\frac{\Delta y_n}{2}\right)\frac{(T_W-T_P)}{\Delta x_w}+k\left(\frac{\Delta y_n}{2}\right)\frac{(T_E-T_P)}{\Delta x_e}+k\left(\frac{\Delta x_w+\Delta x_e}{2}\right)\frac{(T_N-T_P)}{\Delta y_n}$$
$$+\dot{Q}_{gen}\left(\frac{\Delta x_w+\Delta x_e}{2}\right)\left(\frac{\Delta y_n}{2}\right)$$
$$=h_c\left(\frac{\Delta x_w+\Delta x_e}{2}\right)(T_P-T_\infty)$$

Collecting the terms corresponding to the nodal temperatures, the above equation can be written as

$$a_{Ec}T_E+a_{Wc}T_W+a_{Nc}T_N+a_{Pc}T_P=b_{Pc} \tag{3.58}$$

where $a_{Ec}=\left(\frac{k\Delta y_n}{2\Delta x_e}\right)$, $a_{Wc}=\left(\frac{k\Delta y_n}{2\Delta x_w}\right)$, $a_{Pc}=\left(-\frac{k\Delta y_n}{2\Delta x_e}-\frac{k\Delta y_n}{2\Delta x_w}-\frac{k\Delta x_w}{2\Delta y_n}-\frac{k\Delta x_e}{2\Delta y_n}-\frac{h_c\Delta x_w}{2}-\frac{h_c\Delta x_e}{2}\right)$, $b_{Pc}=-\dot{Q}_{gen}\left(\frac{\Delta x_w+\Delta x_e}{2}\right)\left(\frac{\Delta y_n}{2}\right)-h_cT_\infty\left(\frac{\Delta x_w+\Delta x_e}{2}\right)$

Heat Flux Boundary Condition on Side BC (i=M, j=2 to $N-1$*)*
In order to obtain the equation at T_P, the following energy balance is performed after satisfying the heat flux boundary condition (Figure 3.14c).

$$kA_w\frac{(T_W-T_P)}{\Delta x_w}+kA_n\frac{(T_N-T_P)}{\Delta y_n}+kA_s\frac{(T_S-T_P)}{\Delta y_s}+\dot{Q}_{gen}A_P=q_PA_q$$

Substitution of the areas in the above equation yields the following equation.

$$k\left(\frac{\Delta y_n+\Delta y_s}{2}\right)\frac{(T_W-T_P)}{\Delta x_w}+k\left(\frac{\Delta x_w}{2}\right)\frac{(T_N-T_P)}{\Delta y_n}+k\left(\frac{\Delta x_w}{2}\right)\frac{(T_S-T_P)}{\Delta y_s}$$
$$+\dot{Q}_{gen}\left(\frac{\Delta y_n+\Delta y_s}{2}\right)\left(\frac{\Delta x_w}{2}\right)$$
$$=q_P\left(\frac{\Delta y_n+\Delta y_s}{2}\right)$$

Collecting the terms for the nodal temperatures, we get

$$a_{W_q}T_W+a_{N_q}T_N+a_{S_q}T_S+a_{P_q}T_P=b_{P_q} \tag{3.59}$$

where $a_{W_q}=k\left(\frac{\Delta y_n+\Delta y_s}{2\Delta x_w}\right)$, $a_{N_q}=k\left(\frac{\Delta x_w}{2\Delta y_n}\right)$, $a_{S_q}=k\left(\frac{\Delta x_w}{2\Delta y_s}\right)$, $b_{P_q}=q_P\left(\frac{\Delta y_n+\Delta y_s}{2}\right)-\dot{Q}_{gen}\left(\frac{\Delta y_n+\Delta y_s}{2}\right)\left(\frac{\Delta x_w}{2}\right)$

Adiabatic Boundary Condition on Side CD ($j=N$, $i=2$ to $M-1$)

The top side of the computational domain *CD* is subjected to the adiabatic boundary condition. The finite volume equation for the boundary node *P* can be developed by carrying out an energy balance over the control volume (Figure 3.14d) surrounding the node *P*.

$$\left(\text{Heat conducted from node } W \text{ to } P\right)$$
$$+\left(\text{Heat conducted from node } E \text{ to } P\right)$$
$$+\left(\text{Heat conducted from node } S \text{ to } P\right)$$
$$+\left(\text{Heat generated within the control volume around } P\right)=0$$

$$kA_w\frac{(T_W-T_P)}{\Delta x_w}+kA_e\frac{(T_E-T_P)}{\Delta x_e}+kA_s\frac{(T_S-T_P)}{\Delta y_s}+\dot{Q}_{gen}A_P=0$$

After collecting the coefficients of nodal temperatures, the above equation can be written as

$$a_{Wa}T_W+a_{Ea}T_E+a_{Sa}T_S+a_{Pa}T_P=b_{Pa} \tag{3.60}$$

where $a_{Wa}=k\left(\frac{\Delta y_s}{2\Delta x_w}\right)$, $a_{Ea}=k\left(\frac{\Delta y_s}{2\Delta x_e}\right)$, $a_{Sa}=k\left(\frac{\Delta x_w+\Delta x_e}{2\Delta y_s}\right)$, $a_{Pa}=k\left(-\frac{\Delta y_s}{2\Delta x_w}-\frac{\Delta y_s}{2\Delta x_e}-\frac{\Delta x_w+\Delta x_e}{2\Delta y_s}\right)$ and $b_{Pa}=-\dot{Q}_{gen}\left(\frac{\Delta x_w\Delta y_s}{2}\right)$

3.2.4.4 Corner Nodes

The finite volume equations to determine the equation for corner nodes is discussed. While developing these equations care must be taken to consider the boundary conditions at the adjacent boundary sides of the domain.

Corner Node (A)

Grid point A is formed by the adjacent boundary sides *AB* and *DA*, with *AB* subjected to the convective boundary condition and *DA* to the Dirichlet boundary condition (Figure 3.15a). Energy balance around the quarter control volume of node A gives rise to the following equation.

$$kA_n\frac{(T_N-T_P)}{\Delta y_n}+kA_e\frac{(T_E-T_P)}{\Delta x_e}+\dot{Q}_{gen}A_P=h_cA_c(T_P-T_\infty)$$

$$k\left(\frac{\Delta x_w}{2}\right)\frac{(T_N-T_P)}{\Delta y_n}+k\left(\frac{\Delta y_n}{2}\right)\frac{(T_E-T_P)}{\Delta x_e}+\dot{Q}_{gen}\left(\frac{\Delta y_n}{2}\right)\left(\frac{\Delta x_e}{2}\right)=h_cA_c(T_P-T_\infty)$$

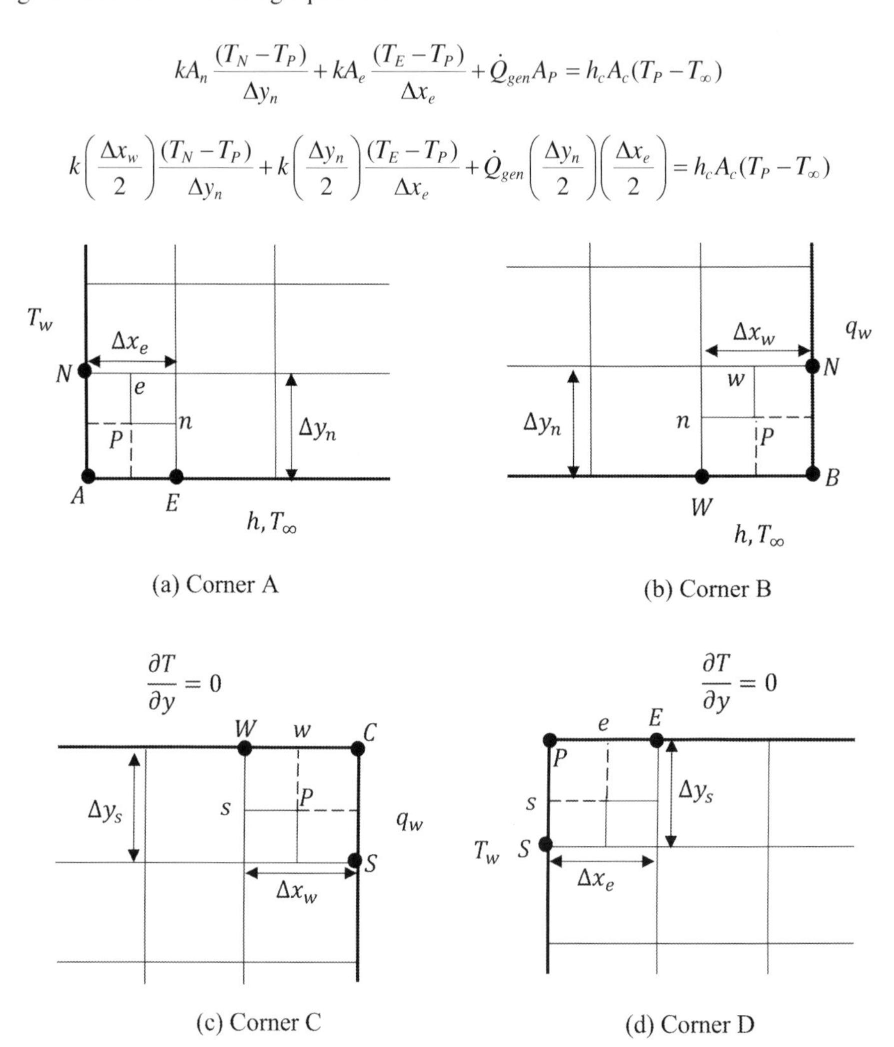

FIGURE 3.15 Control volumes for corner nodes (a) corner A, (b) corner B, (c) corner C and (d) corner D.

The final nodal equation with the coefficients can be written as

$$a_{EA}T_E + a_{NA}T_N + a_{PA}T_P = b_{PA} \tag{3.61}$$

where $a_{EA} = k\left(\dfrac{\Delta y_n}{2\Delta x_e}\right)$, $a_{NA} = k\left(\dfrac{\Delta x_e}{2\Delta y_n}\right)$, $a_{PA} = k\left(-\dfrac{\Delta x_e}{2\Delta y_n} - \dfrac{\Delta y_n}{2\Delta x_e} - h_c\dfrac{\Delta x_e}{2}\right)$,

$b_{PA} = -\dot{Q}_{gen}\left(\dfrac{\Delta x_e \Delta y_n}{4}\right) - \left(\dfrac{h_c \Delta x_e T_\infty}{2}\right)$

Corner Node B

The corner B is formed by side AB subjected to the convective boundary condition and side BC with the heat flux boundary condition (Figure 3.15b). Energy balance over the control volume around node B gives the following equation.

$$kA_n\frac{(T_N - T_P)}{\Delta y_n} + kA_w\frac{(T_W - T_P)}{\Delta x_w} + \dot{Q}_{gen}A_P = h_c A_c(T_P - T_\infty) + qA_c$$

$$k\left(\frac{\Delta x_w}{2}\right)\frac{(T_N - T_P)}{\Delta y_n} + k\left(\frac{\Delta y_n}{2}\right)\frac{(T_W - T_P)}{\Delta x_w} + \dot{Q}_{gen}\left(\frac{\Delta y_n}{2}\right)\left(\frac{\Delta x_w}{2}\right)$$

$$= h_c\left(\frac{\Delta x_w}{2}\right)(T_P - T_\infty) + q\left(\frac{\Delta x_w}{2}\right)$$

The final nodal equation can be expressed as

$$a_{WB}T_W + a_{NB}T_N + a_{PB}T_P = b_{PB} \tag{3.62}$$

where $a_{WB} = k\left(\dfrac{\Delta y_n}{2\Delta x_w}\right)$, $a_{NB} = k\left(\dfrac{\Delta x_w}{2\Delta y_n}\right)$, $a_{PB} = k\left(-\dfrac{\Delta x_w}{2\Delta y_n} - \dfrac{\Delta y_n}{2\Delta x_w} - h_c\dfrac{\Delta x_w}{2}\right)$,

$b_{PB} = -\dot{Q}_{gen}\left(\dfrac{\Delta x_w \Delta y_n}{4}\right) - \left(\dfrac{h_c \Delta x_w T_\infty}{2} - q\dfrac{\Delta y_n}{2}\right)$

Corner Node C

The two adjacent boundary sides BC and CD form corner C, wherein the side BC is subjected to heat flux boundary condition and side CD is adiabatic (Figure 3.15c). Energy balance over the control volume surrounding the corner node C gives the following expression.

$$kA_s\frac{(T_S - T_P)}{\Delta y_s} + kA_w\frac{(T_W - T_P)}{\Delta x_w} + \dot{Q}_{gen}A_P = qA_c$$

$$k\left(\frac{\Delta x_w}{2}\right)\frac{(T_S - T_P)}{\Delta y_s} + k\left(\frac{\Delta y_s}{2}\right)\frac{(T_W - T_P)}{\Delta x_w} + \dot{Q}_{gen}\left(\frac{\Delta y_s}{2}\right)\left(\frac{\Delta x_w}{2}\right) = q\left(\frac{\Delta y_s}{2}\right)$$

The above equation can be written in terms of nodal equation with the respective coefficients as below.

$$a_{WC}T_W + a_{SC}T_S + a_{PC}T_P = b_{PC} \tag{3.63}$$

where $a_{WC} = k\left(\frac{\Delta y_s}{2\Delta x_w}\right)$, $a_{SC} = k\left(\frac{\Delta x_w}{2\Delta y_s}\right)$, $a_{PC} = k\left(-\frac{\Delta x_w}{2\Delta y_s} - \frac{\Delta y_s}{2\Delta x_w}\right)$, $b_{PC} = -\dot{Q}_{gen}\left(\frac{\Delta x_w \Delta y_s}{4}\right) - \left(q\frac{\Delta y_s}{2}\right)$

Corner Node D

The boundary sides *CD* and *DA* form the corner node *D*. The side *CD* is subjected to the adiabatic boundary condition whereas the side *DA* is subjected to the Dirichlet boundary condition (Figure 3.15d).

Energy balance over the control volume around the corner *D* gives the following equation.

$$kA_s\frac{(T_S - T_P)}{\Delta y_s} + kA_w\frac{(T_W - T_P)}{\Delta x_w} + \dot{Q}_{gen}A_P = qA_c$$

$$k\left(\frac{\Delta x_e}{2}\right)\frac{(T_S - T_P)}{\Delta y_s} + k\left(\frac{\Delta y_s}{2}\right)\frac{(T_E - T_P)}{\Delta x_e} + \dot{Q}_{gen}\left(\frac{\Delta y_s}{2}\right)\left(\frac{\Delta x_e}{2}\right) = 0$$

The final nodal equations with respective coefficients is written as

$$a_{ED}T_E + a_{SD}T_S + a_{PD}T_P = b_{PD} \tag{3.64}$$

where $a_{ED} = k\left(\frac{\Delta y_s}{2\Delta x_e}\right)$, $a_{SD} = k\left(\frac{\Delta x_w}{2\Delta y_s}\right)$, $a_{PD} = k\left(-\frac{\Delta x_e}{2\Delta y_s} - \frac{\Delta y_s}{2\Delta x_e}\right)$, $b_{PD} = -\dot{Q}_{gen}\left(\frac{\Delta x_e \Delta y_s}{4}\right)$

Now, Equation (3.57) represents the finite volume equation for the interior grid points of the computational domain in which the two-dimensional heat conduction with heat generation problem needs to be solved. The boundary conditions can be implemented using Equations (3.58) to (3.60) along the boundary sides except the corner nodes, whereas Equations (3.61) to (3.64) can be employed to develop the finite volume equations for the corner nodes of the computational domain.

3.2.5 Complex Geometry and Variable Property

The most important advantage of the finite volume method compared to the finite difference method is its ability to handle curved geometry and the variable property of the medium. In this method, the size of control volumes need not be the same throughout the computational domain, and hence, it is easy to compute gradients of the field variables very accurately by considering small-sized control volumes near the boundaries. The distances, Δx_w, Δx_e, Δy_w and Δy_e considered in defining control volumes in the previous section may be equal or may not be equal, thus enabling discretizing a given computational domain using non-uniform control volumes.

3.2.5.1 Complex Geometry

The computational domains that are discretized during modeling and simulation in flow and heat transfer problems are not always of regular geometry. If the geometries are defined using any one of the Cartesian, cylindrical or polar coordinates, then suitable regular geometrical pattern can be used to discretize the domain. The Cartesian coordinate is the easiest one wherein the subdomains can be easily defined using the coordinates. Similarly, in the case of cylindrical geometry or if it is possible to define a subdomain using cylindrical coordinates, then the dimension of the control volumes can be easily defined. In finite difference and finite volume methods, only the dimension of the control volumes, Δx_w, Δx_e, Δy_w and Δy_e have to be specified, and the connectivity between any two adjacent grid points are left behind undefined. That means the connectivity between the grid points due to the complexity of the geometry of the control volume or subdomain is not defined in these methods. However, this feature will not affect in any way the final computed results of the field variables because the nodal equations are obtained after satisfying the conservation principles in each and every control volume in the finite volume method. The curved geometries are generally approximated by linear approximation in small intervals of the curved line. Hence, geometrical discretization is made free from the type of connectivity between two adjacent nodal values of the field variables. The finite volume method is flexible in dealing with curved surfaces because the sides of the control volumes need not be equal. In the case of the finite difference method, there is no such concept of control volume around each grid point, and the differential equation is converted to the difference equation using Taylor's series expansion scheme. In this method, it becomes more challenging to handle curved boundaries, and generally, linear approximation is implemented to approximate the differential equation, and the accuracy of the difference equation can be improved by choosing a difference equation with higher order accuracy. In the finite difference scheme, the size of the discretized domain depends on the elemental distance used in the differential operator in the respective coordinates. However, in the finite volume method there is no such restriction on the size of the differential element corresponding to the differential operator, and it gives the liberty to choose all different sizes of the sides of a control volume in a two-dimensional domain. The geometrical discretization of computational domain is completely free from the approximation of variation of the adjacent nodal values. The finite element method differs from these two methods in one important feature, in that the user can select the type of variation of the field variable by suitably selecting an approximate polynomial function to represent the variation of the variable within the domain. In this method, geometric approximation and variable approximation can either be chosen by the same function or can be different functions. In this way, finite element method becomes a unique numerical method which enables one to approximate any curved geometry using a suitable discretization function.

3.2.5.2 Variable Property

In two-phase flow heat transfer problems, the thermo-physical properties of the working medium are a strong function of temperature and this can be easily implemented with the help of the finite volume method. This can be demonstrated with the help of a fin problem by assuming thermal conductivity of the solid material as a function of temperature. In general, the thermal conductivity can be written as

$$k = k(T) \tag{3.65}$$

The function on the right-hand side of the above equation may be linear or non-linear depending on the type of material. Once the temperature at a particular grid point is known, then the thermal conductivity at that particular node can be computed using the above equation. Let us revisit Equation (3.46), the finite volume equation derived for the fin problem as below.

$$kA\frac{dT}{dx}\Big|_e - kA\frac{dT}{dx}\Big|_w - \int_w^e h_c P(T - T_\infty)dx = 0$$

The above equation can be rewritten by considering variable thermal conductivity as a function of nodal temperature as

$$k_e A\frac{(T_E - T_P)}{\Delta x_e} - k_w A\frac{(T_P - T_W)}{\Delta x_w} - h_c P(T_P - T_\infty)\Delta x_{we} = 0 \tag{3.66}$$

In the above expression, the thermal conductivity is multiplied by the temperature difference across Δx_e and Δx_w. That means the thermal conductivity should be common for two nodal temperatures and hence, the simplest approximation is to consider the average of the thermal conductivity at the two nodal temperatures. That is $k_e = \frac{(k_E + k_P)}{2}$ and $k_w = \frac{(k_W + k_P)}{2}$, where k_E, k_W and k_P indicate the thermal conductivity evaluated at nodal temperatures, T_E, T_W and T_P respectively. Substituting the above expressions in Equation (3.66), we get

$$\frac{(k_E + k_P)}{2}A\frac{(T_E - T_P)}{\Delta x_e} - \frac{(k_W + k_P)}{2}A\frac{(T_P - T_W)}{\Delta x_w} - h_c P(T_P - T_\infty)\Delta x_{we} = 0$$

After collecting all the terms corresponding to the nodal temperatures, we can write

$$a_{Ek}T_E + a_{Wk}T_W + a_{Pk}T_P = b_{Pk} \tag{3.67}$$

where $a_{Ek} = \frac{A(k_E + k_P)}{2\Delta x_e}$, $a_{Wk} = \frac{A(k_W + k_P)}{2\Delta x_w}$, $a_{Pk} = -\left\{\frac{A(k_E + k_P)}{2\Delta x_e} + \frac{A(k_W + k_P)}{2\Delta x_w} + h_c P \Delta x_{we}\right\}$ and $b_{Pk} = h_c P \Delta x_{we} T_\infty$

Equation (3.67) represents the finite volume equation for any interior node of the computational domain of the fin shown in Figure 3.11. The heat flux and convective boundary conditions derived by Equations (3.49) and (3.50) also have to be modified for the respective thermal conductivity, k_e or k_w as the case may be. Let us modify Equation (3.50) for the convective boundary condition with variable thermal conductivity. Rewriting Equation (3.50) as below by defining the thermal conductivity averaged over the nodes E and M, we get

$$a_{Ec,k} T_E + a_{Mc,k} T_M = b_{Mc,k} \tag{3.68}$$

where $a_{Ec,k} = \frac{(k_E + k_P)A}{2\Delta x_M}$, $a_{Mc,k} = -\left(\frac{(k_E + k_P)A}{2\Delta x_M} + h_c A + h_c P \Delta x_f\right)$ and $b_{Mc,k} = -(h_c A + h_c P \Delta x_f) T_\infty$.

Equations (3.67) and (3.68) can be applied to the fin problem described in Figure 3.11. However, the computational procedure involves an iterative scheme to solve the equations for temperature. Figure 3.16 shows the flow chart for the computational algorithm.

As thermal conductivity is a function of temperature, in order to form the coefficient matrix and load vector matrix, an initial temperature is assumed for all the nodes in the computational domain. Then, the matrices are formed and the equations are solved to get the temperature at different node points. The values of these nodal temperatures are compared with those assumed temperatures, and an iterative procedure is followed until convergence is achieved with some acceptable error limit. The procedure discussed to take care of variable thermal conductivity can be extended for any other property such as specific heat capacity C_p, density ρ and viscosity μ or any other fluid property.

3.2.5.3 Variable Area

The varying cross-sectional area of domains can be easily handled by the finite volume method. Variation of area is generally expressed as a function of coordinates. Let us assume the fin problem in which the area of cross-section of the fin varies continuously along the length of the fin with linear variation. If the relation between the area of the cross-section with respect to the x-coordinate is established, then it can be implemented using the finite volume method.

Let

$$A = A(x) \tag{3.69}$$

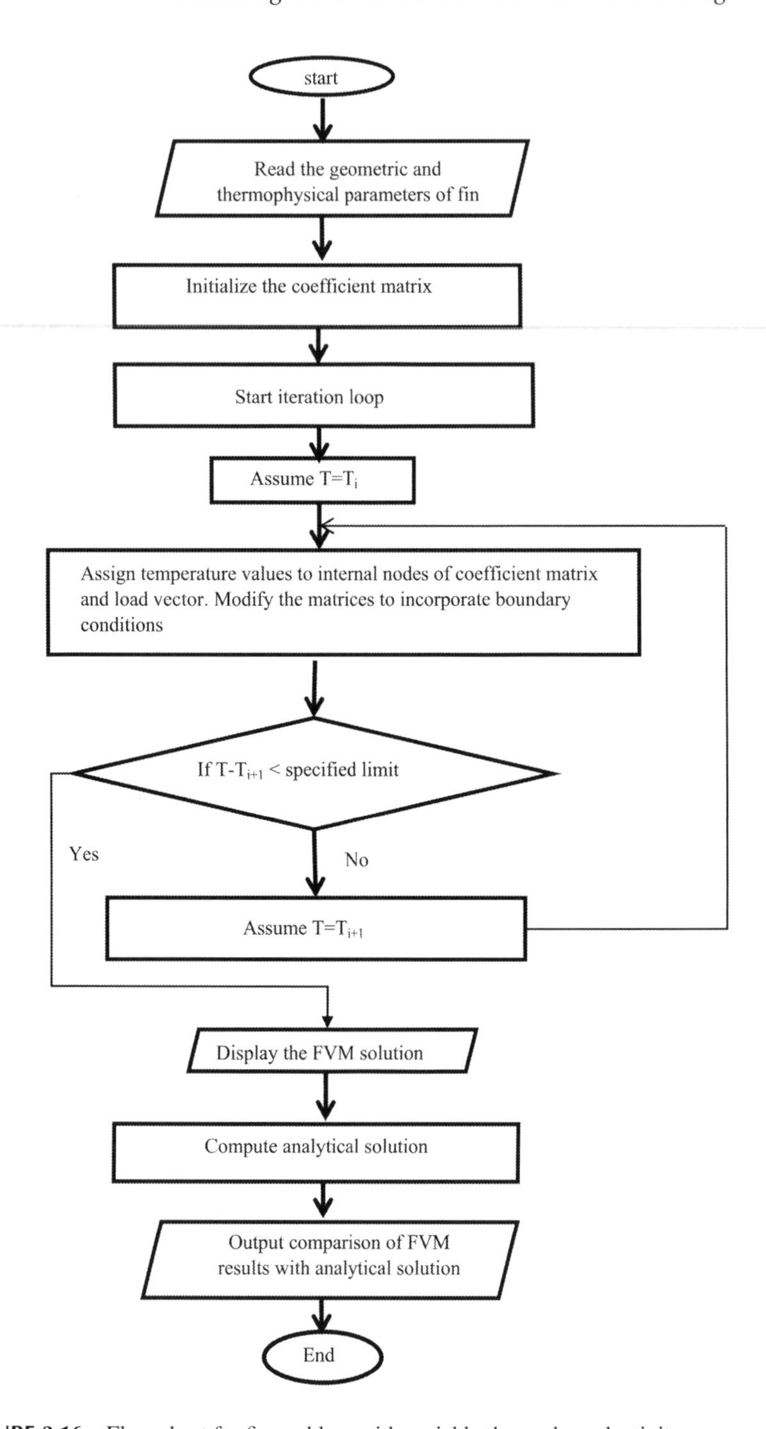

FIGURE 3.16 Flow chart for fin problem with variable thermal conductivity.

Equation (3.46) derived for the fin (Figure 3.11) can be considered as

$$kA\frac{dT}{dx}\Big|_e - kA\frac{dT}{dx}\Big|_w - \int_w^e h_c P(T - T_\infty)dx = 0$$

The above equation can be modified for a varying area of cross-section as

$$kA_e\frac{(T_E - T_P)}{\Delta x_e} - kA_w\frac{(T_P - T_W)}{\Delta x_w} - h_c P(T_P - T_\infty)\Delta x_{we} = 0$$

where $A_e = \frac{(A_E + A_P)}{2}$ and $A_w = \frac{(A_W + A_P)}{2}$, where A_E, A_W and A_P are areas of cross-section at nodes *E*, *W* and *P* respectively. Implementing the variation of area in the finite volume equation, the final equation can be expressed as

$$k\frac{(A_E + A_P)}{2}\frac{(T_E - T_P)}{\Delta x_e} - k\frac{(A_W + A_P)}{2}\frac{(T_P - T_W)}{\Delta x_w} - h_c P(T_P - T_\infty)\Delta x_{we} = 0$$

After collecting the terms for nodal temperatures, the final nodal equation can be written as

$$a_{EA}T_E + a_{WA}T_W + a_{PA}T_P = b_{PA} \tag{3.70}$$

where $a_{EA} = \frac{k(A_E + A_P)}{2\Delta x_e}$, $a_{WA} = \frac{k(A_W + A_P)}{2\Delta x_w}$, $a_{PA} = -\left\{\frac{k(A_E + A_P)}{2\Delta x_e} + \frac{k(A_W + A_P)}{2\Delta x_w} + h_c P\Delta x_{we}\right\}$ and $b_{PA} = h_c P\Delta x_{we}T_\infty$.

The equation for convective boundary condition gets modified as

$$a_{Ec,A}T_E + a_{Mc,A}T_M = -b_{Mc,A} \tag{3.71}$$

where $a_{Ec,A} = \frac{k(A_E + A_P)}{2\Delta x_M}$, $a_{Mc,A} = -\left(\frac{k(A_E + A_P)}{2\Delta x_M} + h_c A_M + h_c P\Delta x_f\right)$ and $b_{Mc,A} = (h_c A_M + h_c P\Delta x_f)T_\infty$ and A_M is the area at the tip of the fin, *M* at x=L.

The computational algorithm for the solution of Equations (3.70) and (3.71) is straightforward, as the area of the cross-section at different control volumes of the computational domain can be computed before evaluating the coefficient matrix and load vector matrix, as the relation is established. Hence, the computational algorithm discussed for the fin problem solved using the finite difference method can be implemented to get the solution for the temperature.

REFERENCES

1. *Finite Difference Methods in Heat Transfer*, M. Necati Ozisik, R. B. Orlande Helcio, Marcelo Jose Colaco and Renato Machado Cotta, 2nd Edition, CRC Press, Boca Raton, FL, USA, 2017.

2. *Fundamentals of Heat and Mass Transfer*, edited by Frank P. Incropera, David P. Dewitt, Theodore L. Bergman and Adrienne S. Lavine, 6th Edition, Wiley, Hoboken, NJ, 2007.
3. *Numerical Heat Transfer and Fluid Flow*, Suhas V. Patankar, Taylor Δ Francis, San Francisco, 1980.

EXERCISE PROBLEMS

Qn: 1 Explain differences between the finite difference method and the finite volume method.

Qn: 2 Explain in detail explicit, implicit and semi-implicit schemes in the finite difference method.

Qn: 3 For a function $f(x) = 4x^2 + 2x$ find $f'(2)$ for different step sizes (a) 0.1 (b) 0.2 (c) 0.5 using the finite difference method and compare errors with analytical solutions.

Qn: 4 Given that $\frac{dy}{dx} = 4y - 3x, y(0) = 2$, using Taylor's series expansion, calculate y for $x = 0.1$.

Qn: 5 The differential equation for the satellite launch is given as $\frac{d^2y}{dt^2} = -g$, where g represents acceleration due to gravity. Obtain the solution using the finite difference method for boundary conditions $y(0) = 0$ and $y(10) = 100.0$.

Qn: 6 The differential equation to calculate deflection in a cantilever beam under uniformly distributed load (Figure P3.1) is given as

$$\frac{d^2y}{dx^2} - \frac{wx(L-x)}{2EI} = 0$$

FIGURE P3.1

where loading intensity (w) is $950\,kN/m$, modulus of elasticity (E) is $200\,\text{GPa}$, moment of inertia (I) is $5 \times 10^7\,\text{mm}^4$ and the length of beam (L) is 2 m. Use a step size of 0.5 m to determine the deflection of beam (y) at $x = 1\,\text{m}$ (Figure P3.2) using the finite difference method.

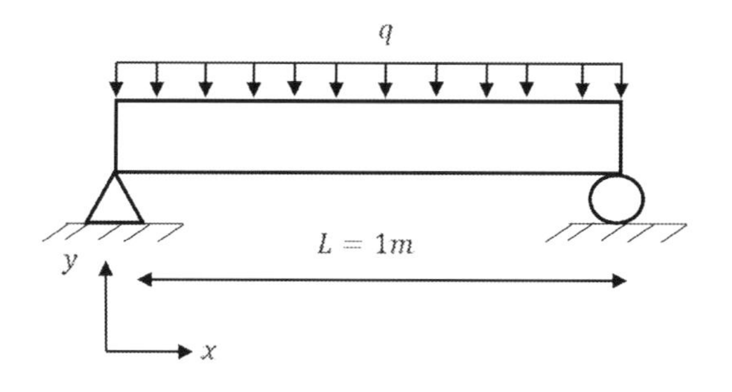

FIGURE P3.2

Qn: 7 A system vibrating without external force and damping is governed by equation

$\frac{d^2u}{dt^2} - \omega^2 u = 0$. For $u(0) = I, u'(0) = 0, t \in (0,T]$, where u represents the displacement of an object under motion, I represents amplitude and T represents the time period of oscillation. For suitable input values of ω, I, T, develop a detailed computational algorithm using the finite difference method for the solution of the governing differential equation.

Qn: 8 Consider a problem of one-dimensional steady state heat conduction without internal heat generation, governed by the following equation:

$$\frac{d}{dx}\left(kA\frac{dT}{dx}\right) = 0$$

The rod (Figure P3.3) is insulated and its ends are maintained at 80°C and 450°C respectively. The thermal conductivity of the rod is 900 W/mK and the cross-sectional area is 0.01 m^2. Find the temperature distribution in the rod using the finite volume method.

$T_1 = 80^o C$ $L = 0.5m$ $T_2 = 450^o C$ x

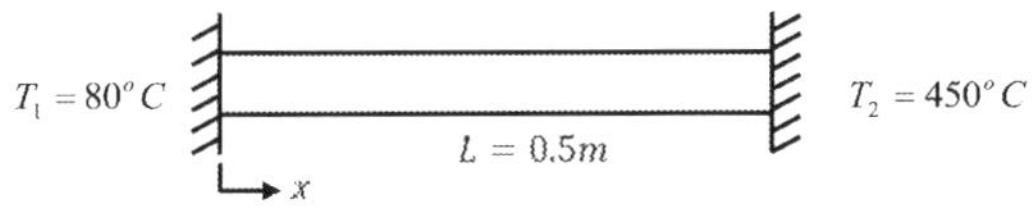

FIGURE P3.3

Qn: 9 Consider a plate of thickness 1.5 cm as shown in Figure P3.4. The faces of the plate are maintained at 150°C and 250°C respectively. The plate has constant thermal conductivity of 0.6 W/m K with uniform internal heat generation of 1 MW/m^3. Find the temperature distribution along the plate thickness using the finite volume method and compare your results with an analytical solution.

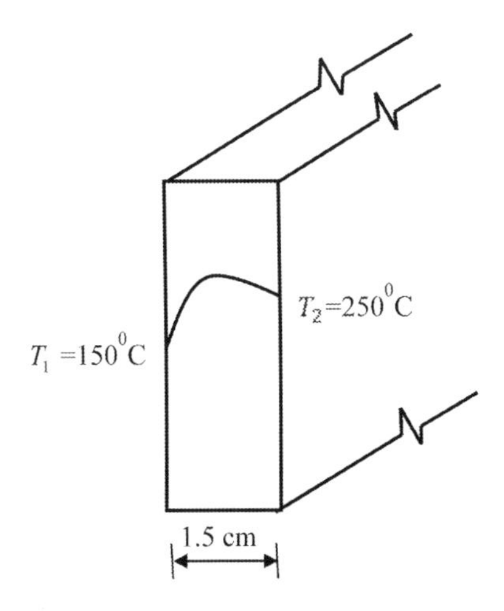

FIGURE P3.4

Qn: 10 Consider one-dimensional heat transfer in a fin which is cooled by convective heat transfer along its length as shown in Figure P3.5. The base temperature of the fin is 120°C whereas its other end is insulated. The fin is exposed to an ambient temperature of 30°C. The governing differential equation is given as:

$$KA\frac{d^2T}{dx^2} - hP\left(T - T_\infty\right) = 0$$

Given that $hP / KA = 30\,\mathrm{m}^{-2}$ and length of the fin is 1 m. Find the temperature distribution along the length of the fin using the finite volume method and compare the results with an analytical solution.

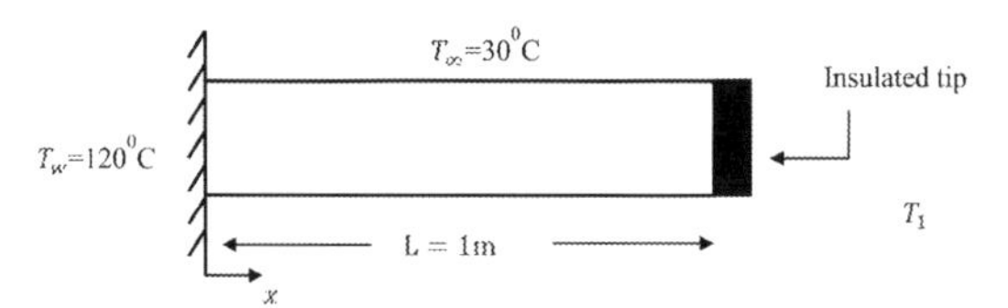

FIGURE P3.5

QUIZ QUESTIONS

Qn: 1 Discrete points in a domain where the solution of the problem is defined are called
(a) nodes (b) cells (c) grids or (d) elements. Tick the correct one.

Qn: 2 The numerical solution of a partial differential equation is said to be consistent when

____________________.

Qn: 3 Discretize the one-dimensional transient heat conduction equation using Crank-Nicholson scheme in FDM.

Qn: 4 Discretize the equation $\frac{d^2\varphi}{dx^2} + \frac{d^2\varphi}{dy^2} = 0$ using central difference method.

Qn: 5 For transient problem the solution using implicit scheme is unconditionally stable: (a) true (b) false. Tick the correct one.

Qn: 6 Continuity equation is based on
(a) mass conversation (b) momentum conservation (c) energy conservation or (d) first law of thermodynamics. Tick the correct one.

Qn: 7 In FVM for the solution of equations without source terms, the surface integrals at each stage are converted into volume integrals: (a) true (b) false. Tick the correct one.

Qn: 8 In FVM, the ____________________ theorem is used for transformation of diffusion terms into a surface based integral.

Qn: 9 Using the finite volume method, discretize the source free unsteady state one-dimensional heat conduction equation by the explicit method.

Qn: 10 In the finite volume method, the stability for the implementation of the Crank-Nicholson scheme is governed by ____________________ number.

4 Finite Element Method

The finite element method is a popularly well-known numerical method in structural analysis and continuum mechanics. This method became very attractive after the availability of computing machines because its solution procedure is based on a matrix form of equations that are solved using computer codes. Among many numerical methods, the finite element method stood apart due to its inherent ability to handle curved or any intricate geometries, which otherwise are difficult to solve using other numerical methods. It follows an integral approach to solve the differential equations derived in flow and heat transfer problems as well as continuum mechanics. The finite element method differs from finite difference and finite volume methods on an important principle, in this method an approximate solution is assumed at the very beginning of the solution procedure. The given differential equations are not altered as done in finite difference and finite volume methods, however, they are integrated over a number of subdomains called elements, to get the final matrix equations. In this method, the given computational domain is discretized into a number of subdomains called elements, resulting in a number of grid points. However, with its basic principle, it establishes a definite relation for the variation of the field variables between two adjacent nodes, which is not the case with the finite difference and finite volume methods. Hence, the characteristics of the finite element method can be described as [1]: (i) it follows an integral approach to get the final algebraic equations as solution for the unknowns and (ii) it employs piecewise smooth functions to approximate the assumed solution for the differential equation. The computational domain is discretized using geometrically well-defined elements so that their geometrical properties related to the elements and nodes can be established. This characteristic of the finite element method makes it easy to discretize any complicated geometry. Domain discretization is accomplished with the help of mesh generation technique, which is a pre-requisite in the solution of differential equations on a given computational domain. This step provides the required geometrical data required for computing matrices in the implementation of the finite element solution procedure. Mesh generation itself is a separate field of study, which deals with different methods to obtain a number of subdomains. The piecewise approximation of the unknown variables makes it possible to obtain non-linear functional variation of the unknowns using a number of linear approximations.

The integral formulation of the finite element method is generally based on two important fundamental theories, (i) variational formulation and (ii) weighted residual method. The variational formulation is based on the principle of minimization of the total potential energy of the system including the forces contributing from the boundaries. In order to obtain the approximate solution of the differential equation, a number of trial functions are assumed with certain unknown parameters. These trial functions are employed to satisfy the minimization of the total potential energy of

DOI: 10.1201/9781003248071-4

the system. The trial function that satisfies the minimum total potential energy of the system is taken as the approximate solution of the differential equation. This formulation is widely used in solid mechanics and structural analysis problems. However, this method cannot be applied to first order derivative of a function. The second method, called the weighted residual method, assumes a solution for the differential equation, which gives rise to a residue, as the assumed solution may not satisfy the given governing equation. Now, reducing the residual to zero will assure the approximate numerical solution of the differential equation. The resulting differential equation with assumed solution is multiplied by a weighting function in the process of making the residue to zero. There are different choices of selecting the weighting functions, depending on which they are classified as collocation method, subdomain method, least square method and Galerkin's method [1]. In the collocation method, a number of points equal to the unknown coefficients in the trial function are selected in the domain as a weighting function, whereas the weighting function is assumed to be equal to one in the subdomain method. The approximate function used to represent the solution of the differential equation itself is assumed as the weighting function in Galerkin's method. When the residue itself is used as the weighting function to get the error function, then the method is called the least square method. Galerkin's weighted residual method is commonly used for the solution of differential equations obtained as conservation equations in flow and heat transfer problems and discussion of this method is elaborated in the following section.

4.1 GALERKIN'S WEIGHTED RESIDUAL METHOD

The Galerkin's weighted residual finite element method is widely used for the solution of flow and heat transfer problems. In this method, the residue obtained by the assumption of approximate solution for the unknown variable of the differential equation is multiplied by the approximate function chosen to represent the variation of the unknowns with respect to the coordinates. This procedure will result in a weak formulation of the given differential equation, which makes it easy to satisfy the continuity of the variable within the domain and also the natural boundary condition. The finite element method differs from finite difference and finite volume method i two important factors: (i) it employs a piecewise approximation method to represent the unknown variable and (ii) the computational domain is discretized using specific elements, well defined by geometrical parameters in the given coordinates. Consider a fin subjected to convective boundary condition at x=L as shown in Figure 4.1(a). The temperature distribution along the length of the fin as predicted by analytical solution (Figure 3.3b) is illustrated in Figure 4.1(b) with a piecewise approximation of the curve. That means the variation between two nodes of an element considered to discretize the computational domain is assumed to be linear and by discretizing the domain using a number of such elements, the same temperature distribution as obtained from analytical solution can be achieved using the Galerkin's weighted residual method. The governing equation for fin is expressed as

$$kA\frac{d^2T}{dx^2} - h_c P(T - T_\infty) = 0 \tag{4.1a}$$

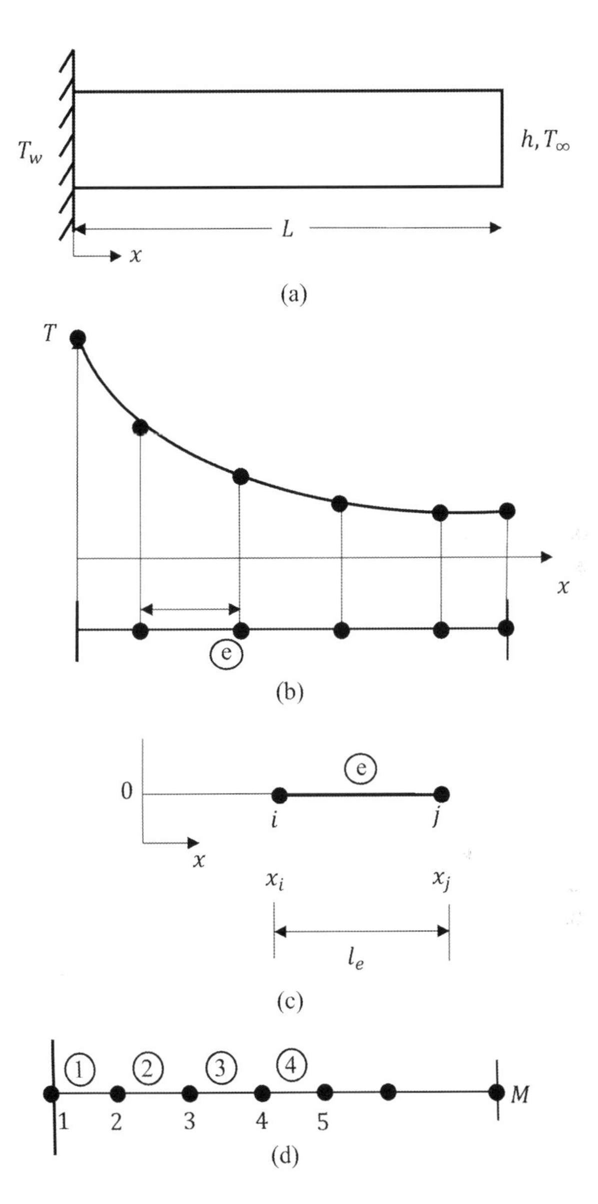

FIGURE 4.1 (a) Schematic diagram of fin, (b) temperature variation along the discretized domain, (c) an element and (d) discretization of computational domain.

Boundary conditions:

$$@\, x = 0, T = T_w \text{ and } @\, x = L, -k\frac{dT}{dx} = h_c\left(T_M - T_\infty\right) \tag{4.1b}$$

Figure 4.1(c) shows the representation of an element, 'e' with nodes 'i' and 'j' with a length of 'l_e' called element length. The fin is discretized using one-dimensional linear elements as shown in Figure 4.1(d), giving rise to a number of elements and nodes. As this element is represented using only two nodes, '*i*' and '*j*', the variation of any attribute within this element is always linear. However, it is possible to discretize the fin using the three node element with quadratic variation of the unknown. Let us start with linear variation with two node elements to solve the fin problem. In the weighted residual method, an approximate solution is assumed for the unknown variable in the governing equation. In the present case, the solution for temperature has to be obtained using Galerkin's weighted residual finite element method. Let us assume linear variation of temperature within an element as represented below.

$$T = \alpha_1 + \alpha_2 x \tag{4.2}$$

Equation (4.2) indicates that the temperature varies linearly between nodes, '*i*' and '*j*' within the element, '*e*'. The constants, α_1, α_2 have to be determined to complete the above expression to obtain temperature at any value of x. Equation (4.2) can be evaluated at nodes, '*i*' and '*j*' to obtain the following equations.

$$T_i = \alpha_1 + \alpha_2 x_i \tag{4.3a}$$

$$T_j = \alpha_1 + \alpha_2 x_j \tag{4.3b}$$

Equation (4.3b)– Equation (4.3a) results in
$T_j - T_i = \alpha_2 (x_j - x_i)$, from which we get

$$\alpha_2 = \frac{T_j - T_i}{(x_j - x_i)} \tag{4.3c}$$

Substituting Equation (4.3c) in Equation (4.3a) will give rise to

$$\alpha_1 = T_i - \frac{(T_j - T_i)}{(x_j - x_i)} x_i \tag{4.3d}$$

Substitution of Equations (4.3c) and (4.3d) in Equation (4.2), we get
$T = \dfrac{\left[(x_j - x)T_i + (x - x_i)T_j\right]}{(x_j - x_i)}$, which can be written as

$T = \left[N_i T_i + N_j T_j\right]$ where N_i, N_j are called shape functions or interpolation functions or basis functions and they are expressed as

$$N_i = \frac{(x_j - x)}{(x_j - x_i)} \text{ and } N_j = \frac{(x - x_i)}{(x_j - x_i)}$$

Now Equation (4.2) can be expressed as

$$T = N_i T_i + N_j T_j = \begin{bmatrix} N_i & N_j \end{bmatrix} \begin{Bmatrix} T_i \\ T_j \end{Bmatrix}$$

$$T = \left[N\right]\left\{T\right\} \tag{4.4}$$

Comparing Equations (4.2) and (4.4), one can appreciate that the function of T with respect to x is transferred to a separate function called the shape function. This means the functional relation of temperature, T with respect to the x-coordinate has been eliminated. Temperatures in the vector $\{T\}$ are only nodal values and do not have any functional relation with x. This is another great advantage of the finite element method, which makes integration of derivatives of temperature much easier. It is interesting to note that if temperatures and coordinates at nodes, 'i' and 'j' are known, then the temperature at any location of x within the above nodes can be computed using Equation (4.4). For this reason, $[N]$ is also called interpolation function. Finite difference and finite volume method do not possess such qualities. However, this advantage of the finite element method is obtained at the cost of a pre-processing procedure called *mesh generation*, which is a unique characteristic of the finite element method. The important characteristics of shape functions are, (i) the summation of shape functions at all the nodes is always equal to one and (ii) the shape function has a value of unity at its own node and zero at other nodes of the element.

The first order derivative of temperature is used to determine heat flux conducted through the solid, hence the derivative of Equation (4.4) can be evaluated as $\frac{dT}{dx}=\frac{d[N]}{dx}\{T\}=\frac{d[N_i \quad N_j]}{dx}\{T\}$. The temperature vector has come out of the derivative term because, only the shape function $[N]$ contains the functions to be differentiated with respect to x. This is another advantage of the finite element method and the above procedure holds good for integration also. Now differentiating the shape functions N_i, N_j with respect to x and expressing $x_j - x_i = l_e$, we get

$$\begin{aligned}\frac{dT}{dx}&=\frac{d[N]}{dx}\{T\}=\frac{d[N_i \quad N_j]}{dx}\{T\}\\ &=\begin{bmatrix}-\frac{1}{l_e} & \frac{1}{l_e}\end{bmatrix}\begin{Bmatrix}T_i\\T_j\end{Bmatrix}\end{aligned} \tag{4.5}$$

$=\left[-\frac{T_i}{l_e}+\frac{T_j}{l_e}\right]=\left[\frac{T_j-T_i}{l_e}\right]$ which is equivalent to Equation (4.3c), α_2, the temperature gradient as obtained from Equation (4.2). Now, the Galerkin's weighted residual method can be applied to solve Equation (4.1a), for which the shape function itself is employed as the weighting function. A close look at the nature of shape function defined by Equation (4.4) indicates that the number of shape functions defined at the nodes is equal to the number of unknown constants used to represent the assumed temperature profile as solution. Equation (4.2) implies the assumed temperature profile is linear and two constants are required to define the expression of temperature in terms of the x-coordinate. Hence, the shape function $[N]$ consists of two shape functions at two nodes, 'i' and 'j', that means when this shape function is used as the weighting function, then it will produce two equations that are required to compute the two constants. For this reason, the shape function $[N]$ is applied in transpose form so that at each node the given residue is multiplied with the respective shape function.

Let us adopt temperature, $\tilde{T}(x)$ as the approximate solution for Equation (4.1a) and is presumed to follow the profile given by Equation (4.2). Then Equation (4.1a) is reduced to

$$kA\frac{d^2\tilde{T}}{dx^2} - h_c P(\tilde{T} - T_\infty) = R \tag{4.6}$$

As the assumed temperature $\tilde{T}(x)$ is not the exact solution or at least the expected approximate solution, then Equation (4.1a) will not be satisfied, that means the right-hand side of the expression will not be equal to zero, instead it will give rise to a residue, which may be positive or negative. Now, the application of Galerkin's weighted residual method to Equation (4.6) with the shape function as the weighting function will yield the following equation.

$\int_\Omega [N]^T R dx = 0$ and after substituting for R from Equation (4.6), we get the following equation.

$$\int_\Omega [N]^T \left(kA\frac{d^2\tilde{T}}{dx^2} - h_c P(\tilde{T} - T_\infty) \right) dx = 0 \tag{4.7}$$

In Equation (4.7), Ω implies the domain over which the integration is carried out. In the case of one-dimensional computational domain, like the one considered in the present problem, it indicates the domain in the x-coordinate. In the case of the two-dimensional domain, it is the x-y space over which the integration is performed. The boundary surface is represented by the symbol Γ. For the purpose of convenience and the fact that the temperature solution being attempted using the finite element method itself is referred to as T, here afterwards $\tilde{T}$, the assumed temperature shown in Equation (4.7) will be written as T.

Consider the term

$\frac{d}{dx}\left([N]^T \frac{dT}{dx}\right) = [N]^T \frac{d^2T}{dx^2} + \frac{dT}{dx}\frac{d[N]^T}{dx}$, from which we obtain

$kA[N]^T \frac{d^2T}{dx^2} = kA\frac{d}{dx}\left([N]^T \frac{dT}{dx}\right) - kA\frac{dT}{dx}\frac{d[N]^T}{dx}$ and hence the integration of the first term in Equation (4.6) can be expressed as

$$kA\int_\Omega [N]^T \frac{d^2T}{dx^2} dx = kA\int_\Gamma \left([N]^T \frac{dT}{dx}\right) d\Gamma - kA\int_\Omega \frac{dT}{dx}\frac{d[N]^T}{dx} dx$$

Application of Green's theorem to the above integral converts the first term on the right-hand side of the above equation to boundary integral. Now, making use of Equation (4.4), the above equation can be expressed as

$$kA\int_{\Omega}[N]^T \frac{d^2T}{dx^2}dx = kA\int_{\Gamma}[N]^T \frac{dT}{dn}d\Gamma - kA\int_{\Omega}\frac{d[N]^T}{dx}\frac{d[N]}{dx}dx\{T\} \tag{4.8}$$

The first term on the right-hand side of the above equation corresponds to the natural boundary condition of the problem considered and its dimension is reduced by one level due to the principle of Green's theorem. Hence, it contributes only to the boundaries. In the case of the one-dimensional domain, it contributes at the end points, in the two-dimensional domain, lines, and in the case of the three-dimensional domain, surfaces. In the present fin problem, it will be considered only while implementing the convective boundary condition considered at $x=L$. From the basic principle of integral formulation of the finite element method, the integration of the governing equation and Green's theorem, naturally produces this particular term, once again highlighting the superiority of the finite element method compared to other numerical methods. For example, in the case of finite difference and finite volume methods, the derivative boundary conditions have to be enforced on the respective boundaries by making use of the basic principle of the respective methods. However, the finite element method is completely free from such procedures to handle the derivative boundary condition. This means, the integral formulation of Galerkin's weighted residual method inherently takes care of the natural boundary condition. In the course of proceeding for integration of other terms in Equation (4.8), the first integral corresponding to the boundaries will not be considered, and generally they are considered at the time of incorporating the boundary conditions. The second term on the right-hand side of Equation (4.8) has to be evaluated.

Now, consider the integration of convective term in Equation (4.7) as shown below.

$$\int_{\Omega}[N]^T h_c P(T - T_\infty)dx$$

$\int_{\Omega}[N]^T h_c PTdx - \int_{\Omega}[N]^T h_c PT_\infty dx$. Rewriting the temperature using Equation (4.4), we get

$$\int_{\Omega}[N]^T[N]h_c Pdx\{T\} - \int_{\Omega}[N]^T h_c PT_\infty dx = 0 \tag{4.9}$$

Substituting Equations (4.8) and (4.9) in Equation (4.7), the final form of the governing equation after the implementation of Galerkin's method can be written as $-kA\int_{\Omega}\frac{d[N]^T}{dx}\frac{d[N]}{dx}dx\{T\} - \int_{\Omega}[N]^T[N]h_c Pdx\{T\} + \int_{\Omega}[N]^T h_c PT_\infty dx = 0$. After collecting the terms of temperature vector together and taking out the negative sign throughout the equation, we get

$$kA\int_{\Omega}\frac{d[N]^T}{dx}\frac{d[N]}{dx}dx\{T\}+\int_{\Omega}[N]^T[N]h_cPdx\{T\}=h_cPT_\infty\int_{\Omega}[N]^T dx \quad (4.10)$$

The above equation shows that the terms multiplied by temperature vector is always has the product of shape function and its transpose either the function or its derivative. The load term is accompanied by terms without the field variable, T but multiplied only by the transpose of the shape function. This has already given an indication that the left-hand side terms will contribute a square matrix for an element multiplied by the temperature vector and the right-hand side term, the load vector, constituting the final equation in the form, $Ax=b$. The first term on the left-hand side of Equation (4.10) corresponds to the conduction term in the governing equation, that means the second order derivative of temperature in the conduction term is reduced to gradients of shape functions due to the integration during the application of the Galerkin's method. Similarly, in the case of convection term, the second term, the product of transpose of shape function and shape function becomes the coefficient for the temperature vector. In the solution of second order partial differential equations with convective heat transfer using the finite element analysis, generally the following convention is adopted.

$$[K]^{(e)}=\int_V[B]^T[D][B]dV+h\int_S[N]^T[N]dS \quad (4.10a)$$

The first term in the above equation corresponds to volume integral to take care of conduction heat transfer, wherein the matrix *[D]* takes care of the material property. If the material is anisotropic, then accordingly only the diagonal elements will have different thermal conductivity in the respective direction. For isotropic material, it simply represents constant thermal conductivity. As convective heat transfer is a surface phenomenon, the respective convective term is integrated over surface area and the coefficient consists of the shape functions only because convective heat transfer is directly proportional to the temperature difference.

4.1.1 Integration of Shape Functions

In the finite element method, the given computational domain is discretized into a number of elements with specified geometric dimensions. Integration of shape functions or their products has to be carried out as a result of implementation of the Galerkin's weighted residual where the shape function used to represent the temperature profile itself is used. In order to make the integrals much simpler and easier, local and natural coordinate systems are used to derive expressions that can be used as integration formulae [1]. Generally, the size of the element is specified in the process of mesh generation. For example, in the case of one-dimensional linear element, the following expression can be employed for integration of shape functions in terms of the length of the element.

$$L\int_0^1 l_1^a l_2^b dl_2 = \frac{a!b!}{(a+b+1)!}L \tag{4.11}$$

where l_1 and l_2 represent the shape functions, N_i and N_j respectively and 'a' and 'b' indicate the powers of these shape functions. Now, expanding the shape function terms in Equation (4.10) and substituting Equation (4.5) for the first term, we get the following equation.

$$kA\int_{l_i}^{l_j}\begin{bmatrix}-\dfrac{1}{l_e}\\ \dfrac{1}{l_e}\end{bmatrix}\begin{bmatrix}-\dfrac{1}{l_e} & \dfrac{1}{l_e}\end{bmatrix}dx\{T\}+h_cP\int_{l_i}^{l_j}\begin{bmatrix}N_i\\ N_j\end{bmatrix}\begin{bmatrix}N_i & N_j\end{bmatrix}dx\{T\}=h_cPT_\infty\int_{l_i}^{l_j}\begin{bmatrix}N_i\\ N_j\end{bmatrix}dx$$

$$\frac{kA}{l_e^2}\begin{bmatrix}1 & -1\\ -1 & 1\end{bmatrix}\int_{l_i}^{l_j}dx\{T\}+h_cP\int_{l_i}^{l_j}\begin{bmatrix}N_i^2 & N_iN_j\\ N_iN_j & N_j^2\end{bmatrix}dx\{T\}=h_cPT_\infty\int_{l_i}^{l_j}\begin{bmatrix}N_i\\ N_j\end{bmatrix}dx \tag{4.12}$$

Using line integral formula given by Equation (4.11), we get the following results.

$\int_{l_i}^{l_j}N_i^2dx = l_e\int_0^1 l_1^2 dl_1 = \frac{2!0!}{(2+0+1)!}l_e = \frac{l_e}{3}$. Similarly, the product of shape functions can be evaluated as

$\int_{l_i}^{l_j}N_iN_jdx = \frac{l_e}{6}$, $\int_{l_i}^{l_j}N_idx = \frac{l_e}{2}$ and $\int_{l_i}^{l_j}N_jdx = \frac{l_e}{2}$. Now, substituting these integrals in Equation (4.12), we get the following expression.

$$\frac{kA}{l_e}\begin{bmatrix}1 & -1\\ -1 & 1\end{bmatrix}\begin{Bmatrix}T_i\\ T_j\end{Bmatrix}+\frac{h_cPl_e}{6}\begin{bmatrix}2 & 1\\ 1 & 2\end{bmatrix}\begin{Bmatrix}T_i\\ T_j\end{Bmatrix}=\frac{h_cPT_\infty l_e}{2}\begin{Bmatrix}1\\ 1\end{Bmatrix} \tag{4.13}$$

Equation (4.13) is the final elemental equation for a generalized element, 'e' with length l_e and nodal temperatures, T_i and T_j. Now, the procedure for evaluating such element matrices has to be repeated to all the elements representing the discretized computational domain. Figure 4.1(d) shows the discretized computational domain into a number of elements. An element number is marked using a circle above the domain and the node numbers by numerals below the domain against the node points. An element in the computational domain is a representative of the characteristic of how the conservation principle in that element is satisfied by the implementation of the finite element solution procedure. This representative behavior holds good for the entire computational domain. Now, in order to get the solution of the field variable for the entire computational domain, all the elements have to be connected together geometrically as well as in the conservation principle sense. Geometrical continuity of the discretized domain in terms of elements is achieved by mesh generation technique, which provides the information about the total number of elements, number of grid points and their coordinates and element-nodal connectivity. In element-nodal connectivity, the connectivity of nodes between the elements is taken care of. The details of mesh generation using transfinite interpolation technique (TFI) will be discussed in the last section of this chapter. Referring to Figure 4.1(d), it is clear that element (1) consists of two nodes, 1 and 2, similarly element 2, with nodes 2 and 3 and so on. Node 2 is common between elements 1 and 2. Generation of such geometrical connectivity details is an important characteristic of the finite element method. Though it is very simple for the one-dimensional domain, it involves a detailed procedure for the case of two- and three-dimensional domains. Of course, there are many software packages exclusively available now, just for generating mesh details of a given computational domain. Documentation of accurate mesh details is important to avoid errors in final computations.

The matrix form of equations as expressed by Equation (4.13) can be easily generated for all the elements with the help of a computer program. The computational domain shown in Figure 4.1(d) is discretized using two node elements and every element in the domain is designated by two nodes.

Then all the element level matrices are assembled to get the final global matrix which will have a number of equations equal to the number of node points and will be solved after modifying the matrices to include the effect of boundary conditions at both ends of the fin. Hence, each element has to be identified by its own nodes depending upon the element number and this is done as part of mesh generation. The logic of programming should be able to add one element after another element into the final global matrix. For this purpose, each element is identified by its local node using '*i*', '*j*' notations and they remain the same for all the elements irrespective of the element number. As it was highlighted earlier, the finite element method starts from an element that represents the characteristic of the whole domain in confirmation with the conservation principle. From Equation (4.13) it is clear that all the elements of the computational domain contribute the same value to the coefficient matrix and the load vector for constant properties and element length. In the finite element method, the coefficient matrix of the field variable is called stiffness matrix *[K]* and the load vector as *{f}*. These matrices are identified at element level as $\left[K\right]^{(e)}$ and $\left\{f\right\}^{(e)}$ and global matrices as *[K]* and *{f}*.

$$\left[K\right]^{(e)} = \frac{kA}{l_e}\begin{bmatrix} 1 & -1 \\ -1 & 1 \end{bmatrix} + \frac{h_c P l_e}{6}\begin{bmatrix} 2 & 1 \\ 1 & 2 \end{bmatrix} \text{ and } \left\{f\right\}^{(e)} = \frac{h_c P T_\infty l_e}{2}\begin{Bmatrix} 1 \\ 1 \end{Bmatrix}$$

The coding and definition of various variables associated with generation of element level stiffness matrix and load vector and assembly of these matrices is shown in Figure 4.2.

Assuming constant thermal conductivity and area of cross-section and element length, l_e, it can be shown that

$$\left[K\right]^{(1)} = \left[K\right]^{(2)} = \left[K\right]^{(3)} = \left[K\right]^{(4)} = \ldots.\left[K\right]^{(M)} \text{ and}$$

$$\left\{f\right\}^{(1)} = \left\{f\right\}^{(2)} = \left\{f\right\}^{(3)} = \left\{f\right\}^{(4)} = \ldots.\left\{f\right\}^{(M)}$$

Computer code 4.1 – Formation of element matrices and global matrices

Define matrices

estiff (2, 2) – element level stiffness matrix for one-dimensional linear element

estiffg (idm, idm) – global stiffness matrix

bload (2) – element level load vector

bloadg (idm) – global load vector

I (idm, 2) – element nodal connectivity array

Define constants

idm – array size, nne – number of nodes per element, nelem – number of elements

$aa1 = \frac{kA}{l_e}$, $aa2 = \frac{h_c P l_e}{6}$ and $aa3 = \frac{h_c P l_e T_\infty}{2}$

Define element matrices

$estiff(1,1) = aa1 + 2 * aa2$, $estiff(1,2) = -aa1 + aa2$, $estiff(2,1) = estiff(1,2)$ and $estiff(2,2) = estiff(1,1)$

$bload(1) = aa3$ and $bload(2) = bload(1)$.

Assembly of matrices

```
do ielem = 1, nelem                                   Do this for all elements
    do I = 1, nne                                     Do it for the row
    Inode = I (ielem, i)                              Designate the global node number in row
  do j = 1, nne                                       Do it for the column
   knode = I (ielem, j)                               Designate the global node number in column
estiffg (Inode, knode) = estiffg (Inode, knode) +     Global stiffness matrix
                         estiff (i, j)
  enddo                                               End the column loop
 bloadg (Inode) = bloadg (Inode) + bload (i)          Do it for global load vector
   enddo                                              End the row loop
enddo                                                 End the element loop
```

FIGURE 4.2 Computer code for formation of element and global matrices.

Assembly of these element level matrices will result in the global matrices. Now, for assembling these element level matrices, the information on element-nodal connectivity of all the elements have to be known and they are stored in an array, which becomes handy for coding. The assembly of element level matrices provides the final global stiffness matrix *[K]* and global load vector *{f}*. The final form of the finite element formulation of the fin problem, before including the boundary conditions can be expressed as

$$\underbrace{[K]}_{M\times M}\underbrace{\{T\}}_{M\times 1}=\underbrace{\{f\}}_{M\times 1} \tag{4.14}$$

The above matrix is of the form $Ax=b$, which can be solved using an equation solver.

4.1.2 Boundary Conditions

The effect of boundary conditions on the stiffness and load vector matrices will be included after obtaining the global matrices because these effects will alter the coefficient values only at the boundary element and nodes. The procedure followed to include convective and Dirichlet boundary conditions in the finite element solution procedure is discussed below.

4.1.2.1 Convective Boundary Condition

The global stiffness and global vector matrices have to be modified for the effect of boundary conditions specified at $x=0$ and $x=L$. Let us discuss the procedure to incorporate the convective boundary condition at $x=L$ as given below.

$-kA\dfrac{dT}{dx}=h_cA(T-T_\infty)$, where A refers to the cross-sectional area. The first term on the right-hand side of Equation (4.8) refers to the left-hand side of the above expression except the minus sign. It has to be recalled that Equation (4.10) was multiplied by minus sign throughout the equation because it was common for all the terms in the equation. This amounts to the fact that the first term on the right-hand side of Equation (4.8) is also multiplied by minus sign and becomes equal to the left-hand side of the above expression for the boundary condition at $x=L$. The finite element form of the boundary term can now be evaluated as

$$\begin{aligned}
-\int[N]^T kA\frac{dT}{dx}d\Gamma &= \int[N]^T h_cA(T-T_\infty)d\Gamma\\
&= h_cA\int[N]^T T d\Gamma - h_cAT_\infty\int[N]^T d\Gamma\\
&= h_cA\int[N]^T[N]d\Gamma\{T\} - h_cAT_\infty\int[N]^T d\Gamma\\
&= h_cA\int\begin{bmatrix}N_i\\N_j\end{bmatrix}\begin{bmatrix}N_i & N_j\end{bmatrix}d\Gamma\{T\} - h_cAT_\infty\int\begin{bmatrix}N_i\\N_j\end{bmatrix}d\Gamma
\end{aligned} \tag{4.15}$$

At the last element, the boundary condition is applied at the boundary node 'j' and hence in the above equation, N_i does not contribute and hence it is made equal to zero, to get the following expression.

$$= h_c A \int \begin{bmatrix} 0 \\ N_j \end{bmatrix} \begin{bmatrix} 0 & N_j \end{bmatrix} d\Gamma \{T\} - h_c A T_\infty \int \begin{bmatrix} 0 \\ N_j \end{bmatrix} d\Gamma$$

At the boundary node at $x=L$, integration has to be performed only at a point. As the convection at $x=L$ end takes place through the cross-sectional area of the fin, the integration at node 'j' gives rise to the cross-sectional area, A itself. Now, the above expression takes the following from.

$= h_c A \begin{bmatrix} 0 & 0 \\ 0 & 1 \end{bmatrix} \{T\} - h_c A T_\infty \begin{Bmatrix} 0 \\ 1 \end{Bmatrix}$. With this, the expression for the stiffness matrix and load vector for the last element can be written as

$$\left[K\right]^{(M)} = \frac{kA}{l_e} \begin{bmatrix} 1 & -1 \\ -1 & 1 \end{bmatrix} + \frac{h_c P l_e}{6} \begin{bmatrix} 2 & 1 \\ 1 & 2 \end{bmatrix} + h_c A \begin{bmatrix} 0 & 0 \\ 0 & 1 \end{bmatrix} \text{ and } \{f\}^{(M)} = \frac{h_c P T_\infty l_e}{2} \begin{Bmatrix} 1 \\ 1 \end{Bmatrix} + h_c A T_\infty \begin{Bmatrix} 0 \\ 1 \end{Bmatrix}$$

The computer code for inclusion of convective and Dirichlet boundary conditions is elaborated in Figure 4.3.

4.1.2.2 Dirichlet Boundary Condition

When the unknown variable is known at certain boundary nodes, then such boundary condition is called the Dirichlet boundary condition. The final global matrices after incorporating the natural boundary conditions have to be modified to include the effect of known variables at the boundary nodes [2]. Figure 4.3 shows the computer code for the inclusion of convective and Dirichlet boundary conditions for the fin problem. When the unknown variables at certain boundary nodes are known, then these values can be substituted in the final matrix which is of the form $Ax=b$, as shown by Equation (4.14). In order to make the matrix of the form $Ax=b$ consistent, after multiplying the respective coefficient by the known temperatures, these equations are simplified to get the effect of known temperature. That means, the coefficient of the diagonal entry of the A matrix becomes equal to 1.0 with the respective known temperature becoming the load vector of that particular equation, with all the remaining coefficients in that equation equated to zero. This procedure makes the equation consistent, without reducing the number of equations to be solved.

4.1.3 Example Problem: Fin (Computer Code fin _ FEM.for)

Using the finite element formulation discussed through Equations (4.13) to (4.15) and with the help of computer code shown in Figures 4.2 and 4.3, a computer program in FORTRAN has been developed to obtain temperature at all the nodes of the computational domain. Figure 4.4 shows the results obtained for the fin problem using Galerkin's finite element method. Initially the simulation results for temperature distribution were obtained using only five elements as shown in Figure 4.4(a) and the

```
Define matrices

estiffg (idm, idm) – global stiffness matrix

bloadg (idm) – global load vector

Define constants

idm – array size, nnode – number of nodes, nelem – number of elements,

nbt – number of Dirichlet boundary nodes, ibt (idm) – array with Dirichlet boundary nodes

bt (idm) – value of temperature at Dirichlet boundary nodes

aa4 = h_c A , aa5 = h_c A T_∞

Include convective boundary condition

Global stiffness matrix

estiffg (nnode, nnode) = estiffg (nnode, nnode) + aa4

Global load vector

bloadg (nnode) = bloadg (nnode) + aa5

Dirichlet boundary condition

do i = 1, nbt                                      Do this for all Dirichlet boundary nodes
    value = bt(i)                                  Assign Dirichlet boundary nodal temperature
    lnode = ibt (i)                                Designate the Dirichlet boundary node number
  do j = 1, nnode                                  Do it for all the nodes
  bloadg (j) = bloadg (j) – estiffg (j, lnode)*value   Modify the load vector with the known temperature
 estiffg (lnode, j) = 0.0                          Make coefficients zero in all the columns at the respective row
 estiffg (j, nnode) = 0.0                          Make coefficients zero in all the rows at the respective column
  enddo                                            End the loop for all the nodes
estiffg (lnode,lnode) = 1.0                        Make the diagonal coefficient equal to 1.0
 bloadg (lnode) = value                            Assign the temperature value to the load vector
enddo                                              End the loop for Dirichlet boundary nodes
```

FIGURE 4.3 Computer code for inclusion of convective and Dirichlet boundary conditions.

nodal temperatures are joined together by linear approximation as assumed in the finite element formulation. The non-linear behavior of temperature variation along the length of the fin is depicted using piecewise approximation by linear approximation in this figure. Figure 4.4(b) illustrates the results for mesh sensitivity study using 50, 250 and 500 elements. It can be noticed that all the results obtained with the above number of elements coincide with each other and the finite element method could predict the non-linear variation accurately even with two node elements of linear approximation. Figure 4.4(c) shows the validation results with analytical

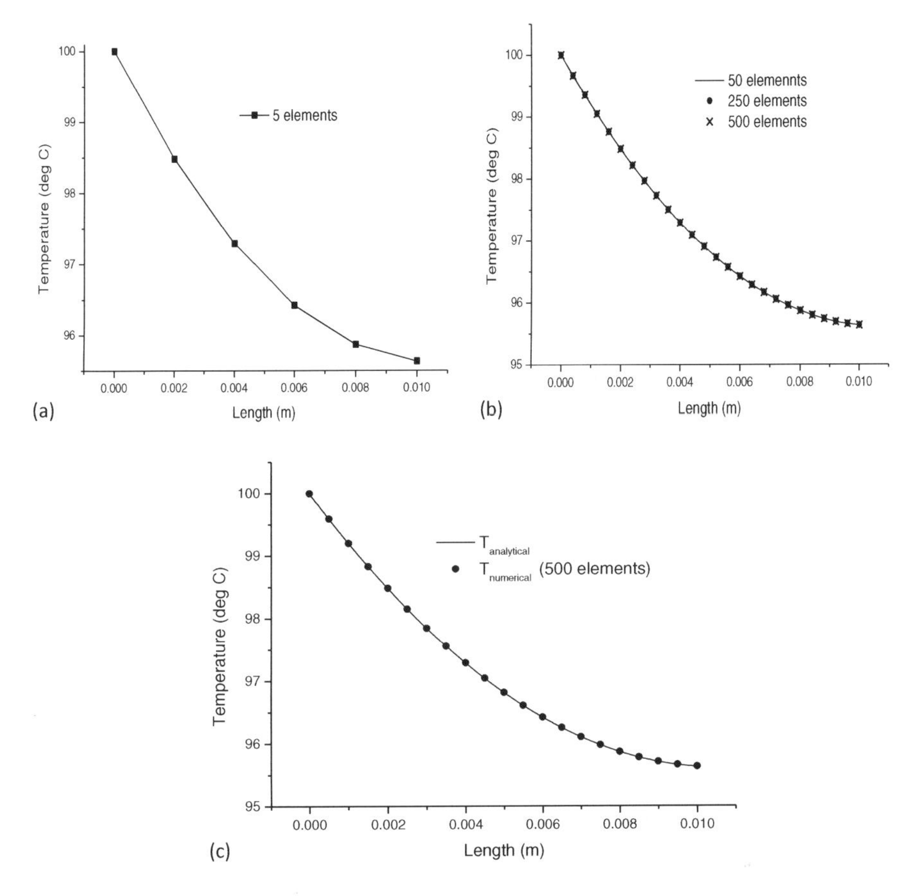

FIGURE 4.4 (a) Temperature results for five elements, (b) mesh sensitivity results and (c) validation results.

solution, which clearly indicates that the finite element procedure could simulate the temperature variation exactly as the analytical solution.

Once the validation of the computer code is accomplished, the program can be used to obtain simulation results. Figures 4.5(a), 4.5(b) and 4.5(c) depict the effect of thermal conductivity, heat transfer coefficient and ambient temperature respectively on temperature distribution along the length of the fin. As thermal conductivity increases, the temperature difference between the two ends of the fin keeps decreasing as shown in Figure 4.5(a). With higher thermal conductivity, smaller temperature differences will cause the required heat conduction along the fin. An increase in convective heat transfer coefficient enables higher heat dissipation for the same temperature of the fin at $x=0$ as illustrated in Figure 4.5(b). The increased ambient temperature reduces its ability to absorb heat from the fin, thus maintains a higher temperature along the length of the rod as demonstrated in Figure 4.5(c).

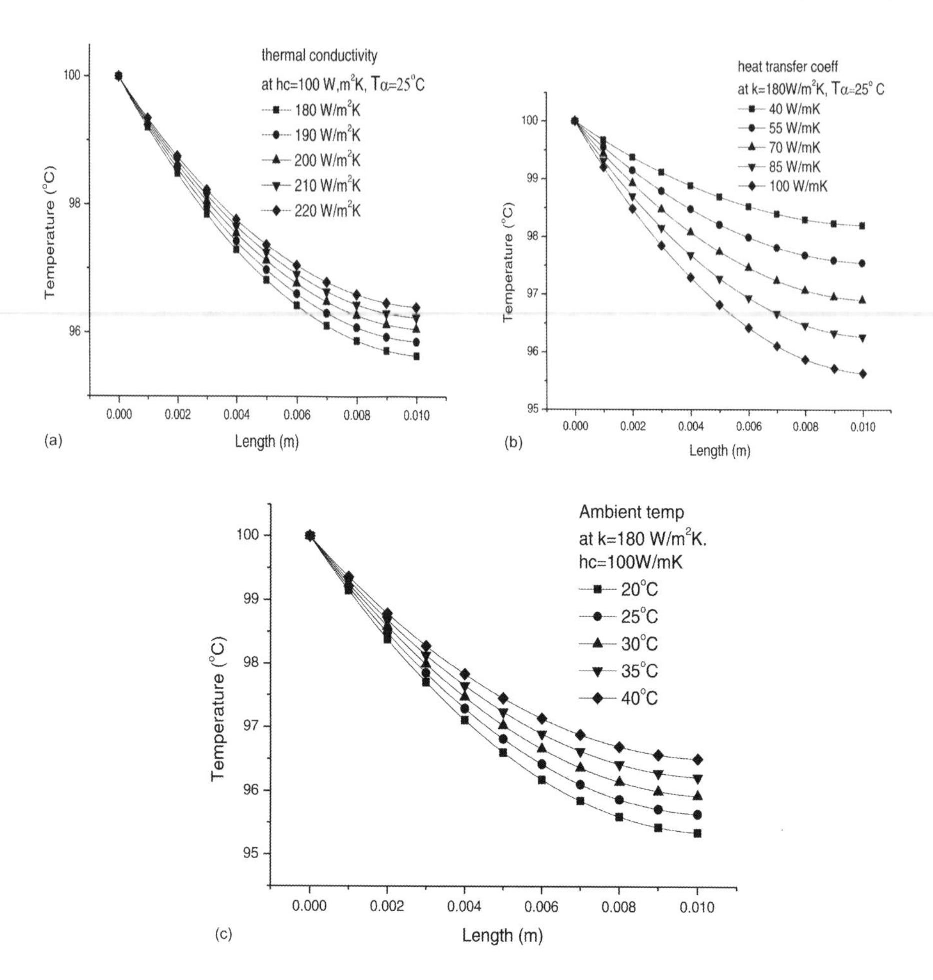

FIGURE 4.5 (a) Effect of thermal conductivity, (b) heat transfer coefficient and (c) ambient temperature on temperature distribution.

Thus, the finite element solution obtained by the implementation of Galerkin's weighted residual method is capable of predicting the underlying physics for the change of performance parameters of the fin.

4.2 DOMAIN DISCRETIZATION AND ISOPARAMETRIC FORMULATION

4.2.1 Domain Discretization

From the very basic principle of the finite element method, the computational domain has to be discretized into a number of elements, to which the Galerkin's weighted residual method is applied to obtain the matrix equations. As the type of equation used to represent the piecewise approximation of the solution determines the variation of the variable within an element, it is important to follow a systematic method to discretize the computational domain. In the finite element methodology, discretization of the computational domain is called the *pre-processing* phase of the whole computational procedure. Based on the significant variation of the variables in different coordinate directions of a given geometry of the system, the problem under investigation can be classified as one-dimensional, two- and three-dimensional. In the case of a fin problem, the temperature gradient which causes the heat conduction through the solid is significant only in one dimension, hence it is treated as a one-dimensional conduction-convection problem. Accordingly, the computational domain, the replica of the real geometry of the system, needs to be discretized using some standard characteristic elements. The accuracy of the finite element method is decided based on the fact that how the results converge between two meshes with a refinement of meshes. The finite element solution is an approximate solution, which deviates from the exact solution. However, some criteria have to be established to confirm that the solution obtained is independent of the size of the mesh. This is achieved by means of mesh sensitivity analysis, wherein with refinement of mesh, the error between the solutions obtained by two consecutive meshes should keep decreasing. In numerical simulation procedure, generally results with error less than 1% is assumed to be an acceptable solution. This mesh refinement can be performed either using *h-refinement* or *p-refinement*. In the *h-refinement* scheme, the size of the mesh is reduced, in other words, the number of elements is increased to test the convergence of the simulation results. In the case of *p-refinement*, a higher order polynomial is chosen as an interpolation function to get the best mesh which will result in a converged solution [3]. The mesh obtained using the discretization method can be classified as structured mesh or unstructured mesh. Unstructured mesh is preferable to represent irregular geometry compared to the structured mesh.

4.2.2 Isoparametric Formulation

Galerkin's weighted residual finite element method is an integral method and it has been already noticed that during the solution procedure, the integration of the differential form of the field variables, is reduced to the integration of shape functions or its derivatives at element level. When a computational domain is discretized into a number of elements, depending on the position of the element in the domain, the limits of integration of the coordinate directions will vary. If the global coordinates are used for the integration of these shape functions, then the computation of integration becomes highly cumbersome due to varying limits of integration for different elements due to fixed single origin. This issue can be easily resolved by defining local coordinates whose coordinates may vary from 0 to 1 or length of element. In a local coordinate system, the origin of the coordinate of each element gets shifted to maintain the same limits of integration [1]. Natural coordinate systems with limits of –1 and+1 are the most preferable one because Gaussian quadrature formulae can be implemented for integration. Implementation of Gaussian quadrature formulae converts integration into summation, and hence integration of shape functions or their derivatives can be easily programmed. As the limits of integration vary only from –1 to+1, the integration can be implemented for any element within the domain and the actual value of the integrand is obtained by means of coordinate transformation. Figures 4.6(a) and 4.6(b) show one-dimensional domain discretized using global coordinates and natural coordinates respectively. In global coordinates the origin is fixed at 0 and the distances of all the nodes in the computational domain are measured from this origin. However, in the case of the natural coordinate system shown in Figure 4.6(b), the origin is located at the center of the element with coordinates of the end nodes of the elements always varying from –1 to+1. The natural coordinate system which is the basis for the isoparametric formulation will be discussed in this textbook and computer programming is also explained for formulations obtained using isoparametric formulation. Implementation of the finite element method revolves around two important concepts, one is the continuous piecewise approximation of the field variable and the other is discretization of the computational domain into the required number of elements. In natural coordinate systems, other than the field variables, the coordinates of the geometry themselves can be modeled. However, one should always remember that the final solution of the field variables is required in global coordinates. This necessitates the transformation of one coordinate to another coordinate. Let us understand this concept from a one-dimensional coordinate system. Figure 4.6(a) shows an element of length, l_e in x-coordinate system, with x_i and x_j as coordinates at the nodes 'i' and 'j'. The shape functions at the respective nodes are

$N_i(x) = \dfrac{x_j - x}{x_j - x_i}$ and $N_j(x) = \dfrac{x - x_i}{x_j - x_i}$. Now consider a natural coordinate system whose origin is located at the center of the element, or at $\dfrac{l_e}{2}$ from nodes 'i' or 'j' with $\xi = -1$ at node 'i' and $\xi = +1$ at node 'j'. This makes the length of the element as 2 in ξ-coordinate. It has to be observed that the origin for the x-coordinate system, called the global coordinate system starts somewhere, well before the node 'i' and hence with the location of elements in the right-hand direction of the given element, the coordinates keep increasing in the x-direction, thus the limits of integration of various matrices across each element varies continuously. However, in the case of ξ-coordinate, the origin is always located at the center of the element and hence the limits of integration over an element remain constant from –1 to+1. As the origin for ξ-coordinate is located always at the center of the element, the ξ axis need not be parallel to x-coordinate. The length of the element in x-coordinate system has always length dimension whereas in the case of ξ- coordinate, called the natural coordinate system, the length of the element is non-dimensional. Hence, this characteristic makes it possible even to represent the x-coordinate in terms of ξ- coordinate, thus enabling us to achieve coordinate transformation or mapping. As the ξ- coordinate is always located at the center of an element and need not be parallel to the surface to be transformed into natural coordinates, it is possible to represent any complex geometry in terms of these natural coordinates. Later it will be shown that the integration of shape functions or derivatives in terms of these natural coordinates can be easily achieved by Gaussian integration scheme, making integration into summation.

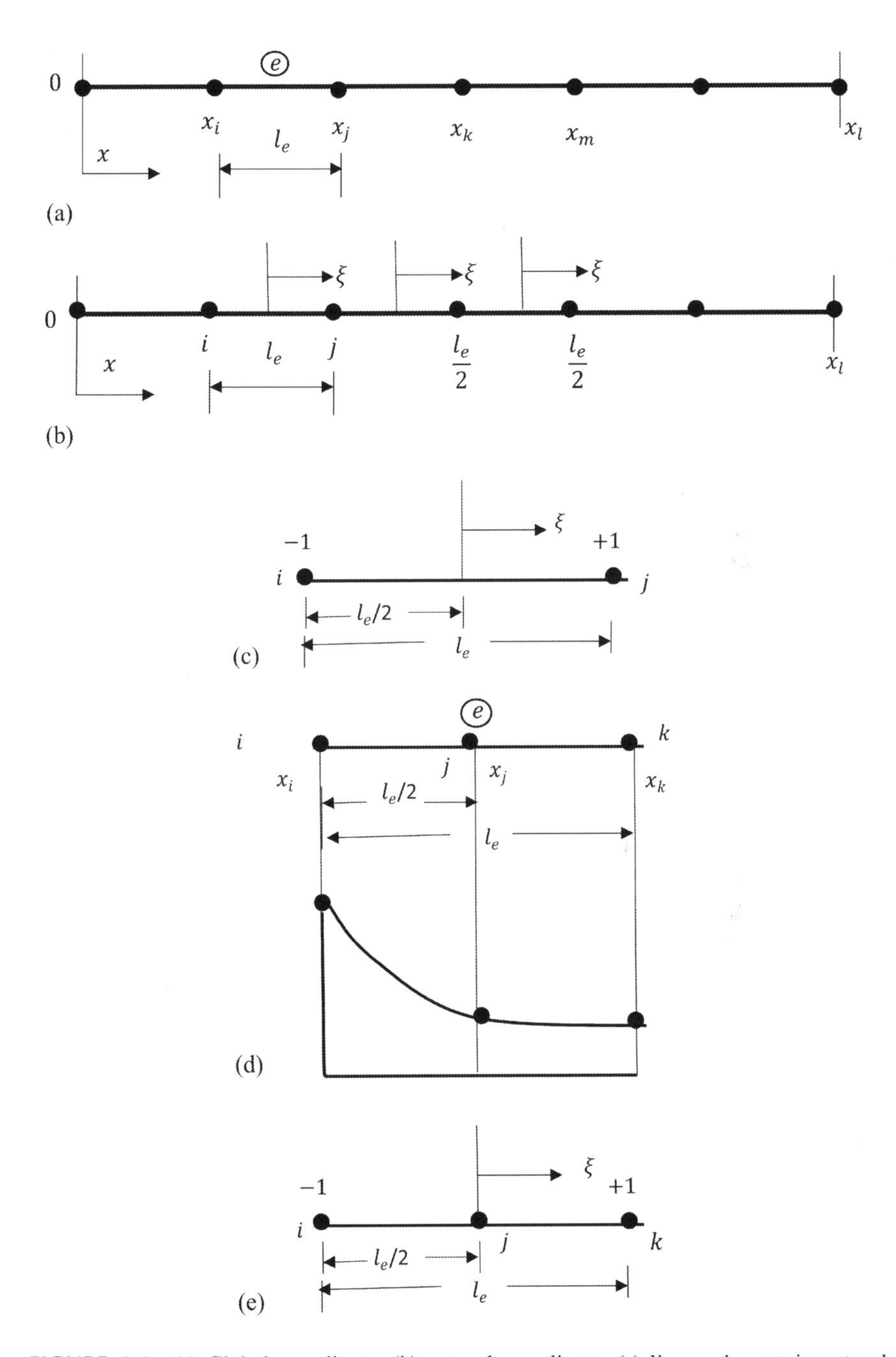

FIGURE 4.6 (a) Global coordinate, (b) natural coordinate, (c) linear element in natural coordinate, (d) quadratic element in global coordinate and (e) quadratic element in natural coordinate.

The main objective of using these natural coordinate systems is that integration of elements could be carried out locally within an element with the knowledge of the geometric information about the elements and this will reduce the intricacy associated with integration using global coordinate systems. Thus, the idea is to transform the integration of shape functions and derivatives to natural coordinate systems. This can be performed only after establishing the relation between the global Cartesian coordinate system and the natural coordinate system. The basic principle involved in coordinate transformation can be understood initially considering only one-dimensional coordinate, that is establishing a relation between x-coordinate and ξ- coordinate. Referring to Figure 4.6(c), the relation between x-coordinate and ξ- coordinate has to be understood. The origin of ξ- coordinate is located at the center of the element and let us represent this origin as x_0 and it can be expressed as

$$x_0 = x_i + \frac{(x_j - x_i)}{2}$$

Thus, $x_0 = \frac{(x_i + x_j)}{2}$. As ξ- coordinate is non-dimensional, the scaling factor used to represent any x distance is $\frac{l_e}{2}$ as the origin of ξ- coordinate is located from both the nodes 'i' and 'j' at this length.

Thus, $\xi = \frac{(x - x_0)}{l_e/2}$ and from this expression, x can be related to ξ as, $x = \frac{(x_i + x_j)}{2} + \xi\left(\frac{l_e}{2}\right)$.

The relation between x- and ξ- coordinates can be either assumed as a linear function or non-linear function. Let us assume a linear mapping between the elements represented by x-coordinate with nodes, 'i' and 'j' having origin left of node 'i' and an element in ξ- coordinate as depicted in Figure 4.6(c).

That is,

$$x = \alpha_1 + \alpha_2 \xi \tag{4.16a}$$

This equation can be solved by substituting the values of x and ξ at nodes, 'i' and 'j' as

$$x_i = \alpha_1 + \alpha_2 \xi_i \tag{4.16b}$$

$$x_j = \alpha_1 + \alpha_2 \xi_j \tag{4.16c}$$

Solving Equations (4.16b) and (4.16c) for constants, α_1 and α_2, we get $\alpha_1 = \frac{(x_i + x_j)}{2}$ and $\alpha_2 = \frac{(x_j - x_i)}{2}$. After substituting these constants in Equation (4.16a), we get

$$x = \frac{1}{2}(1 - \xi)x_i + \frac{1}{2}(1 + \xi)x_j \tag{4.16d}$$

which can be expressed as

$$x = N_i(\xi)x_i + N_j(\xi)x_j \tag{4.16e}$$

where $N_i(\xi) = \frac{1}{2}(1-\xi)$ and $N_j(\xi) = \frac{1}{2}(1+\xi)$.

It can be found that

$$\sum_{i=1}^{n} N^{(e)}{}_i = 1 \tag{4.16f}$$

One should notice that the shape function is always a non-dimensional entity whether it is written in global coordinates or natural coordinates. It can be tested for the basic characteristics of shape functions; at its own node, the above shape functions become unified, and in the other node, it is zero. As was discussed earlier, the discretization of domain is carried out with the help of standard elements, which are well defined in the literature [3]. Now, Equation (4.16e) can be written as

$$x = \sum_{i=1}^{n} N^{(e)}{}_i(\xi)x_i$$

Like the field variables, the coordinates also can be represented using a smooth continuous piecewise approximation using the natural coordinate system. Field variables are also represented using some piecewise approximate functions. When these two approximate functions are selected as the same, then such formulation is called isoparametric formulation. In other words, in isoparametric formulation, the same interpolation function used for approximating the variable is employed for approximating the geometry of the computational domain. In the solution of flow and heat transfer problems using the finite element method, it is required to compute the first order derivatives of field variables, which are reduced to first order derivatives of the shape functions.

Let us consider the first order derivative of shape functions relating x- and ξ- coordinates as

$\frac{dN_i}{dx} = \frac{dN_i}{d\xi}\frac{d\xi}{dx}$, wherein N_i represents all the shape functions for a given element.

Of course, in the present case, it is two-noded linear element. The above expression can be rewritten as

$$\frac{dN_i}{dx} = J^{-1}\frac{dN_i}{d\xi} \tag{4.17}$$

where J is called the Jacobian matrix, $J = \frac{dx}{d\xi}$.

Differentiate Equation (4.16e) with respect to ξ,

$$\frac{dx}{d\xi} = \frac{dN_i}{d\xi}x_i + \frac{dN_j}{d\xi}x_j$$

$$= -\frac{1}{2}x_i + \frac{1}{2}x_j = \frac{1}{2}\left(x_j - x_i\right)$$

$= \frac{l_e}{2} = J$. For any function, the coordinate transformation can be written as

$\int_{x_i}^{x_j} F(x)dx = \int_{-1}^{+1} F(\xi)Jd\xi$. For one-dimensional element this can be written as

$= \frac{l_e}{2}\int_{-1}^{+1} F(\xi)d\xi$. Further this type of integration can be carried out approximately using Gaussian integration technique as

$\int_{-1}^{+1} F(\xi)d\xi \approx \sum_{i=1}^{ngp} W_i F(\xi_i)$, where '*ngp*' indicates the number of Gaussian integration points and W_i are the integration weights. The accuracy of the above integration can be improved by selecting a higher number of sample points. The Gauss sample points and corresponding weights are available for one-dimensional and two-dimensional integration in standard textbooks [1, 2].

4.3 DISCRETIZATION OF ONE-DIMENSIONAL DOMAIN

Many steady state and transient one-dimensional problems can be solved using Galerkin's weighted residual technique either assuming linear or quadratic function for piecewise approximation of the field variables. The method of linear approximation of the field variable has already been discussed in the previous section in detail. It is observed that approximating the field variable with a function gives rise to the concept of shape function or interpolation function, which enables the decoupling of the functional dependence of the nodal temperatures through shape functions. Once the temperatures are designated as nodal temperatures, those temperatures are not functions of the coordinate being referred because the nodal temperatures are defined at specified nodes whose coordinates are well defined, hence the temperature becomes free from functional dependence. For most of the problems encountered in flow and heat transfer analysis, linear one-dimensional elements make a good approximation for the field variables. However, the accuracy of approximation can be improved by selecting a quadratic variation as

$$T = \alpha_1 + \alpha_2 x + \alpha_3 x^2 \tag{4.18}$$

Referring to Figure 4.6(d), it can be noticed that the temperature variation is expressed using three nodes resulting in quadratic approximation. Equation (4.18) can be written for all the three nodes, 'i', 'j' and 'k' and the resulting three equations could be solved to obtain the constants, α_1, α_2 and α_3 to provide the final expression for T in terms of shape functions and respective nodal temperatures. However, generally in order to avoid the cumbersome algebra involved in the above-mentioned procedure, the shape functions are obtained using Lagrangian interpolation polynomial, resulting in the following shape functions at all three nodes.

$$N_i(x) = \frac{(x - x_j)(x - x_k)}{(x_i - x_j)(x_i - x_k)}$$

$$N_j(x) = \frac{(x - x_i)(x - x_k)}{(x_j - x_i)(x_j - x_k)}$$

$N_k(x) = \dfrac{(x - x_i)(x - x_j)}{(x_k - x_i)(x_k - x_j)}$. After substituting the nodal values for x-coordinate in the denominator and assuming the node 'j' is located at the center of the element, the above shape functions can be rewritten as

$$N_i(x) = \frac{2}{l_e^2}(x - x_j)(x - x_k) \tag{4.19a}$$

$$N_j(x) = \frac{-4}{l_e^2}(x - x_i)(x - x_k) \tag{4.19b}$$

$$N_k(x) = \frac{2}{l_e^2}(x - x_i)(x - x_j) \tag{4.19c}$$

Hence,

$$T(x) = N_i T_i + N_j T_j + N_k T_k \tag{4.20}$$

It is understood that application of Galerkin's weighted residual method for the solution of second order derivatives results in first order derivative of shape functions. As the shape function involves product of coordinate terms, the integration of the shape functions or their derivatives will be highly challenging involving a lot of algebra. Hence, on the lines of discussion of natural coordinate system for one-dimensional linear elements, the concept can be extended for quadratic elements as well. Let us represent the spatial coordinate x as a quadratic function of ξ as follows (Refer Figure 4.6(e)).

$$x = \alpha_1 + \alpha_2 \xi + \alpha_3 \xi^2 \tag{4.21}$$

The above equation is written at all the three nodes as

$$x_i = \alpha_1 + \alpha_2 \xi_i + \alpha_3 \xi_i^2 \tag{4.22a}$$

$$x_j = \alpha_1 + \alpha_2 \xi_j + \alpha_3 \xi_j^2 \tag{4.22b}$$

$$x_k = \alpha_1 + \alpha_2 \xi_k + \alpha_3 \xi_k^2 \tag{4.22c}$$

After substituting the values of ξ at different nodes of the element, the above equations are solved to obtain the constants and then Equation (4.21) can be finally written as

$$x = \frac{\xi}{2}(\xi - 1)x_i + (1+\xi)(1-\xi)x_j + \frac{\xi}{2}(1+\xi)x_k \tag{4.23}$$

The details of integration of element matrices using isoparametric formulation will be discussed in the following section.

4.4 DISCRETIZATION OF TWO-DIMENSIONAL DOMAIN

Many problems modeled in flow and heat transfer fields are two-dimensional in nature. The most commonly used or the elements with the minimum number of nodes are the triangular elements. Figure 4.7(a) shows a two-dimensional domain discretized using triangular elements along with a single element. The variation of field variable in a triangular element can be represented as

$$T(x,y) = \alpha_1 + \alpha_2 x + \alpha_3 y \tag{4.24}$$

which clearly indicates linear variation within the element. Writing the above expression at all the three nodes, 'i', 'j' and 'k', we get the following equations.

$$T_i(x,y) = \alpha_1 + \alpha_2 x_i + \alpha_3 y_i \tag{4.25a}$$

$$T_j(x,y) = \alpha_1 + \alpha_2 x_j + \alpha_3 y_j \tag{4.25b}$$

$$T_k(x,y) = \alpha_1 + \alpha_2 x_k + \alpha_3 y_k \tag{4.25c}$$

The detailed derivations to obtain the expressions for the shape functions can be obtained from other reference [1]. Equation (4.24) can be expressed using the shape functions as

$$T(x,y) = N_i T_i + N_j T_j + N_k T_k \tag{4.26}$$

where $N_i(x,y) = \dfrac{1}{2A}(a_i + b_i x + c_i y)$

$$N_j(x,y) = \frac{1}{2A}(a_j + b_j x + c_j y)$$

$N_k(x,y) = \dfrac{1}{2A}(a_k + b_k x + c_k y)$ and

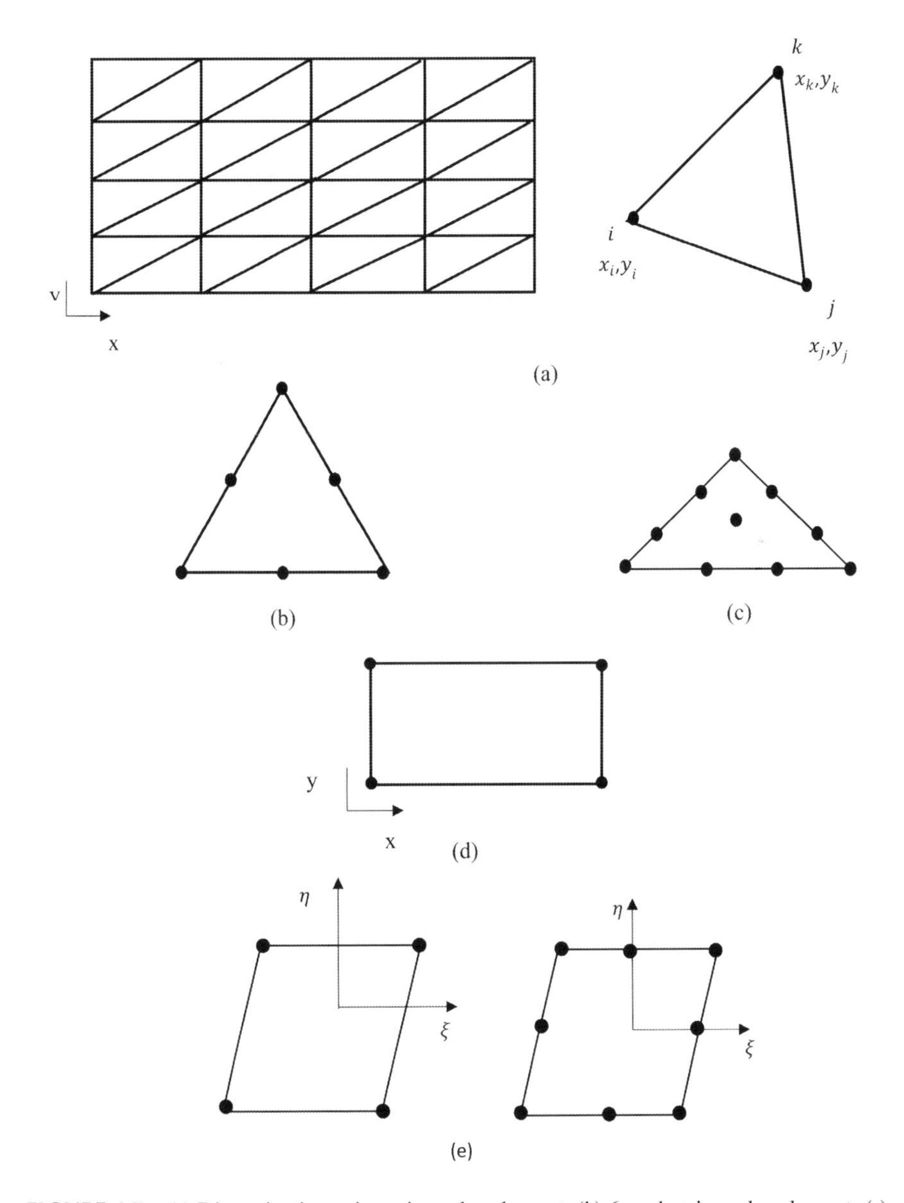

FIGURE 4.7 (a) Discretization using triangular element, (b) 6-node triangular element, (c) 10-node triangular element and (d) rectangular element.

$$a_i = x_j y_k - x_k y_j;\ b_i = y_j - y_k;\ c_i = x_k - x_j$$

$$a_j = x_k y_i - x_i y_k;\ b_j = y_k - y_i;\ c_j = x_i - x_k$$

$$a_k = x_i y_j - x_j y_i;\ b_k = y_i - y_j;\ c_k = x_j - x_i \text{ and } 2A = \det\begin{vmatrix} 1 & x_i & y_i \\ 1 & x_j & y_j \\ 1 & x_k & y_k \end{vmatrix}.$$

Figures 4.7(b) and 4.7(c) show 6-node and 10-node higher order triangular elements respectively. The natural coordinate system can be extended to triangular elements also to obtain the interpolation functions to represent both the spatial coordinates and field variables. However, isoparametric formulation for the triangular element is not very popular compared to the quadrilateral element.

4.4.1 Rectangular and Quadrilateral Elements

The most commonly used elements to discretize two-dimensional domains are the rectangular elements and quadrilateral elements. Rectangular elements are generally used to discretize regular two-dimensional geometries and they become the basis for the development of quadrilateral elements. Referring to Figure 4.7(d), the variation of field variable using rectangular element can be represented as

$$T(x,y) = \alpha_1 + \alpha_2 x + \alpha_3 y + \alpha_4 xy \tag{4.27}$$

Representing the algebraic equation for the field variable becomes more complex with an increase in the number of nodes of the element. Hence, the normalized coordinate system, also called the natural coordinate system is employed to obtain the shape functions. As a first step, the coordinate transformation between x-y system and ξ- η system will be derived before proceeding to obtain the field variables. The derivation of shape functions in ξ- η coordinates can be achieved either using Lagrange polynomials or using the rectangular coordinate system and normalized coordinate system and the details of the derivations can be obtained from textbooks on the finite element method [1]. Four node quadrilateral and 8-node quadrilateral elements as shown in Figures 4.7(e) and 4.7(f) can also be employed for generating the mesh.

The main objective of this textbook is to focus on providing guidelines in developing computational algorithms and computer codes for modeling and simulation of flow and heat transfer problems. Hence, details of the above derivations are not discussed here. Let us represent the x and y coordinates in terms of the natural coordinates as follows.

$$x(\xi,\eta) = \alpha_1 + \alpha_2\xi + \alpha_3\eta + \alpha_4\xi\eta \tag{4.28a}$$

$$y(\xi,\eta) = \alpha_5 + \alpha_6\xi + \alpha_7\eta + \alpha_8\xi\eta \tag{4.28b}$$

Making use of the values of x and y coordinates at the respective nodes and the corresponding values of ξ- η coordinates, the constants in the above equations can be obtained to get the following form of equations.

$$x(\xi,\eta) = N_i x_i + N_j x_j + N_k x_k + N_m x_m \tag{4.29a}$$

$$y(\xi,\eta) = N_i y_i + N_j y_j + N_k y_k + N_m y_m \tag{4.29b}$$

where

$$N_i(\xi,\eta)=\frac{1}{4}(1-\xi)(1-\eta) \tag{4.30a}$$

$$N_j(\xi,\eta)=\frac{1}{4}(1+\xi)(1-\eta) \tag{4.30b}$$

$$N_k(\xi,\eta)=\frac{1}{4}(1+\xi)(1+\eta) \tag{4.30c}$$

$$N_m(\xi,\eta)=\frac{1}{4}(1-\xi)(1+\eta) \tag{4.30d}$$

The actual quadrilateral element shown in Figure 4.8(a) defined with global coordinates can be mapped into the parent element shown in Figure 4.8(b), defined using ξ-η coordinates in terms of the above-defined shape functions.

In finite element solution of differential equations, most of the time it is required to integrate the first order derivative of shape functions with respect to the global coordinates, x, y and z. Hence, the main objective of mapping the actual element to the element defined by natural coordinates, is not just finding the shape functions, but also to evaluate the derivatives and find a relation between the derivatives of the global coordinates and the natural coordinates. As it is easy to integrate the derivatives of functions represented in ξ-η coordinates, it is essential to establish the relation between the derivatives. Let us start this with the derivatives of the coordinates x and y defined by Equation (4.29). Instead of finding the actual derivatives of the shape functions given by Equation (4.30), it is good enough to represent the derivatives of the shape functions as it is. While developing the computer code, the shape functions and their derivatives at all the nodes will be defined and evaluated at different Gauss points during integration. There are two coordinates, x and y which need to be differentiated with respect to ξ-η coordinates, thus giving rise to the following four partial derivatives.

$$\frac{\partial x}{\partial \xi}=\frac{\partial N_i}{\partial \xi}x_i+\frac{\partial N_j}{\partial \xi}x_j+\frac{\partial N_k}{\partial \xi}x_k+\frac{\partial N_m}{\partial \xi}x_m \tag{4.31a}$$

$$\frac{\partial x}{\partial \eta}=\frac{\partial N_i}{\partial \eta}x_i+\frac{\partial N_j}{\partial \eta}x_j+\frac{\partial N_k}{\partial \eta}x_k+\frac{\partial N_m}{\partial \eta}x_m \tag{4.31b}$$

$$\frac{\partial y}{\partial \xi}=\frac{\partial N_i}{\partial \xi}y_i+\frac{\partial N_j}{\partial \xi}y_j+\frac{\partial N_k}{\partial \xi}y_k+\frac{\partial N_m}{\partial \xi}y_m \tag{4.31c}$$

$$\frac{\partial y}{\partial \eta}=\frac{\partial N_i}{\partial \eta}y_i+\frac{\partial N_j}{\partial \eta}y_j+\frac{\partial N_k}{\partial \eta}y_k+\frac{\partial N_m}{\partial \eta}y_m \tag{4.31d}$$

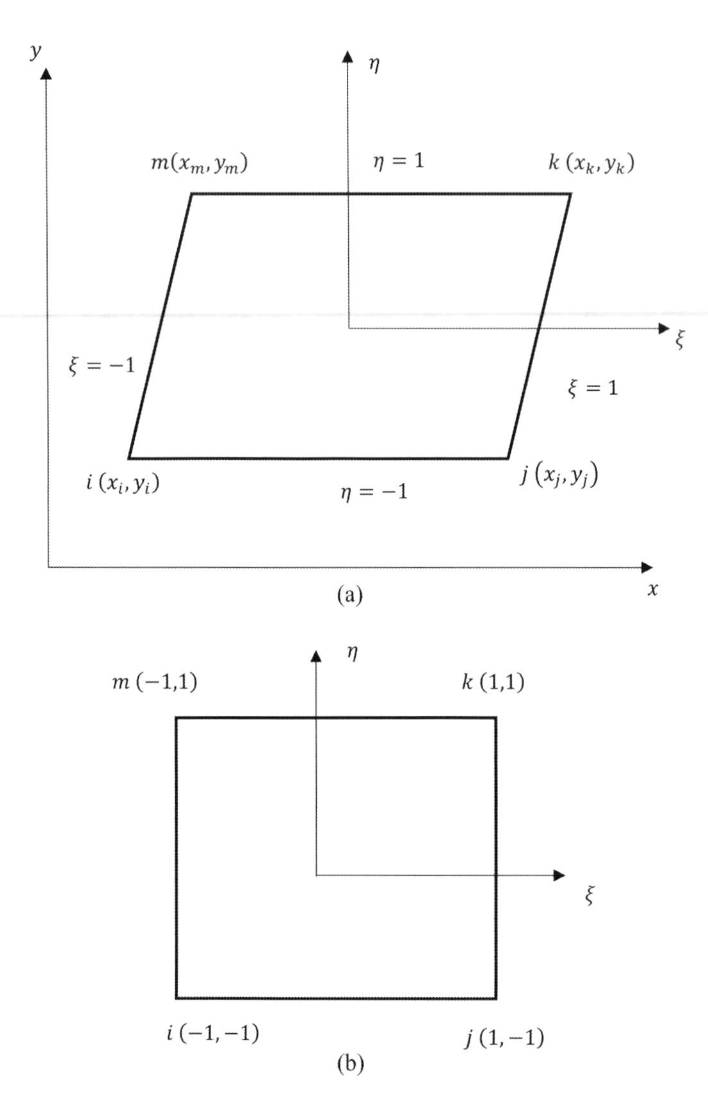

FIGURE 4.8 (a) Quadrilateral physical domain in global coordinates and (b) quadrilateral parent element in natural coordinates.

Also the shape functions N_i is a function of ξ-η coordinates, the chain rule of differentiation gives the following.

$$\frac{\partial N_i}{\partial \xi} = \frac{\partial N_i}{\partial x}\frac{\partial x}{\partial \xi} + \frac{\partial N_i}{\partial y}\frac{\partial y}{\partial \xi}$$

$$\frac{\partial N_i}{\partial \eta} = \frac{\partial N_i}{\partial x}\frac{\partial x}{\partial \eta} + \frac{\partial N_i}{\partial y}\frac{\partial y}{\partial \eta}$$

The above two equations can be written in matrix form as

$$\begin{Bmatrix} \dfrac{\partial N_i}{\partial \xi} \\ \dfrac{\partial N_i}{\partial \eta} \end{Bmatrix} = \begin{bmatrix} \dfrac{\partial x}{\partial \xi} & \dfrac{\partial y}{\partial \xi} \\ \dfrac{\partial x}{\partial \eta} & \dfrac{\partial y}{\partial \eta} \end{bmatrix} \begin{Bmatrix} \dfrac{\partial N_i}{\partial x} \\ \dfrac{\partial N_i}{\partial y} \end{Bmatrix} \tag{4.32}$$

The vectors in the above equation represent the derivatives with respect to global and natural coordinates connected by the coordinate mapping matrix, called the Jacobian matrix defined as

$[J] = \begin{bmatrix} \dfrac{\partial x}{\partial \xi} & \dfrac{\partial y}{\partial \xi} \\ \dfrac{\partial x}{\partial \eta} & \dfrac{\partial y}{\partial \eta} \end{bmatrix}$ and hence Equation (4.32) can be written as

$\begin{Bmatrix} \dfrac{\partial N_i}{\partial \xi} \\ \dfrac{\partial N_i}{\partial \eta} \end{Bmatrix} = [J] \begin{Bmatrix} \dfrac{\partial N_i}{\partial x} \\ \dfrac{\partial N_i}{\partial y} \end{Bmatrix}$. However, it is required to find the global derivatives of the shape functions and from the above equation, it can be written as

$$\begin{Bmatrix} \dfrac{\partial N_i}{\partial x} \\ \dfrac{\partial N_i}{\partial y} \end{Bmatrix} = [J]^{-1} \begin{Bmatrix} \dfrac{\partial N_i}{\partial \xi} \\ \dfrac{\partial N_i}{\partial \eta} \end{Bmatrix} \tag{4.33}$$

As the Jacobian in the above equation appears in the denominator, it can never take a value equal to zero, in which case it will lead to an indeterminate problem. In other words, for the existence of mapping between the global and natural coordinates, the Jacobian should be always non-zero. The Jacobian matrix can be evaluated using Equations (4.31a) to (4.31d) as

$$[J] = \begin{bmatrix} \dfrac{\partial x}{\partial \xi} & \dfrac{\partial y}{\partial \xi} \\ \dfrac{\partial x}{\partial \eta} & \dfrac{\partial y}{\partial \eta} \end{bmatrix} = \begin{bmatrix} \dfrac{\partial N_i}{\partial \xi} & \dfrac{\partial N_j}{\partial \xi} & \dfrac{\partial N_k}{\partial \xi} & \dfrac{\partial N_m}{\partial \xi} \\ \dfrac{\partial N_i}{\partial \eta} & \dfrac{\partial N_j}{\partial \eta} & \dfrac{\partial N_k}{\partial \eta} & \dfrac{\partial N_m}{\partial \eta} \end{bmatrix} \begin{bmatrix} x_i & y_i \\ x_j & y_j \\ x_k & y_k \\ x_m & y_m \end{bmatrix} \tag{4.34}$$

It is worth observing that for evaluating both the global derivatives and Jacobian matrix, derivatives of shape functions in natural coordinates are required as noted from Equations (4.33) and (4.34). Hence, in the process of integration of shape function derivatives in global coordinates, the derivatives of shape functions in natural coordinates take care of both coordinate transformation and evaluation of the Jacobian matrix. For a given quadrilateral element with global coordinates of all the four nodes specified, the integration of shape function derivatives in natural coordinates is achieved by considering a number of Gauss points. Numerical integration of shape functions and their derivatives is reduced to summation at all the selected number of Gauss points. Expanding the product of two matrices in equation (4.34) gives rise to

$$\left[J\right]=\begin{bmatrix} \sum_{p=i}^{m}\frac{\partial N_p}{\partial \xi}x_p & \sum_{p=i}^{m}\frac{\partial N_p}{\partial \xi}y_p \\ \sum_{p=i}^{m}\frac{\partial N_p}{\partial \eta}x_p & \sum_{p=i}^{m}\frac{\partial N_p}{\partial \eta}y_p \end{bmatrix}$$ and the determinant of *[J]* is computed as

$$\left|J\right|=\left[\left(\sum_{p=i}^{m}\frac{\partial N_p}{\partial \xi}x_p\right)\left(\sum_{p=i}^{m}\frac{\partial N_p}{\partial \eta}y_p\right)-\left(\sum_{p=i}^{m}\frac{\partial N_p}{\partial \eta}x_p\right)\left(\sum_{p=i}^{m}\frac{\partial N_p}{\partial \xi}y_p\right)\right] \tag{4.35}$$

Thus, integration of any function in x, y-coordinates can be represented in terms of ξ-η coordinates as

$$\int_A F(x,y)dxdy=\int_{-1}^{+1}\int_{-1}^{+1}F(\xi,\eta)\left|J\right|d\xi d\eta$$

The detailed application of a natural coordinate system for the integration of global derivatives obtained as a result of application of Galerkin's weighted residual method to heat transfer problems will be discussed in the next chapter.

4.5 DISCRETIZATION OF THREE-DIMENSIONAL DOMAIN

Numerical simulation of three-dimensional problems in heat transfer and flow analysis are more easily handled by using commercial software than writing own code except for certain specialized applications that involve coupled and non-linear differential equations. Irrespective of the type of numerical method employed such as finite difference, finite volume and the finite element method, the commercial software has occupied the simulation space compared to computer codes developed by academics. Many real life problems with complex geometries can be analyzed using software packages such as ANSYS, OpenFOAM, COMSOL Multiphysics etc. just to mention a few among a number of popular software. However, knowledge about the basics in solving three-dimensional problems and associated computational algorithms will help the users to understand the working of the software and better equipped with knowledge to interpret the simulation results. The most challenging part of solution of three-dimensional problems is domain discretization and book keeping of all the mesh details. There are free open source codes such as Gmsh, ICEM, snappyHexMesh, cfMesh, Salome etc. widely used for generating three-dimensional meshes. Four node tetrahedral elements (Figure 4.9a) are the basic element that can be used to discretize three-dimensional computational domain. Making use of an approach

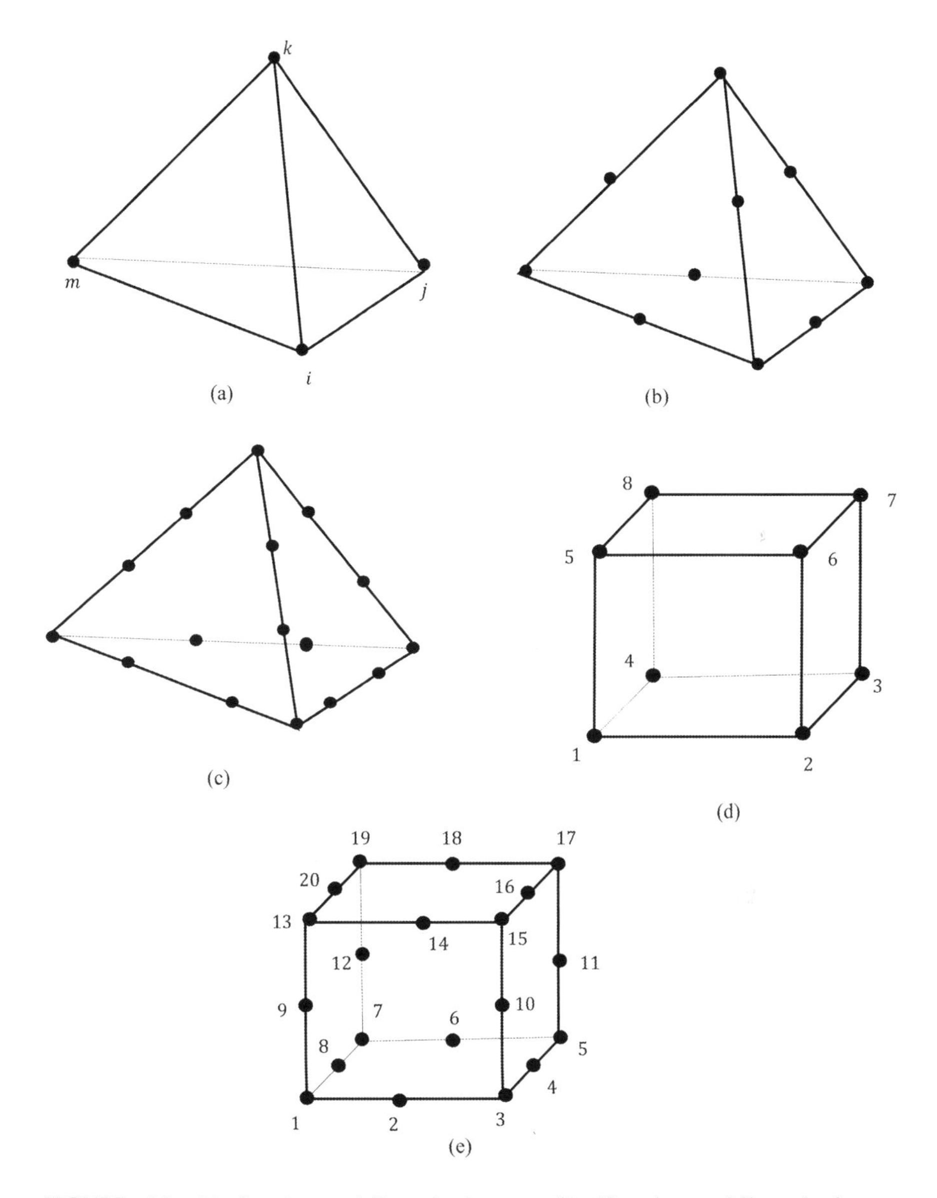

FIGURE 4.9 (a) 4-node quadrilateral element, (b) 10-node quadrilateral element, (c) 16-node quadrilateral element, (d) 8-node brick element and (e) 20-node brick element.

based on area coordinates employed for triangular elements [3], volume coordinates are defined to derive an interpolation function to determine the variation of field variable at any point X within the element. Assuming linear piecewise variation of the field variable, $\phi(x, y, z)$within the element, we can write

$$\phi(x, y, z) = N_i\phi_i + N_j\phi_j + N_k\phi_k + N_m\phi_m \tag{4.36}$$

where

$$N_i(x,y,z) = \frac{1}{6V}\left(a_i + b_i x + c_i y + d_i z\right) \tag{4.37a}$$

$$N_j(x,y,z) = \frac{1}{6V}\left(a_j + b_j x + c_j y + d_j z\right) \tag{4.37b}$$

$$N_k(x,y,z) = \frac{1}{6V}\left(a_k + b_k x + c_k y + d_k z\right) \tag{4.37c}$$

$$N_m(x,y,z) = \frac{1}{6V}\left(a_m + b_m x + c_m y + d_m z\right) \tag{4.37d}$$

and

$6V = \begin{vmatrix} 1 & x_i & y_i & z_i \\ 1 & x_j & y_j & z_j \\ 1 & x_k & y_k & z_k \\ 1 & x_m & y_m & z_m \end{vmatrix}$ and the various coefficients used in Equation (4.37) can be evaluated as

$$a_i = \begin{vmatrix} x_j & y_j & z_j \\ x_k & y_k & z_k \\ x_m & y_m & z_m \end{vmatrix},\ b_i = \begin{vmatrix} 1 & y_j & z_j \\ 1 & y_k & z_k \\ 1 & y_m & z_m \end{vmatrix},\ c_i = \begin{vmatrix} 1 & x_j & z_j \\ 1 & x_k & z_k \\ 1 & x_m & z_m \end{vmatrix} \text{ and } d_i = \begin{vmatrix} 1 & x_j & y_j \\ 1 & x_k & y_k \\ 1 & x_m & y_m \end{vmatrix}$$

Similarly, the other constants can be obtained. It may be required to integrate shape functions and this can be achieved in volume coordinates as follows.

$$\int_V N^a{}_i N_j^b N_k^c N_m^d dV = \frac{a!b!c!d!}{\left(a+b+c+d+3\right)!}(6V) \tag{4.38}$$

Non-linear variation also can be considered along all the edges of the basic tetrahedral element as shown in Figures 4.9(b) and 4.9(c). Quadratic variation gives rise to 10-node tetrahedral elements (Figure 4.9(b)) whereas the cubic approximation gives rise to 20-node elements (Figure 4.9(c)). However, it is difficult to achieve geometric isotropy in tetrahedral elements and hence, their use is highly limited. The next most common type of element used for three-dimensional domain discretization is brick elements. The basic element with linear variation consists of 8-nodes as depicted in Figure 4.9(d) in Cartesian global coordinates. The piecewise approximation of field variable within the element can be represented as

$$\phi(x,y,z) = \alpha_1 + \alpha_2 x + \alpha_3 y + \alpha_4 z + \alpha_5 xy + \alpha_6 xz + \alpha_7 yz + \alpha_8 xyz \tag{4.39}$$

Application of natural coordinate system in all the three Cartesian directions with limits of –1 and +1 gives rise to the following interpolation expression.

$x(\xi,\eta,\varsigma)=\sum_{i=1}^{8}N_i(\xi,\eta,\varsigma)x_i,\quad y(\xi,\eta,\varsigma)=\sum_{i=1}^{8}N_i(\xi,\eta,\varsigma)y_i$ and $z(\xi,\eta,\varsigma)=\sum_{i=1}^{8}N_i$ $(\xi,\eta,\varsigma)z_i$

in which

$$N_1(\xi,\eta,\varsigma)=\frac{1}{8}(1-\xi)(1-\eta)(1+\varsigma)$$

$$N_2(\xi,\eta,\varsigma)=\frac{1}{8}(1+\xi)(1-\eta)(1+\varsigma)$$

$$N_3(\xi,\eta,\varsigma)=\frac{1}{8}(1+\xi)(1+\eta)(1+\varsigma)$$

$$N_4(\xi,\eta,\varsigma)=\frac{1}{8}(1-\xi)(1+\eta)(1+\varsigma)$$

$$N_5(\xi,\eta,\varsigma)=\frac{1}{8}(1-\xi)(1-\eta)(1-\varsigma)$$

$$N_6(\xi,\eta,\varsigma)=\frac{1}{8}(1+\xi)(1-\eta)(1-\varsigma)$$

$$N_7(\xi,\eta,\varsigma)=\frac{1}{8}(1+\xi)(1+\eta)(1-\varsigma)$$

$$N_8(\xi,\eta,\varsigma)=\frac{1}{8}(1-\xi)(1+\eta)(1-\varsigma) \tag{4.40}$$

The derivatives of x, y and z-coordinates are determined as

$\frac{\partial x}{\partial \xi}=\sum_{i=1}^{8}\frac{\partial N_i}{\partial \xi}x_i,\ \frac{\partial y}{\partial \xi}=\sum_{i=1}^{8}\frac{\partial N_i}{\partial \xi}y_i$ and $\frac{\partial z}{\partial \xi}=\sum_{i=1}^{8}\frac{\partial N_i}{\partial \xi}z_i$

$\frac{\partial x}{\partial \eta}=\sum_{i=1}^{8}\frac{\partial N_i}{\partial \eta}x_i,\ \frac{\partial y}{\partial \eta}=\sum_{i=1}^{8}\frac{\partial N_i}{\partial \eta}y_i$ and $\frac{\partial z}{\partial \eta}=\sum_{i=1}^{8}\frac{\partial N_i}{\partial \eta}z_i$

$\frac{\partial x}{\partial \varsigma}=\sum_{i=1}^{8}\frac{\partial N_i}{\partial \varsigma}x_i,\ \frac{\partial y}{\partial \varsigma}=\sum_{i=1}^{8}\frac{\partial N_i}{\partial \varsigma}y_i$ and $\frac{\partial z}{\partial \varsigma}=\sum_{i=1}^{8}\frac{\partial N_i}{\partial \varsigma}z_i$. Then the Jacobian matrix can be obtained as

$$[J]=\begin{bmatrix} \frac{\partial x}{\partial \xi} & \frac{\partial y}{\partial \xi} & \frac{\partial z}{\partial \xi} \\ \frac{\partial x}{\partial \eta} & \frac{\partial y}{\partial \eta} & \frac{\partial z}{\partial \eta} \\ \frac{\partial x}{\partial \varsigma} & \frac{\partial y}{\partial \varsigma} & \frac{\partial z}{\partial \varsigma} \end{bmatrix}=\begin{bmatrix} \frac{\partial N_1}{\partial \xi} & \frac{\partial N_2}{\partial \xi} & .. & \frac{\partial N_8}{\partial \xi} \\ \frac{\partial N_1}{\partial \eta} & \frac{\partial N_2}{\partial \eta} & .. & \frac{\partial N_8}{\partial \eta} \\ \frac{\partial N_1}{\partial \varsigma} & \frac{\partial N_2}{\partial \varsigma} & .. & \frac{\partial N_8}{\partial \varsigma} \end{bmatrix}\begin{bmatrix} x_1 & y_1 & z_1 \\ x_2 & y_2 & z_2 \\ . & . & . \\ . & . & . \\ x_8 & y_8 & z_8 \end{bmatrix} \tag{4.41}$$

The integration of functions in global coordinates can be performed using the natural coordinate system as

$$\int_V F(x,y,z)dxdydz=\int_{-1}^{+1}\int_{-1}^{+1}\int_{-1}^{+1} F(\xi,\eta,\varsigma)\left|J\right| d\xi d\eta d\varsigma. \tag{4.42}$$

When piecewise approximation within the element is assumed as quadratic, then 20-node brick element is obtained as shown in Figure 4.9(e). Interpolation equations based on natural coordinates can be developed on the lines of the steps followed in the derivation of shape functions for 8-node brick element.

Thus far the discussion was focused on isoparametric formulation in which both the field variables and element geometry are defined using the same type of element. That means the mapping between the global and natural coordinates is achieved using the same parent element used to represent the field variable. This rule need not be the same in all cases. When the number of nodes considered for approximating element geometry is less than those used for approximating the field variables, then it is called *subparametric* formulation and similarly when the number of nodes for approximating the element geometry is higher than those used for the field variables, then it is called *superparametric* formulation. In general, isoparametric formulation is preferably employed in most of the heat transfer and flow problems. Once again, it is worth recalling the very purpose of employing natural coordinate systems is to represent piecewise approximation of either geometry or field variable in the finite element method, which makes the integration easily programmable. The finite element method is an integral formulation, in which the application of Galerkin's method always results in integration of shape functions or their derivatives. Use of global coordinates during these integrations results in tedious ways of bookkeeping of nodal coordinates of all the elements because these coordinates are defined using a fixed origin. However, in natural coordinates system, the center of each element is considered as the origin to define the coordinates of the nodes of the element and the coordinates always vary from –1 to +1, which enables integration as summation through Gaussian quadrature. Discretization of computational domain in the finite element method is achieved through a systematic procedure called mesh generation.

4.6 MESH GENERATION

The basic steps of a solution procedure start from an element and then it is extended for the entire computational domain. The finite element method is distinct in the characteristic that the defined element becomes the representative of the complete solution procedure for the entire computational domain. The finite element principle satisfies the underlying conservation principle of the representative governing equations in differential form in each element. Though generating grid details in a computational domain is common for any type of numerical method such as finite difference or finite volume method, in the finite element method, the solution procedure starts from the definition of an element. Mesh generation consists of (i) discretization of the computational domain using some standard procedure, (ii) establishing the required connectivity between the elements so that the continuity of field variables being solved is satisfied and (iii) generating mesh data in the form of coordinates of vertices of each element and data on nodal-element connectivity. A computational domain is a representative or imaginary domain in space considered as a replica of the real geometry of the system under investigation. The final objective of modeling and simulation is to find out or predict the performance of a real system which can be defined using its geometrical topology with the help of some coordinate systems. Hence, care must be taken to generate a computational domain that represents geometrically a similar structure of the real system. It is well known that most of the real systems are complex in geometry, however, it is not impossible to geometrically define their details. It is needless to say that some systems are very regular in their shape, which can be represented using regular geometric surfaces of solids. In mesh generation techniques, the method of mesh generation differs based on the simplicity of the geometry, and in a major classification, they are categorized as structured and unstructured meshes. Generating mesh details using regular geometries which can be defined using well-established geometrical topologies such as rectangle, quadrilateral or brick shape etc., is called structured mesh generation. The given computational domain can be easily divided into a number of such well-defined geometries and hence, the inter connectivity of the vertices of each element also can be conveniently documented to convert it finally into a computer code. It becomes simple to generate conformal mapping between the parent element and real elements in structured mesh generation. Algebraic methods such as transfinite interpolation (TFI) technique, partial differential equations (PDE) technique etc. are some of the methods to generate structured mesh. For generating mesh details for two-dimensional domain using structured mesh, quadrilateral elements are commonly used whereas hexahedral elements are employed for three-dimensional domain. Coding and bookkeeping of all element details of structured mesh is simple compared to unstructured mesh.

The concept of the finite element method became very popular in computational science mainly because of its ability to handle complex geometries of real systems, without sacrificing the details and accuracy. Any given complex geometry can be discretized using unstructured mesh generation technique. In order to accomplish the discretization of a complex geometry without satisfying the connectivity between all the elements, triangular elements are employed due to their ability to replicate any intricate shape in a given geometry. The size of each element also plays a major role in defining mesh details of a computational domain. In structured mesh also, the size of the elements need not be the same and uniform throughout the computational domain. In flow and heat transfer problems, it is always required to determine the gradients of field variables at the boundaries of the domain. Hence, in order to accurately calculate the gradients near the boundary with simple linear elements, it is important to generate finer elements at the boundaries and this is called mesh refinement. In the case of unstructured mesh, the basic principle behind this mesh generation is, that the grading of the elements should be easily achieved so that the curved or any intricate boundaries are meshed accurately. The commonly employed technique is called Delauny triangulation, which has the ability to change the size of the triangular elements depending on the location in the given computational domain. As the boundaries of sharp curved geometry is approached, the size of the triangle is also reduced without satisfying the continuity requirement between the elements. In unstructured meshing, triangular elements are employed for the two-dimensional domain, whereas tetrahedral elements are used for three-dimensional domains. Due to the nature of the varying size of the triangles within the domain in different directions, the vertices of all the triangles are not positioned in an ordered pattern, and this poses the challenge for coding such meshing details. The description of different mesh generation techniques and the mathematics behind all these techniques are discussed in detail in reference [4]. Discussion on mesh generation is included in this chapter in the context of analysis of finite element method. As many example problems discussed in this textbook employ transfinite interpolation (TFI) technique, some details about this mesh generation method will be discussed in the following section.

4.7 TRANSFINITE INTERPOLATION TECHNIQUE (TFI)

The well-established structured meshing technique using the algebraic method is the transfinite interpolation technique, which is easy to understand and code. This technique can generate structured mesh based on an interpolation scheme conforming to specified boundaries. It makes use of multivariate interpolation technique using Boolean sum of univariate interpolation procedure in each of the computational coordinate direction. The mapping is done from a regular-shaped parent domain to the given arbitrary domain. The basic principle revolves around generating interpolation function for the boundaries or the sides of the given geometry using unidirectional interpolation function that satisfies one-to-one mapping between the parent geometry and the physical geometry. The parent geometry is always considered as unit square for two-dimensional geometries. Let us assume the parent geometry is represented by ξ and η coordinates with two vertical sides, $g(0,\eta)$, $g(1,\eta)$ and two horizontal sides, $g(\xi,0)$ and $g(\xi,1)$ as shown in Figure 4.10(a). Now, consider the physical geometry, ABCD depicted in Figure 4.10(b), which needs to be discretized into a number of sub-domains giving rise to grid points. The physical domain ABCD is represented using global coordinates, x and y.

Final mesh generation should provide information on the x and y coordinates of all the grid points in the physical domain ABCD. As it is difficult to discretize the irregular physical domain, the principle of mapping between the parent element to the physical domain is adopted [5]. The parent element is represented by coordinates (0,0), (1, 0), (1, 1) and 0, 1) respectively corresponding to the corners, A, B, C and D

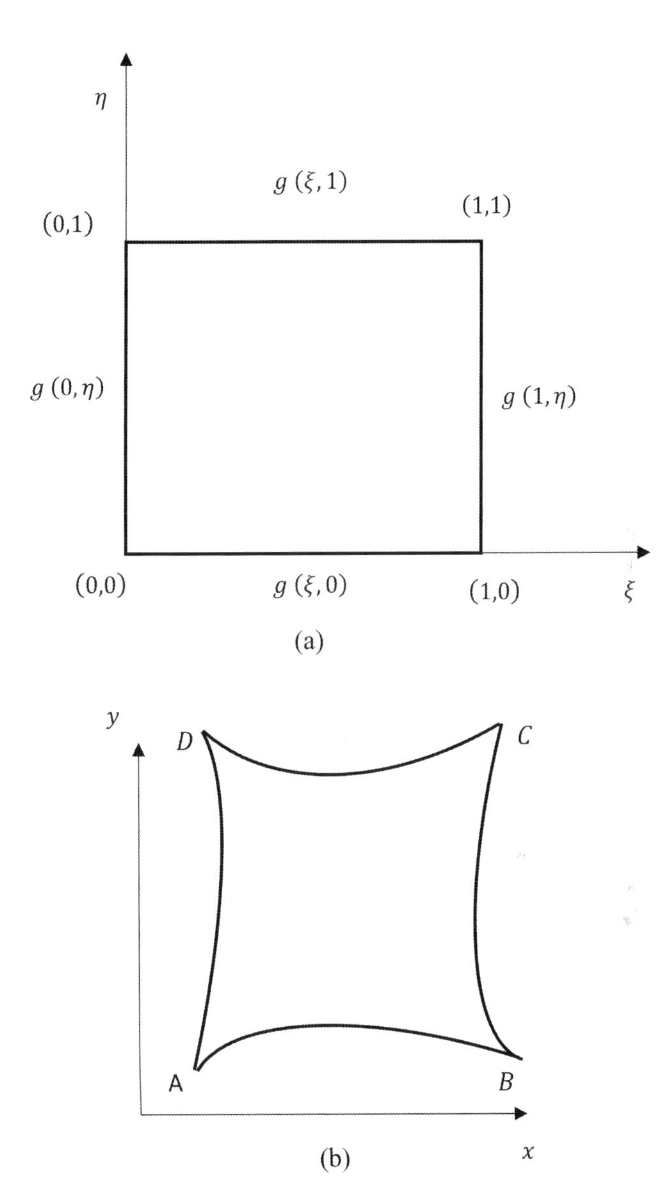

FIGURE 4.10 (a) Square parent element in natural coordinates and (b) quadrilateral physical domain in global coordinates.

of the physical domain. Using unidirectional linear interpolation along AB, one can write the interpolation function in the ξ- direction for total divisions, M from i=1 at different levels of η varying from j=1 to N, as

$$g(\xi_i, \eta_j) = (1 - \xi_i) g(0, \eta_j) + \xi_i g(1, \eta_j) \tag{4.43a}$$

Similar expression can be obtained for interpolation along the η direction as

$$g(\xi_i,\eta_j)=(1-\eta_j)g(\xi_i,0)+\eta_j g(\xi_i,1) \tag{4.43b}$$

where ξ_i varies as $0\le\dfrac{i-1}{M-1}\le 1$ and η_j varies as $0\le\dfrac{j-1}{N-1}\le 1$. As only linear interpolation has been employed, the mapping of ABCD will produce only the linear variation between the vortices. Hence, the generalized projectors that map points in the parent element to the physical element for both the directions can be written as

$$U_\xi(\xi,\eta)=(1-\xi)g(0,\eta)+\xi g(1,\eta) \tag{4.44a}$$

$$V_\eta(\xi,\eta)=(1-\eta)g(\xi,0)+\eta g(\xi,1) \tag{4.44b}$$

The sides represented by the sides in the parent element, $\eta=0$, $\eta=1$, $\xi=0$ and $\xi=1$ can be mapped to the straight edges of the respective sides of the physical domain, AB, CD, DA and BC respectively. However, the entire the curved sides of the geometry ABCD can be mapped using the parent element with composite mapping using Boolean sum of transformations, U_ξ and V_η, denoted by $U_\xi\oplus V_\eta$ and this can be expanded as

$$U_\xi\oplus V_\eta=U_\xi+V_\eta-U_\xi V_\eta \tag{4.45}$$

The product term in the above equation can be obtained as

$$U_\xi V_\eta=U_\xi\left[(1-\eta)g(\xi,0)+\eta g(\xi,1)\right]$$

In the above expression, U_ξ has to be evaluated at both $\xi=0$ and $\xi=1$. The expansion of the above expression produces the following equation.

$$U_\xi V_\eta=(1-\xi)\left[(1-\eta)g(0,0)+\eta g(0,1)\right]+\xi\left[(1-\eta)g(1,0)+\eta g(1,1)\right]$$

$$=(1-\xi)(1-\eta)g(0,0)+(1-\xi)\eta g(0,1)+\xi(1-\eta)g(1,0)+\xi\eta g(1,1) \tag{4.46}$$

After substituting Equations (4.44) and (4.46) in Equation (4.45), we get the final mapping function as

$$U_\xi\oplus V_\eta=(1-\xi)g(0,\eta)+\xi g(1,\eta)+(1-\eta)g(\xi,0)+\eta g(\xi,1)-$$

$$(1-\xi)(1-\eta)g(0,0)+(1-\xi)\eta g(0,1)+\xi(1-\eta)g(1,0)+\xi\eta g(1,1) \tag{4.47}$$

Equation (4.47) represents the basic transformation principle used in transfinite interpolation mesh generation technique. Initially discrete values of ξ_i and η_j will be generated using the expressions $0\le\xi_i=\dfrac{i-1}{M-1}\le 1$ and $0\le\eta_j=\dfrac{j-1}{N-1}\le 1$ to generate the mesh in the domain ABCD using the parent element. Equation (4.47) can be used to generate the x and y coordinates of the physical domain ABCD by replacing the functions $\boldsymbol{g}$ in terms of the respective coordinates. It has to be noted that the bottom and top sides of the domain will be discretized using the discrete values of ξ_i, whereas the left and right sides are discretized using the discrete values of η_j. As the interpolation functions are generated between the vertical and horizontal sides of the parent and physical domains, it becomes easy to identify these sides with generalized notations such as left, right, bottom and top. All these four sides will have x and y coordinates, for example, for the bottom side, the y-coordinate is zero, however, the x-coordinate varies as discrete intervals in terms of ξ_i and the bottom side is represented as $g(\xi,0)$. Similarly, the vortices of the geometry have to be identified. Let us adopt the following notations to represent all four sides and four vortices of the physical geometry.

$g(0,\eta)$ = left side, $ul(\eta)-x$ coordinate; $vl(\eta)-y$ coordinate;

$g(1,\eta)$ = right side, $ur(\eta)-x$ coordinate; $vr(\eta)-y$ coordinate

$g(\xi,0)$ = bottom side, $ub(\xi)-x$ coordinate; $vb(\xi)-y$ coordinate;

$g(\xi,1)$ = top side, $ut(\xi)-x$ coordinate; $vt(\xi)-y$ coordinate

x_{00} – x-coordinate of bottom left corner; y_{00} – y-coordinate

x_{10} – x-coordinate of bottom right corner; y_{10} – y-coordinate

x_{11} – x-coordinate of right top corner; y_{11} – y-coordinate

x_{01} – x-coordinate of left top corner; y_{01} – y-coordinate

Now making use of Equation (4.47), the x and y coordinates of grid points generated within the physical domain, expressed by the above corner coordinates can be written by the following equations.

$$x(\xi,\eta)=(1-\xi)ul(\eta)+\xi ur(\eta)+(1-\eta)ub(\xi)+\eta ut(\xi)-(1-\eta)(1-\xi)x_{00}-(1-\xi)\eta x_{01}-(1-\eta)\xi x_{10}-\xi\eta x_{11} \quad (4.48a)$$

$$y(\xi,\eta)=(1-\xi)vl(\eta)+\xi vr(\eta)+(1-\eta)vb(\xi)+\eta vt(\xi)-(1-\eta)(1-\xi)y_{00}-(1-\xi)\eta y_{01}-(1-\eta)\xi y_{10}-\xi\eta y_{11} \quad (4.48b)$$

The consistency conditions at all the four corner nodes of the physical domain have to be satisfied with the following constraints.

$$ub(0) = ul(0); ub(1) = ur(0); ur(1) = ut(1); ul(1) = ut(0) \tag{4.49a}$$

$$vb(0) = vl(0); vb(1) = vr(0); vr(1) = vt(1); vl(1) = vt(0) \tag{4.49b}$$

Using Equations (4.48) and (4.49), the transfinite interpolation grids can be obtained for any given physical domain whose dimensions are specified in terms of x and y coordinates. Another important consistency condition to be satisfied is the number of divisions for both the horizontal sides should be the same and similarly for the two vertical sides. Mesh generation using the transfinite interpolation method involves three important steps as described below.

(i) Computing the divisions on all the four sides of the parent element with unit coordinates. Making divisions along the vertical and horizontal boundary sides using the specified coordinates of the vortices and the number of divisions in x and y coordinates.
(ii) Checking for the consistency conditions with respect to the x and y coordinates of all the four vortices.
(iii) Computing the x and y coordinates of all the grid points generated by interpolation using Equation (4.48).

The above computational procedure assumes constant divisions along the x and y-coordinates, however, a non-uniform grid can be generated using TFI technique. As it has been explained, the TFI technique computes the interpolation of grid points within the computational domain by making use of the divisions made on the boundaries. For generating non-uniform grids along both x- and y-directions, initially the non-uniform divisions on the respective boundaries need to be generated, and then it interpolates the divisions within the computational domain following the same pattern of variation of the size of the grids on the boundaries. In non-uniform grid generation also, the parent element in $\xi - \eta$ coordinates will have only unit square, and the mapping functions generated using Equations (4.48) takes care of the required interpolation within the computational domain. A computer code, ***Tfi_2D.for*** in FORTRAN has been developed to generate mesh in two-dimensional domain using TFI technique. Figure 4.11(a) shows the parent element in $\xi - \eta$ coordinates with unit square. A simple uniform TFI grid with straight edges is depicted in Figure 4.11(b), whereas TFI grid for curved boundaries is shown in Figure 4.11(c). It is also possible to generate non-uniform grid closer to the boundaries of the domain. In order to calculate the gradients of field variables at the boundaries in many flow and heat transfer problems, non-uniform mesh is preferred and Figure 4.11(d) shows one such grid generated using TFI technique.

It has to be noticed that the grid spacing keeps increasing from the boundaries towards the center of the domain along all the four boundaries. Hence, a symmetry can be observed along the x and y-directions in the computational domain about increasing grid spacing from the boundaries. It is essential to maintain the dimensional symmetry of the domain as well while generating non-uniform mesh. Generally, an exponential function is used for controlling the size of the grid spacing from the boundary up to half the dimension of the domain in each direction.

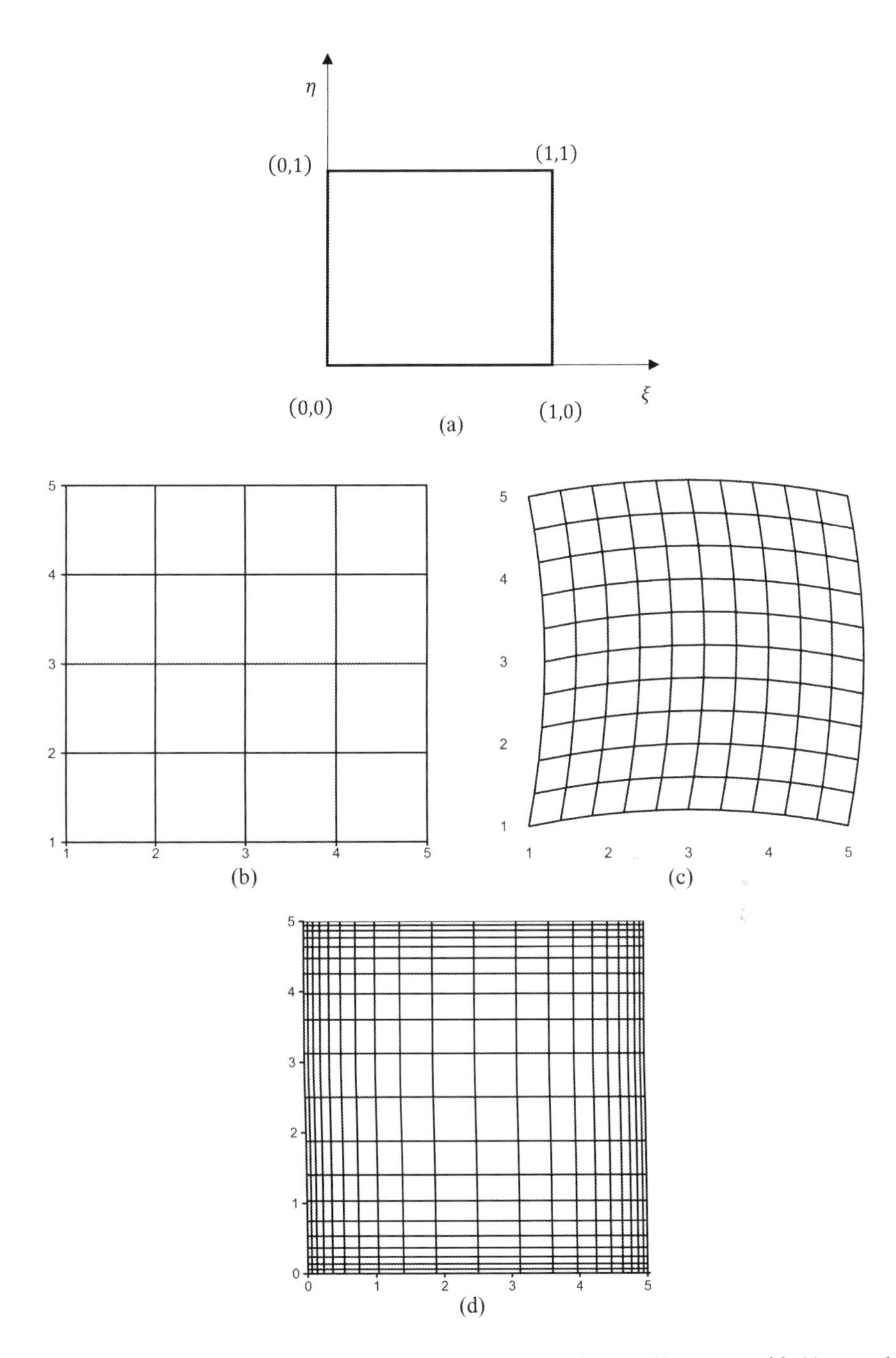

FIGURE 4.11 (a) Square parent element in natural coordinates, (b) square grid, (c) curved grid and (d) non-uniform grid.

4.7.1 Multi-Block TFI Grid Generation

Many flow and heat transfer problems involve geometries other than regular shape, hence it is essential to develop algorithms for mesh generation of irregular shapes. Figure 4.12 depicts one such geometry of square section with a square hole at the center symmetric about both x and y-axes. Though it is a simple square, the central hole makes it a little challenging, demanding discretizing the domain into a number of blocks marked as 1, 2, 3, 4, 5, 6, 7 and 8 as shown in Figure 4.12(a). The final mesh required as the computational domain is shown in Figure 4.12(b).

The computational algorithm remains the same, except that all the computational procedures followed for one sub-domain will be repeated for the eight sub-domains. Due to the positioning of the various sub-domains in the overall geometry, there is a constraint on the number of divisions that can be made in the x and y-directions. For example, sub-domains 1, 4 and 6 should have the same number of divisions in the x-direction, similarly between 2 and 7 and the last three, 3, 5 and 8. Following the same logic, the number of y-divisions will remain constant for the three set of sub-domains, 1, 2, 3, 4, 5 and 6, 7 and 8. The algorithm for programming multi-block TFI mesh generation can be explained as below.

(i) Read total number of sub-domains.
(ii) Start the do loop for all the sub-domains – input data.
    Read number of divisions in x- and y- directions for each sub-domain
    Read x and y coordinates of all the four corners of each sub-domain
  End the do loop – input data
(iii) Compute the number of rows and columns for each sub-domain.
(iv) Start the do loop for the sub-domains – divisions on boundary sides.
    Compute the divisions on unit square in ξ and η coordinates
    Compute x and y divisions on all the boundary sides
    Compute any non-uniform divisions on the specified boundary sides
  End the do loop – Divisions on boundary sides
(v) Start the do loop for all the sub-domains – x and y coordinates of interior nodes of physical domain.
    Start the loop for row – for all sub-domains in y-direction
    Start the loop for column – for all sub-domains in x-direction
    Compute x and y coordinates of all the grid points in the interior domain using Equations (4.48(a) and 4.48(b))
    End the loop – for all sub-domains in x-direction
    End the loop – for all sub-domains in y-direction
  End the do loop for all the sub-domains – x and y coordinates of interior nodes of physical domain
(vi) Stop and output the results for making grid view using graphing software.

4.7.2 Three-Dimensional TFI Meshing

The principle of transfinite interpolation technique discussed in the above section for generating x and y coordinates of all the interior nodes of a physical domain

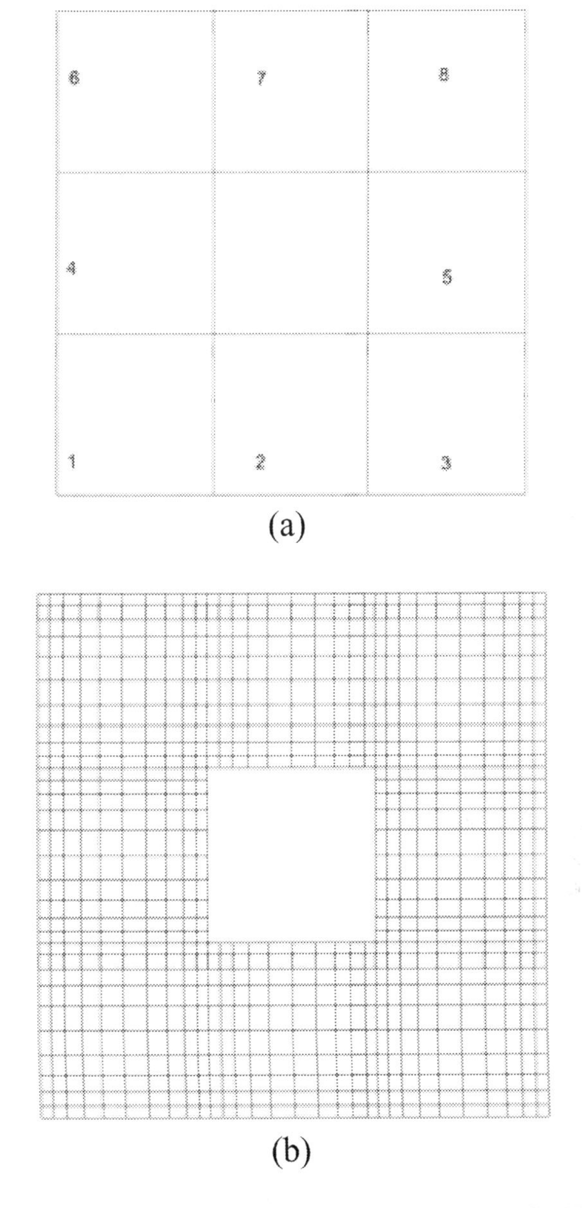

FIGURE 4.12 Multi-block TFI mesh, (a) discretion of blocks and (b) final TFI mesh.

in two-dimension can be extended for three-dimensional geometries. Figure 4.13(a) shows the parent element defined using natural coordinates, ξ, η and ς in order to discretize the physical domain ABCDEFGH shown in Figure 4.13(b) defined using x, y and z-coordinates.

Following the principle used in the definition of linear interpolation functions as defined by Equations (4.43), the linear basis functions can be described for all the

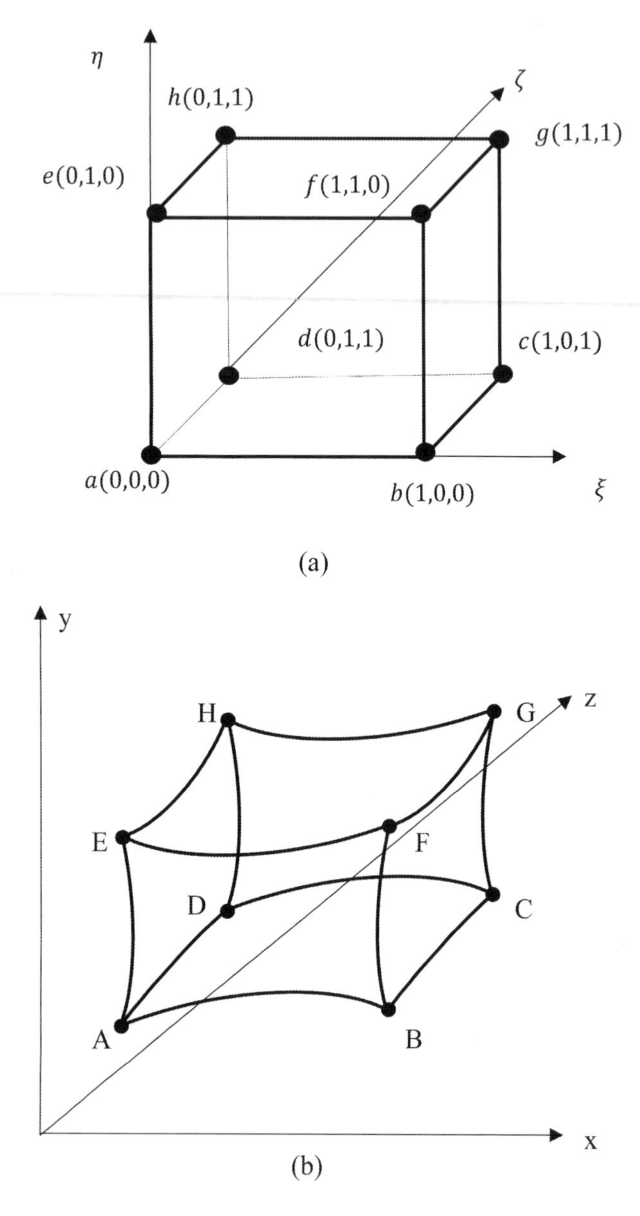

FIGURE 4.13 (a) Unit size cubic element in natural coordinates and (b) parallelepiped with curved surfaces

three ξ, η and ς directions for divisions, i varying from 1 to P, j varying from 1 to Q and k varying from 1 to R as below.

$$g(\xi_i,\eta_j,\varsigma_k)=(1-\xi_i)g(0,\eta_j,\varsigma_k)+\xi_i g(1,\eta_j,\varsigma_k) \tag{4.50a}$$

$$g(\xi_i,\eta_j,\varsigma_k)=(1-\eta_i)g(\xi_i,0,\varsigma_k)+\eta_i g(\xi_i,1,\varsigma_k) \tag{4.50b}$$

$$g(\xi_i,\eta_j,\varsigma_k)=(1-\varsigma_i)g(\xi_i,\eta_j,0)+\varsigma_i g(\xi_i,\eta_j,1) \tag{4.50c}$$

where ξ_i varies as $0 \le \frac{i-1}{P-1} \le 1$ and η_j varies as $0 \le \frac{j-1}{Q-1} \le 1$ and $0 \le \frac{k-1}{R-1} \le 1$. A cube contains twelve straight edges and each set of four edges are parallel to all the three ξ, η and ς- directions. The corresponding mapping projectors pair can be obtained by taking the product of

$U_\xi V_\eta$ for four edges parallel to ς as mapping on the corresponding four edges on $g(0,0,\varsigma)$.
$U_\xi W_\varsigma$ for four edges parallel to η as mapping on the corresponding four edges on $g(0,\eta,0)$.
$V_\eta W_\varsigma$ for four edges parallel to ξ as mapping on the corresponding four edges on $g(\xi,0,0)$.

The above bilinear transformation tensor product can be evaluated as

$$U_\xi V_\eta=(1-\xi)(1-\eta)g(0,0,\varsigma)+(1-\xi)\eta g(0,1,\varsigma)+\xi(1-\eta)g(1,0,\varsigma)+\xi\eta g(1,1,\varsigma) \tag{4.51a}$$

$$U_\xi W_\varsigma=(1-\xi)(1-\varsigma)g(0,\eta,0)+(1-\xi)\varsigma g(0,\eta,1)+\xi(1-\varsigma)g(1,\eta,0)+\xi\varsigma g(1,\eta,1) \tag{4.51b}$$

$$V_\eta W_\varsigma=(1-\eta)(1-\varsigma)g(\xi,0,0)+(1-\eta)\varsigma(\xi,0,1)+\eta(1-\varsigma)g(\xi,1,0)+\eta\varsigma(\xi,1,1) \tag{4.51c}$$

As it was implemented for two-dimensional mapping, the composite mapping of physical domain ABCDEFGH can be mapped using the Boolean sum of transformations, U_ξ, V_η and W_ς as follows.

$$U_\xi \oplus V_\eta \oplus W_\varsigma = U_\xi + V_\eta + W_\varsigma - U_\xi V_\eta - U_\xi W_\varsigma - V_\eta W_\varsigma + U_\xi V_\eta W_\varsigma \tag{4.52}$$

The mapping transformation functions can be written for generalized x, y and z coordinates as below [4].

$$U_\xi = U(\xi,\eta,\varsigma)=(1-\xi_i)X(0,\eta_j,\varsigma_k)+\xi_i X(1,\eta_j,\varsigma_k) \tag{4.53a}$$

$$V(\xi,\eta,\varsigma)=(1-\eta_i)X(\xi_i,0,\varsigma_k)+\eta_i X(\xi_i,1,\varsigma_k) \tag{4.53b}$$

$$W(\xi,\eta,\varsigma)=(1-\varsigma_i)X(\xi_i,\eta_i,0)+\varsigma_i X(\xi_i,\eta_i,1) \tag{4.53c}$$

$$\begin{aligned} UV(\xi_i,\eta_j,\varsigma_k) &= (1-\xi_i)(1-\eta_j)X(0,0,\varsigma_k)+(1-\xi_i)\eta_j X(0,1,\varsigma_k) \\ &+\xi_i(1-\eta_j)X(1,0,\varsigma_k)+\xi_i\eta_j X(1,1,\varsigma_k) \end{aligned} \tag{4.53d}$$

$$\begin{aligned} UW(\xi_i,\eta_j,\varsigma_k) &= (1-\xi_i)(1-\varsigma_k)X(0,\eta_j,0)+(1-\xi_i)\varsigma_k X(0,\eta_j,1) \\ &+\xi_i(1-\varsigma_k)g(1,\eta_j,0)+\xi_i\varsigma_k X(1,\eta_j,1) \end{aligned} \tag{4.53e}$$

$$VW(\xi_i,\eta_j,\varsigma_k) = (1-\eta_j)(1-\varsigma_k)X(\xi_i,0,0) + (1-\eta_j)\varsigma_k X(\xi_i,0,1) \\ +\eta_j(1-\varsigma_k)X(\xi_i,1,0) + \eta_j\varsigma_k X(\xi_i,1,1) \tag{4.53f}$$

$$UVW(\xi_i,\eta_j,\varsigma_k) = (1-\xi_i)(1-\eta_j)(1-\varsigma_k)X(0,0,0) + (1-\xi_i)(1-\eta_j)\varsigma_k X(0,0,1) \\ +(1-\xi_i)\eta_j(1-\varsigma_k)X(0,1,0) + (1-\xi_i)\eta_j\varsigma_k X(0,1,1) \\ +\xi_i\eta_j(1-\varsigma_k)X(1,1,0) + \xi_i\eta_j\varsigma_k X(1,1,1) + \tag{4.53g}$$

Using Equation (4.53), the x, y and z-coordinates of all the interior grid points of the physical computational domain ABCDEFGH can be generated using the following expression.

$$UVW(\xi_i,\eta_j,\varsigma_k) = U(\xi_i,\eta_j,\varsigma_k) + V(\xi_i,\eta_j,\varsigma_k) + W(\xi_i,\eta_j,\varsigma_k) \\ -UV(\xi_i,\eta_j,\varsigma_k) - UW(\xi_i,\eta_j,\varsigma_k) - VW(\xi_i,\eta_j,\varsigma_k) \\ +UVW(\xi_i,\eta_j,\varsigma_k) \tag{4.54}$$

As discussed in Equation (4.49), the consistency conditions for all the eight vortices of the parent domain have to be described. The computational algorithm follows all the steps discussed for the two-dimensional TFI mesh generation procedure, however, all the six boundary surfaces and twelve edges of the physical domain have to be discretized using the expressions given by Equations (4.53) and (4.54). Though it is more tedious to develop computer code for a three-dimensional domain compared to a two-dimensional domain, the code is developed only once, and it can be modified for further improvements. A computer code, ***Tfi_3D.for*** in FORTRAN has been developed to generate grid points in the three-dimensional computational domain using TFI technique. Figure 4.14(a) shows a simple rectangular prism discretized using TFI technique to generate x, y and z-coordinates of all the interior grid points. As it is well known, all the real life problems are not of regular geometries. Let us consider ducting of air-conditioning system in big auditoriums or cinema halls. These ducts may have to pass through a number of column structures in the building. Using the multi-block concept, any type of three-dimensional domain can be discretized using TFI technique, as was discussed for two-dimensional domain in Figure 4.12. Figures 4.14(b) and 4.14(c) show similar types of geometries with holes which have to be discretized using the multi-block principle.

The given domain will be divided into a number of sub-domains and they are numbered in sequence except the sub-domains which have to be left without meshing. Then the TFI technique will be applied to each sub-domain using the same formulation, however, care must be taken to assure continuity and common edges and surfaces of the sub-domains that have to be meshed. Node numbering of all the interior grid points of the sub-domains is also highly challenging for which suitable algorithm has to be developed. The very purpose of using TFI technique

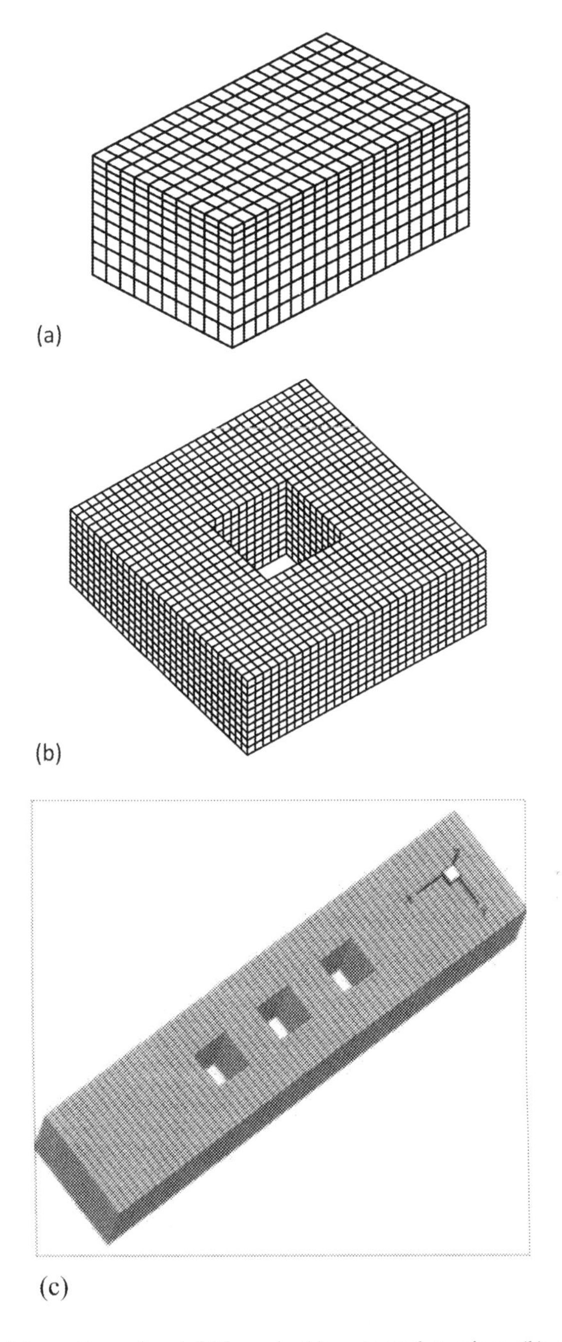

FIGURE 4.14 Three-dimensional TFI mesh, (a) rectangular prism, (b) square duct with square hole and (c) long duct with three square holes.

is to discretize the given physical domain into a number of grid points with their coordinates and in the finite element method, other than the spatial coordinates, the element-nodal connectivity data also has to be generated for the simulation program.

4.8 TIME-DEPENDENT PROBLEMS

Many real problems are transient or time-dependent problems, with an important characteristic of variation of the variables under consideration with time. In a way, all the problems that we observe in the universe are time dependent and the concept of steady state conditions are hypothetical situations with certain assumptions on the negligible time transients. It is the strength of these time transients that determine the problem as transient or steady state. That means certain situations are named as steady state conditions because the transient variation of the parameters are well within the acceptable limits. However, these limits are neither constants nor fixed for any given situation. For example, a voltage stabilizer unit is designed to give output voltage at constant value with permissible allowance of variation. These variations may be insignificant for certain applications and may become very important for some other situations. In engineering analysis, the level of acceptable transients is decided purely based on the expected performance of the system, the cost involved to maintain these levels and other factors that may influence the performance of the system itself. Time transients may vary from a few milliseconds during combustion in heat engines to a few seconds in thermostat controls, a few hours during heat treatment of metals and a few years in decay of nuclear waste materials. In flow and heat transfer problems, only first order time derivatives are obtained while modeling transient problems. However, time derivatives of second order also find some engineering applications in other fields of applications. In this textbook, only the transients due to first order time derivatives will be discussed. To start with let us try to understand the nature of transient problems by considering one-dimensional transient conduction problem, which is described as

$$\rho C_P \frac{\partial T}{\partial t} = k \frac{\partial^2 T}{\partial x^2} \tag{4.55}$$

It is explained in conduction heat transfer analysis that the above equation represents the rate of change of energy of the control volume with time is equal to the net heat conducted through the medium. Under the assumption of steady state condition, the rate of change of energy of the control volume with time becomes equal to zero and hence whatever heat enters the control volume, the same amount of heat comes out of the control volume. Any realistic problem should satisfy the statement of energy balance through the boundary conditions for the energy equation shown in Equation (4.55). If the left-hand side of the above equation becomes negative, then the right-hand side also should be negative. Unless there is a sufficient amount of energy entering the solid and being conducted, the energy of control volume cannot increase with time. Assuming Equation (4.55) is applied to the realistic situation after satisfying the underlying physics, now the time variation has to be numerically modeled. Unlike the spatial coordinates which have a fixed origin, the time coordinate does not have a fixed origin, for this reason, the initial condition is introduced as a boundary condition for the time coordinate. Any time at which the analysis or process is initiated to monitor the transient behavior of the system is called initial condition. In transient problems, solutions for the field variables are approached from a starting time to a particular end time, at which the process is assumed to come to an end. In most of the transient problems, it is the history of variation of the field variables during the transients that are important than the results at the end of the total process time. That means the evolution of variables at different time intervals are required in numerical solutions. When experiments are conducted during annealing of metal products, parameters of importance are measured at a fixed interval of time, and the progression of change of variables is plotted to understand the underlying physics. However, in the case of a numerical solution, like spatial variables, time is also a variable and hence it has to be discretized into a number of intervals in the time domain for the total time period of the process under consideration. In finite difference and finite volume methods, generally the time derivative is approximated using the difference equation. In the case of the finite element method, a separate shape function for time can be defined just like the shape functions described for the spatial coordinates. However, this method is not preferred in common in finite element solution procedure; instead the conventional finite difference method is employed. Now, consider the application of Galerkin's weighted residual method to Equation (4.55).

$$\int_{\Omega}[N]^T\left(\rho C_P\frac{\partial T}{\partial t}-k\frac{\partial^2 T}{\partial x^2}\right)dx=0$$

$$\rho C_P\int_{\Omega}[N]^T\frac{\partial T}{\partial t}dx-\int_{\Omega}[N]^T k\frac{\partial^2 T}{\partial x^2}dx=0$$

$$\rho C_P\int_{\Omega}[N]^T[N]dx\frac{\partial\{T\}}{\partial t}-k\int_{\Omega}\frac{\partial[N]^T}{\partial x}\frac{\partial[N]}{\partial x}dx=0$$

The above equation can be written in matrix form for one element as

$$[C]^{(e)}\frac{d\{T\}}{dt}+[K]^{(e)}\{T\}=\{f\}^{(e)} \tag{4.56}$$

In the above equation, the partial derivative notation for time derivative is replaced with ordinary derivative because the time discretization will be dealt with using the mean value theorem. Consider temperature variation during time interval Δt discretized between time levels, m and n as shown in Figure 4.15.

Using the finite difference method, the temperature gradient at time t can be expressed as

$$\frac{dT(t)}{dt}=\frac{T(n)-T(m)}{(n-m)} \tag{4.57}$$

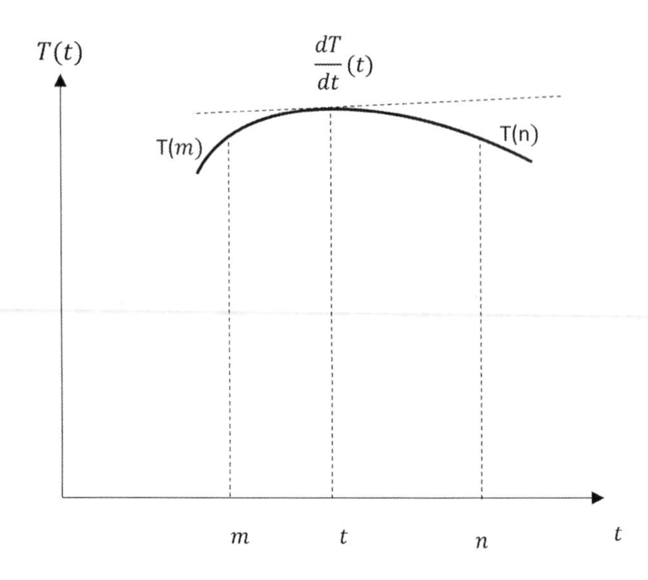

FIGURE 4.15 Determination of temperature gradient with time using mean value theorem principle.

However, our main aim is to represent T at any value of t lying between m and n. This can be determined by defining temperature at m within the time interval $(t-m)$ using Equation (4.57) and using mean value theorem as

$$T(m) = T(t) - (t-m)\frac{dT(t)}{dt}$$

That is, $T(t) = T(m) + (t-m)\frac{dT(t)}{dt}$

$T(t) = T(m) + (t-m)\frac{T(n)-T(m)}{(n-m)}$. In the above equation, the ratio $\frac{(t-m)}{(n-m)}$ determines the evaluation of *T(t)* in terms of difference temperatures between time levels m and n. Hence, calling this ratio as $\Theta = \frac{(t-m)}{(n-m)}$, we get the final expression as

$$T(t) = T(m) + \Theta\left[T(n) - T(m)\right]$$

$$T(t) = (1-\Theta)T(m) + \Theta T(n) \tag{4.58}$$

Hence, the temperature gradient in Equation (4.56) can be expressed as

$$\frac{d\{T\}}{dt} = \frac{\{T\}_n - \{T\}_m}{\Delta t} \tag{4.59}$$

where

$$\{T\} = (1-\Theta)\{T\}_m + \Theta\{T\}_n \tag{4.60}$$

The effect of temperature levels on temperature variation at any time t can be extended to the load vector as

$$\{F\} = (1-\Theta)\{F\}_m + \Theta\{F\}_n \tag{4.61}$$

The expression for temperature gradient and generalized temperature vector as given by Equations (4.59) and (4.60) respectively can be substituted in Equation (4.56) and after collecting the terms for time levels m and n together, the final equation is expressed as

$$\begin{aligned}\left(\left[C\right]^{(e)} + \Theta\Delta t\left[K\right]^{(e)}\right)\{T\}_n^{(e)} &= \left(\left[C\right]^{(e)} + (1-\Theta)\Delta t\left[K\right]^{(e)}\right)\{T\}_m^{(e)} \\ &\quad + \Delta t\left((1-\Theta)\{F\}_m^{(e)} + \Theta\{F\}_n^{(e)}\right)\end{aligned} \tag{4.62}$$

The above equation enables us to compute the variation of temperature in the computational domain at the end of every specified Δt. Equation (4.62) is of the form $Ax=b$, in which A corresponds to the coefficient matrix with temperature vector at time level, n, x is the unknown temperature vector $\{T\}$ at time level, n and b is the load vector obtained as a combination of load contributions from the boundary conditions $\{F\}$ at different time levels and due to the temperature at time level, m with the coefficient matrix. After deriving the final equation for computations, it is worth analyzing the meaning of Θ in transient analysis.

We know,

$\Theta = \dfrac{(t-m)}{(n-m)}$. It is easy to visualize that at $t=m$, $\Theta=0$ and substitution of this value in Equation (4.62) makes the contribution of the stiffness matrix zero in the coefficient matrix for temperature at level, n and makes its full contribution for stiffness matrix at time level, m on the right-hand side of Equation (4.62). At this condition, the temperature gradient averaging is focused at time level, m. Equation (4.62) provides solution for temperature in one-dimensional domain at time level, n using the values of temperature and load vector at time level, m. When the temperature evolution with time takes place from time level, m to n with a time interval of Δt, the time variation in temperature of all the domains is attributed to the effect of heat conduction within the spatial domain at the previous time level, m. That means the effect of conduction within the spatial domain at time level, n is ignored. In a physical sense, or at least by intuition, this looks oversimplified. With $\Theta=0$, the time marching of temperature evolution with time at the end of every time step, Δt can be computed by knowing only the temperature distributions within the domain at the previous time step. In a computational algorithm for computing time marching solution, an outermost time loop is introduced to carry out the computations of all the data using the previous time level. Such a scheme is called *explicit time marching scheme*. This scheme results in a set of simultaneous equations which can be easily solved to get the time marching solution for temperature, however, if mass lumping technique is employed, then there is no need to solve the simultaneous equations.

When $t=n$, $\Theta=1$ and substitution of this value in Equation (4.62) makes the stiffness matrix contribution on the right-hand side of the equation equal to zero, that means the effect of heat conduction within the domain at time level, m is ignored and that effect is considered only at the current time level, n. The effect of temperature at the previous time level contributes only to the capacitance matrix, of course, this will be the case for any value of Θ. It is assumed that the spatial temperature distributions at time level, m does not influence the temperature variation at the current time level, n. That means, both the effect of rate of change of variation of thermal energy within the control volume over Δt and conduction within the domain during this time interval, contribute to the temperature history computed at the current time level, n. Avoiding the effect of previous temperature effect on conduction is implicit in this scheme, hence, it is called *implicit time marching scheme*. Implementation of implicit scheme necessitates the use of an equation solver to get solution for temperature for every time interval. In a realistic situation, suppose a heated rod is cooled down in an environment, it will be challenging to follow the time history and spatial distribution of temperature accurately. Even with the help of a number of temperature sensors embedded at a small interval of spatial distance along the length of the rod, it may be difficult to notice the temperature variation during the time lapse at all the sensors and the type of conduction that has taken place between any two consecutive sensors. Thus, cooling of the rod problem cannot be either attributed to an explicit scheme or an implicit scheme. However, the concept of explicit and implicit is the consequence of discretizing the time derivative between two time levels across a Δt and is nothing to do with the actual physics that is taking place during cooling of the rod. As it has been discussed earlier, modeling means representing the physical phenomena using certain conservation principle and representing it in mathematical equations, hence, the mathematical approximation of the conservation principle sometimes introduces certain challenges. Time discretization methodology discussed till now is one such challenge, which is not able to accurately bring the realistic situation into mathematical form.

It is well known that conduction takes place at all time levels in a transient heat conduction problem and the challenge is how to predict the temperature evolution without much sacrificing the underlying physics behind the problem. The only foolproof method available to achieve this is seeking the help of experimental data for verification of the mathematical model. However, to continue the discussion on explicit and implicit methods, it is always possible to take a middle path between these two schemes. In a way, the implicit scheme looks more accurate compared to the explicit scheme because it takes care of temperature evolution at the current time level by considering all the effects at that time level itself. In the case of an explicit scheme, as it ignores temperature variation due to heat conduction at the current time level and depends only on the effect at the previous time level, it may not be accurate compared to the implicit scheme. In order to make a realistic scheme, it is better to relax the implicit scheme, to take care of the conduction effect at the previous time level. That means, some percentage weightage for this effect can be attributed while computing the temperature at the current time level. When the midway is considered, it looks more realistic compared to both the explicit and implicit schemes. That is for $t = \frac{\Delta t}{2}$, the effect of both explicit and implicit scheme can be taken into account for temperature evolutions at time level, n. It has to be recalled that for explicit scheme, $t=m$ and for implicit scheme, $t=n$ and the time levels, m and n are measured from the origin, 0 of the time scale. Hence for $t=\frac{\Delta t}{2}$, the computation of Θ becomes as

$\Theta=\frac{t-m}{n-m}=\frac{\Delta t/2-m}{n-m}=\frac{\left(m+\frac{n-m}{2}\right)-m}{n-m}=\frac{1}{2}$. Please note m is added to compute the mid-point of $(n-m)$ from the origin of the time coordinate. When $\Theta=\frac{1}{2}$ is substituted in Equation (4.62), there is equal contribution of the stiffness matrix at both time levels, m and n, resulting in a scheme called *semi-implicit time marching scheme*. Equation (4.62) is rewritten for the semi-implicit scheme as follows.

$$\left(\left[C\right]^{(e)}+0.5\Delta t\left[K\right]^{(e)}\right)\left\{T\right\}_n^{(e)}=\left(\left[C\right]^{(e)}+0.5\Delta t\left[K\right]^{(e)}\right)\left\{T\right\}_m^{(e)} \\ +\Delta t\left(0.5\left\{F\right\}_m^{(e)}+0.5\left\{F\right\}_n^{(e)}\right) \tag{4.63}$$

In most of the simulation works, the effect of Θ on the load vector is ignored unless the boundary conditions are time dependent. The semi-implicit scheme is also called Crank-Nicholson scheme, which is very commonly used for simulation of time-dependent problems. Now, the time marching schemes can be identified with the value of t and Θ as below.

@ $t=m$, $\Theta=0$ Fully explicit scheme also known as forward difference scheme
@ $t=n$, $\Theta=1$ Fully implicit scheme – backward difference scheme
@ $t=\frac{\Delta t}{2}$, $\Theta=0.5$ Semi-implicit scheme – Crank-Nicholson scheme

Sometimes, $\Theta=(2/3)$ is considered and is called Galerkin's method. While dealing with time-dependent problems, the selection of a suitable time marching scheme is crucial in order to obtain realistic and physically meaningful simulation results.

4.8.1 Stability Conditions

Time discretization differs from spatial discretization in that the final simulation results at the end of total transients purely depend on the results obtained on the entire domain at the end of each time step. In the case of spatial discretization, once computational results are obtained over the entire domain, that becomes final at that instant, except the effect of neighboring grid points, which is already taken care by the specific numerical method employed. However, in the case of transient results, either the final results or the history of field variables with time, depend on the characteristic of results, that means, how far they are close to the expected realistic results. If some unrealistic results are computed at the initial time steps and allowed the computation to proceed without notice, then the final results will not provide the required output from the simulation. Sometimes, such results are attributed to what is called *stability* of transient numerical simulation. As the time derivative is discretized generally using finite difference technique, a mathematical expression can be derived to understand this condition of *numerical stability*. For this purpose, the one-dimensional transient conduction equation given by Equation (4.55) is discretized using finite difference method for all the three time marching scheme and the stability condition for the respective scheme is discussed below.

For simplicity Equation (4.55) is written as

$$\frac{\partial T}{\partial t} = \alpha \frac{\partial^2 T}{\partial x^2} \tag{4.64}$$

where $\alpha = \dfrac{k}{\rho C_p}$ is thermal diffusivity of the solid.

Application of finite difference method to Equation (4.64) gives rise to the following difference equation.

$$\frac{T^{n+1} - T^n}{\Delta t} = \alpha \frac{\left(T_{i+1}^{\beta} + T_{i-1}^{\beta} - 2T_i^{\beta}\right)}{(\Delta x^2)}$$

$$T_i^{n+1} = T_i^n + \left(\frac{\alpha \Delta t}{\Delta x^2}\right)\left(T_{i+1}^{\beta} + T_{i-1}^{\beta} - 2T_i^{\beta}\right) \tag{4.65}$$

In the above equation the time level for the diffusion term on the right-hand side of the equation is indicated by a generalized notation β because this will change depending on the type of time marching scheme chosen.

4.8.1.1 Explicit Scheme

Equation (4.65) is rewritten for explicit scheme using forward difference scheme as

$$T_i^{n+1} = T_i^n + \left(\frac{\alpha \Delta t}{\Delta x^2}\right)\left(T_{i+1}^{n} + T_{i-1}^{n} - 2T_i^{n}\right) \tag{4.66}$$

The temperature distribution over the domain at time level *(n+1)* after time interval of Δt is dependent only on the temperature of the domain at the previous time level *(n)*.

Rewriting Equation (4.66) with respective coefficients of various nodal temperatures results in the following equation.

$$T_i^{n+1} = \left(\frac{\alpha \Delta t}{\Delta x^2}\right) T_{i-1}^n + \left(1 - 2\frac{\alpha \Delta t}{\Delta x^2}\right) T_i^n + \left(\frac{\alpha \Delta t}{\Delta x^2}\right) T_{i+1}^n \tag{4.67}$$

In the above equation consider the term, $\left(\dfrac{\alpha \Delta t}{\Delta x^2}\right)$, called Fourier number, which is the same for all the nodal temperatures except for T_i^n, that is the node at which the temperature is being computed at the current time level, *(n+1)*. When the Fourier number is equal to $\frac{1}{2}$, then the coefficient for T_i^n becomes equal to zero. That means, the coefficient of the node number which had some temperature value at time level *(n)*, now does not contribute anything for temperature evolution at current time level, *(n+1)*. In other words, when the final set of simultaneous equations corresponding to all the nodes in the computational domain are formed, the diagonal node becomes zero. This is unrealistic, because during advancing in time, every node contributes for the evolution of temperature at any given node, however, the contribution may vary depending on its proximity to the node under consideration. In a sense, such a situation is not possible in reality because during one Δt interval, there cannot be any sudden change in the contribution from the same node at the previous time level. For any value of Fourier number greater than $\frac{1}{2}$, the coefficient for T_i^n becomes negative. This coefficient consists of only thermo-physical properties of the material and spatial discretization and these values cannot be negative and hence, the resulting situation will lead to negative sign for the temperature, T_i^n. When conduction heat transfer is taking place from node *(i-1)* towards nodes, *(i)* and *(i+1)*, there cannot be sudden drop in temperature at node *(i)* and then increase towards node *(i+1)* and such a condition will violate Second law of Thermodynamics, that is heat is transported from a lower temperature to a higher temperature medium. Hence, this condition also will not lead to simulations for realistic conditions and thus, the Fourier number gives a constraint on its value in order to obtain physically correct simulation results. This leads to a conclusion that Fourier number has to be less than $\frac{1}{2}$ to satisfy the stability condition for the explicit scheme. That is, $Fo = \left(\frac{\alpha \Delta t}{\Delta x^2}\right) < \frac{1}{2}$.

4.8.1.2 Implicit Scheme

Now, Equation (4.65) can be modified for fully implicit time marching scheme as follows.

$$T_i^{n+1} = T_i^n + \left(\frac{\alpha \Delta t}{\Delta x^2}\right)\left(T_{i+1}^{n+1} + T_{i-1}^{n+1} - 2T_i^{n+1}\right)$$

After collecting the coefficients for the nodal temperatures, the above equation can be written as

$$FoT_{i-1}^{n+1} + (2 + Fo)T_i^{n+1} + FoT_{i+1}^{n+1} = T_i^n \tag{4.68}$$

In the above equation, it is clearly noted that there is no restriction on the value of *Fo* in order to obtain stable numerical solutions for temperature evolutions. It is surprising to note that the load vector for the fully implicit scheme comes only from the temperature of the same node at the previous time level, ignoring the contribution from the neighboring nodes. However, in the case of fully explicit scheme, the load vector is contributed from all the neighboring nodes and the node under consideration at the previous time level. This also confirms that the fully implicit scheme is more stable compared to the fully explicit scheme.

4.8.1.3 Semi-Implicit Scheme (Crank-Nicholson Scheme)

The stability condition for semi-implicit scheme can be determined by the application of equal contribution at both time levels and Equation (4.65) gets modified as below.

$T_i^{n+1} = T_i^n + \frac{Fo}{2}\left(T_{i-1}^{n+1} - 2T_i^{n+1} + T_{i+1}^{n+1}\right) + \frac{Fo}{2}\left(T_{i-1}^n - 2T_i^n + T_{i+1}^n\right)$, which can be further simplified as

$$\left(Fo+1\right)T_i^{n+1} - \frac{Fo}{2}\left(T_{i-1}^{n+1} + T_{i+1}^{n+1}\right) = \left(1-Fo\right)T_i^n + \frac{Fo}{2}\left(T_{i-1}^n + FoT_{i+1}^n\right) \quad (4.69)$$

In the above equation when Fo becomes less than 1, then an instability condition similar to the one discussed for fully explicit scheme is encountered. Hence, for semi-implicit scheme (Crank-Nicholson) the stability criteria is specified as

$Fo = \left(\frac{\alpha \Delta t}{\Delta x^2}\right) < 1$. After establishing the stability condition for all the three different time marching schemes, it is clear that fully implicit scheme is found to be highly stable without any constraint on the Fourier number, whereas the fully explicit scheme is found to be the least stable scheme and the semi-implicit scheme falls in between these two schemes. As the stability condition is specified in terms of Fourier number, it is worth analyzing how this number plays a role in numerical computation of time evolution of field variables.

4.8.1.4 Significance of Fourier Number

Fourier number is a combination of thermal diffusivity and spatial and temporal discretization. Spatial discretization is decided on the computational domain which is the replica of the physical domain under simulation, whereas the time domain is an abstract domain, which cannot be defined in a physical sense like the spatial domain. In a way Fourier number enforces a limit between the relative discretized size of the physical domain and time domain, especially when fully implicit and semi-implicit schemes are employed in the computations. Thermal diffusivity relates two phenomena related to heat transfer, one is conduction and the other one, heat storage within the medium. Transient problem itself is the outcome of the condition that the stored thermal energy changes due to the conduction of heat within the medium. The energy conservation principle applied to a transient heat conduction reveals that the rate of change of thermal energy stored within the medium is equal to the net heat conducted out of the medium. However, if the net heat conducted is achieved by the application of heat at the boundary of the medium, without any chance for storage, then the problem becomes steady state conduction. The ratio between the rate at which heat is conducted within the medium to the rate at which the stored thermal energy changes is called the Fourier number. It is the temperature difference between the time interval that decides the rate of change of stored thermal energy, whereas the temperature gradient between the adjacent nodes within the medium at a given time level that dictates the heat conduction. It is impossible to compute temperature evolutions at a given time level without the knowledge of temperature variations within the domain at the previous time level. Temperature distribution in a given computational stencil at the given instant of time *(n+1)* is a consequence of change in energy storage of the medium from the temperature distribution at the previous time level *(n)*. Conduction within the domain at a given instant of time *(n+1)* is due to the existing temperature gradient, rather created by the change in stored thermal energy. Heat conduction is governed by temperature variation between the adjacent nodes separated by Δx for the given thermal conductivity of the medium, whereas it is the size of Δt that determines the change in thermal energy storage of the medium for the given constant properties of the medium. Hence, thermal conductivity, density and specific heat capacity of the medium remain constant with time, only the size of Δx and Δt obtained during discretization, control the quantum of heat conducted and variation in thermal energy storage. The conservation of energy principle dictates that the amount of heat that could be conducted is fixed by the amount of change in thermal energy storage. It is worth remembering that Δx and Δt are our own making and are obtained by the implementation of numerical discretization of spatial and time domains with certain approximations and assumptions. Hence, there is a natural limitation on these two quantities while dealing with fully explicit and semi-implicit time marching schemes. The size of these quantities has to be selected such that the underlying physics and the coupling between conduction and energy storage are not altered artificially. It is well understood now that there is a coupling between the size of spatial discretization and the time step and they have to be selected by satisfying the stability condition, otherwise, the unrealistic temperature evolutions obtained at the earlier time steps will lead to spurious solution or oscillating solution at later time steps. As there is no stability condition to be satisfied with fully implicit scheme, higher time steps can be implemented. Finer grids are preferred at the boundaries in order to accurately compute the flux quantities in flow and heat transfer problems and hence, the minimum spatial discretization size restricts the size of time step. It is always the time step which is decided later after making the mesh size. While checking on mesh sensitivity analysis, the time step is never considered, that means independent of the time step, the selected spatial discretization should be able to provide consistent simulation results. That shows clearly that it is the time step that guides the modeler to select the time marching scheme. Of course, one cannot affirm that the mesh size decides the time step, sometimes, problems that have to be simulated for long durations of time demand higher time step in order to reduce the computational time. For example, heat transfer analysis of thermal energy dissipation from nuclear waste materials have to be simulated for half-life periods of decades. Such problems demand higher time steps, with smaller time steps, it may not be physically possible to carry out the simulations for the entire period, so that the design parameters can be obtained. In general, smaller time steps are used with fully explicit time marching scheme, slightly larger time steps for semi-implicit scheme and higher time intervals for the fully implicit scheme. However, there is no standard rule to decide exactly the time step, though certain mathematical approaches can be followed to compute the correct time step for stable solution. Most of the modelers do not prefer such mathematical formulation for time step for the reason that real-life problems are highly complex in geometry and boundary conditions. The more pragmatic approach is to carry out the simulation with some assumed time step based on either available work or experience and then validate the results with available literature.

4.8.1.5 Alternate Direction Implicit (ADI) Method

The different time marching scheme used in flow and heat transfer problems were discussed by considering transient one-dimensional conduction problem. Whatever principles of time stepping methods discussed for one-dimensional very well hold good for multi-dimensional problems also. However, the implicit method discussed in the above section can be applied with certain modifications so that the diffusion in x and y-directions are considered well. The ADI method is an improvised method that combines the benefits of fully explicit and fully implicit methods. Let us consider a two-dimensional transient conduction problem discretized using the finite difference method as follows.

$$T_i^{n+1} = T_i^n + \left(\frac{\alpha \Delta t}{\Delta x^2}\right)\left(T_{i+1}^{\beta} + T_{i-1}^{\beta} - 2T_i^{\beta}\right) + \left(\frac{\alpha \Delta t}{\Delta y^2}\right)\left(T_{i+1}^{\beta} + T_{i-1}^{\beta} - 2T_i^{\beta}\right) \quad (4.70)$$

Instead of considering either the fully explicit or fully implicit method, a combination of these two methods by considering heat diffusion in both the directions at both the time levels can be implemented. Peaceman-Rachfod [6] suggested a predictor and corrector time marching scheme, by considering both explicit and implicit schemes for both the directions, however, executed in two half time steps. That means the given time step Δt is split into two halves, to carry out predictor and corrector steps in the first and second half time step. During the predictor step, diffusion in the x-direction is considered implicitly, keeping the y-direction diffusion in fully explicit mode and Equation (4.70) modified to this effect is given as below.

$$T_i^{n+\frac{1}{2}} = T_i^n + \left(\frac{\alpha \Delta t}{\Delta x^2}\right)\left(T_{i+1}^{n+\frac{1}{2}} + T_{i-1}^{n+\frac{1}{2}} - 2T_i^{n+\frac{1}{2}}\right) + \left(\frac{\alpha \Delta t}{\Delta y^2}\right)\left(T_{i+1}^{n} + T_{i-1}^{n} - 2T_i^{n}\right) \quad (4.71)$$

Then the next half time step is completed in the corrector step by considering implicit scheme for diffusion in the y-direction and explicit scheme for diffusion in the x-direction as shown in the following equation.

$$T_i^{n+1} = T_i^{n+\frac{1}{2}} + \left(\frac{\alpha \Delta t}{\Delta x^2}\right)\left(T_{i+1}^{n+\frac{1}{2}} + T_{i-1}^{n+\frac{1}{2}} - 2T_i^{n+\frac{1}{2}}\right) + \left(\frac{\alpha \Delta t}{\Delta y^2}\right)\left(T_{i+1}^{n+1} + T_{i-1}^{n+1} - 2T_i^{n+1}\right) \quad (4.72)$$

The fully implicit method considers heat diffusion effect in both the directions simultaneously, without giving any importance to the residual effect of diffusion that has taken place in the previous time level. Though the semi-implicit method takes care of the effect in both time levels, it has missed out the directional effect while considering heat diffusion. The ADI method provides compensation for the directional effect by considering diffusion in both directions both explicitly and implicitly. There are many time stepping schemes that can improve the accuracy of computations in transient problems. In the above method itself, the corrector step can be carried out using a semi-implicit scheme as well or sometimes three-level time stepping scheme such as Adams-Bashforth also can be implemented.

REFERENCES

1. *Applied Finite Element Analysis*, Larry J. Segerlind, 2nd Edition, Wiley, Hoboken, NJ, 1984, TA347.F5S43.
2. *The Finite Element Method in Heat Transfer and Fluid Dynamics*, J. N. Reddy and D. K. Gartling, 3rd Edition, CRC Press, Boca Raton, FL, 2010.
3. *Fundamentals of Finite Element Analysis*, edited by David V. Huttan, McGraw Hill Higher Education, New York, 2004.
4. *Handbook of Grid Generation*, edited by Joe F. Thompson, Bharat K. Sony and Nigel P. Weatherill, Chapter 3 Transfinite Interpolation (TFI) Generation systems, Robert E. Smith, CRC Press, Boca Raton, FL, USA, 1999.
5. *Basic Structured Grid Generation with an Introduction to Unstructured Grid Generation*, M. Farashkhalvat and J. P. Miles, Butterworth Heinemann, Burlington, MA, USA, 2003.
6. Peaceman, D. W. and Rachfod, Jr. H. H. The numerical solution of parabolic and elliptic equations, *Journal of SIAM*, 1955; 3: 28–41.

EXERCISE PROBLEMS

Qn: 1 In a heat transfer problem, it is required to discretize a given one-dimensional computational domain of length 0.2 m using 20 linear elements. Draw a diagram indicating the element, node numbers and element-nodal connectivity. Develop a computer program to execute the above exercise. Repeat this problem for discretization of the above domain using 3-node non-linear elements and a computer code.

Qn: 2 A one-dimensional steady state heat conduction is considered in a rod of length L whose end at $x = 0$ is subjected to Dirichlet boundary condition and the end $x = L$ to heat flux boundary condition. Apply the Galerkin's weighted residual finite element method to solve the governing equation and show the finite element formulation of the heat flux boundary condition.

Qn: 3 Consider steady state three-dimensional heat conduction equation in a cube. Apply isoparametric formulation for coordinate transformation and develop the equations relating the local and global coordinates for an 8-node brick element. Develop a computational algorithm to implement this scheme to discretize the computational domain using 8-node brick elements.

Qn: 4 It is required to model the heat load due to solar radiation into a room which is maintained at a temperature lower than that of the atmosphere. Develop the governing equation assuming the heat entering from the outer side of the wall is dissipated by convection inside the room. Identify the boundary conditions by treating this heat transfer problem as one-dimensional. Apply Galerkin's method to solve the governing equations and reduce the final discretized form of the governing equations in $Ax = b$ form.

Qn: 5 Electronic chip cooling problem can be modeled as a two-dimensional heat conduction problem. Assume the heat to be dissipated from the chip as a uniformly distributed heat source of constant magnitude. The dimensions of the rectangular chip can be represented by dimension $L \times H$ where L indicates the dimension in the x-direction and H the dimension along the y-direction. The computational domain of the chip can be given notation as ABCD, AB being the bottom edge of the domain and BC is the right vertical edge. Develop the relevant governing equation using energy conservation principle. Solve the equation numerically using the finite element method when the bottom edge of the domain is adiabatic for heat transfer, right edge is convecting heat to the atmosphere, top edge is maintained at a constant temperature and left end is subjected to heat flux boundary conditions.

Qn: 6 Using the TFI mesh generation computer code given in the chapter, generate the computational mesh for the computational domain described in Qn: 5 assuming quadrilateral elements. Generate different meshes assuming uniform size grid points and repeat mesh generation assuming non-uniform mesh near the edges of the computational domain.

Qn: 7 The outer side of a concrete wall of 0.15 m thick is exposed to high temperature gaseous medium at 300°C. The rate of heating of the wall has to be estimated for safety purpose. The inner side of the wall is subjected to convective atmosphere with $T_\infty = 28^\circ\text{C}$ and convective heat transfer coefficient, $h_c = 30\ \text{W/m}^2\text{K}$, which is one fifth of the convective heat transfer coefficient experienced at the outer wall. The time rate of change of temperature at the outer and inner surface of the wall has to be monitored continuously with time. Develop the governing equation to model this heat transfer situation as an unsteady problem. Solve the equation using the finite element method.

Qn: 8 Heat at the rate of 2 kW/m^2 is supplied to a hot plate of 10 mm thickness at the bottom of the plate. The top surface of the plate is exposed to an ambient at 25°C with convective heat transfer coefficient of 40 W/m^2 K. Write down the relevant conservation equation and the boundary conditions. Apply the finite element method to obtain numerical solution of temperature variation at the centre point of the plate assuming the hot plate as one-dimensional domain.

Qn: 9 A two-dimensional plane wall of unit thickness with dimension 0.12 m $\times$ 0.05 m is subjected to Dirichlet boundary conditions with temperature equal to 400°C on the left end of the wall and 200°C at the top end of the wall. The bottom side of the wall is subjected to adiabatic boundary condition, whereas the right end of the wall is exposed to an ambient at 30°C and h_c=40 W/m^2 K. Obtain finite element numerical solution using Galerkin's finite element method to estimate the temperature at a location (0.04 m, 0.02 m) in the wall.

Qn: 10 In heat transfer analysis of a heat sink used with an electronic system, the heat sink can be assumed as a rectangular plate of dimension 10 cm × 4 cm, with ignorable thickness for heat transfer. The top and bottom sides of the heat sink are enforced with Dirichlet boundary conditions of 40°C and 60°C respectively. A heat flux of 100 W/m^2 is passing through the left end of the heat sink with convective boundary condition at the right end having free stream temperature of 30°C and convective heat transfer coefficient of 40 W/m^2 K. It is required to compute the temperature variation at the left end of the heat sink with time at a spatial interval of 1 cm. Solve the governing heat transfer equation using Galerkin's finite element method and adopt semi-implicit time marching scheme to determine the temperature with time.

QUIZ QUESTIONS

Qn: 1 Name the important methodologies employed in Finite Element Analysis.

Qn: 2 Given that a one-dimensional element is divided into 3 nodes then order of stiffness matrix is ____________________

Qn: 3 Rayleigh Ritz method is based on weighted residual approach (a) True (b) False.

Qn: 4 The determinant for stiffness matrix is ____________________

Qn: 5 Degree of freedom for a triangular stress element are (a) 3 (b) 4 (c) 5 (d) 6.

Qn: 6 For ____________________ elements the geometry is described by higher order shape functions.

Qn: 7 For one-dimensional heat conduction problem temperature distribution using linear elements is written as
(a) $T=T_1N_1+T_2N_2$ (b) $T=T_1N_1/ T_2N_2$ (c) $T=T_1N_1- T_2N_2$ (d) None

Qn: 8 For a bar of length 10 cm determine the temperature at a distance of 4 cm if the ends of the bar are maintained at 100°C and 150°C respectively. Assume linear variation of temperature.

Qn: 9 Given that in a two-dimensional domain thermal conductivity is constant throughout then Poison equation is written as
(a) $k\nabla^2\varphi + Q = 0$ (b) $\nabla^2\varphi + Q = 0$ (c) $\nabla^2\varphi = 0$ (d) None

Qn: 10 Write the Jacobian matrix for a one-dimensional element given that nodal co-ordinates are $x_i = 1, x_j = 2, x_k = 3$

5 Modeling of Heat Transfer Problems

Application of modeling in heat transfer problems vary based on different conditions such as one-dimensional, two- and three-dimensional problems, type of boundary conditions, steady state, transient and problems with and without heat generation. Heat transfer with heat generation finds wide engineering applications such as electric heating, storage of chemicals, nuclear waste disposal, nuclear reaction in rod bundles etc. The presence of heat generation in a medium influences the temperature distribution to a large extent and it is also possible to control heat loss or gain through the boundaries by suitably controlling the heat generation parameters. Though most of the heat generation problems in the one-dimensional domain can be solved analytically, numerical solution provides more insight into the nature of temperature distribution and heat transfer. Governing equations for conduction with heat generation can be derived using the basic energy conservation principles. In the solution of such problems using numerical methods, the heat generation and the source term contribute to the load vector of the final simultaneous equations. The associated boundary conditions for such problems have to be identified and incorporated in the solution procedure. The simulation parameters are decided based on the initial objectives of the simulation program.

Heat transfer through multi-dimensional solids are discussed in many textbooks with details of numerical methods. As far as mass transfer is concerned, its significance in engineering applications can be understood by modeling and simulating simultaneous heat and mass transfer problems with some applications. Engineering applications such as food processing, drying of grains, wet porous solids, storage of solar energy in solar ponds, evaporative cooling etc. involve simultaneous transport of heat energy and mass flux of certain species. Use of deodorant spray in a room involves mass transfer by virtue of the difference between the concentration of the species in the deodorant and the same species in the ambient air. When tap water in a bucket is left open for a few hours, there is a little drop in temperature of water up to the limit of wet bulb temperature corresponding to the given ambient conditions. This principle is utilized in the design of desert coolers in which spraying water over air flowing through the fills produces a cooling effect. The mass diffusion of water species in water surface is at its maximum concentration and whenever the concentration of water vapor in the ambient is air very low due to seasonal variation, then this difference in species concentration causes the transport of water vapor to the ambient air. Thus the vapor pressure difference between the water surface and ambient air causes evaporation of water taking heat from the water itself. This process is called drying if instead of water surface, the moisture is evaporated from any wet solid, or cloth etc. Drying finds wide engineering applications in food processing,

DOI: 10.1201/9781003248071-5

manufacturing of bricks, tiles and ceramic products used with electronic equipment. There are different stages during a drying process, which can be understood by referring to relevant literature. In the present modeling and simulation exercise in this chapter, a problem related to heat and moisture transport through soil with application to landmine detection will be described in detail.

Landmines are used during wartime to deter entry of enemies into a particular region and are buried under the ground in shallow depth. They are devised such that due to application of certain pressure, they will explode, thus when humans step over the earth under which a landmine is buried, it will explode. After the war, it becomes highly challenging to demine these areas are in order to make them suitable for human living. When metallic landmines are used, they can be detected easily using metal detectors. However, in recent decades, non-metal landmines are used to avoid their identification and deactivation. Such problems are very interesting for an engineer to find out a solution to exactly locate the landmine buried under the ground. As the mines are buried under the ground, the problem can be treated as one that involves heat and mass transfer phenomena. The ground mainly consists of soil along with minerals and moisture, and this moisture content may vary depending on the seasonal variation at a particular location. A novel idea has been proposed [1] to detect the buried landmine under the ground using the concept of *thermal signature*. It is well known that the ground receives solar radiation continuously and conducts heat to the lower level, in a way, the ground can be treated as a good solar energy storage medium. Due to the variation of solar heat during a diurnal cycle, the surface temperature of the ground varies continuously over 24 hours. When a landmine made of specific material is buried under the ground at a certain distance, the region of ground just above the landmine will not receive any heat compared to other adjacent sides of the ground. This characteristic sets up a temperature difference between the region of ground without landmine and the region just above the landmine and the resulting temperature difference is called *thermal signature* of the landmine for the given conditions. Variation of thermal signature during a diurnal cycle becomes a distinctive nature of the mine, ground and ambient (including solar irradiation) combination. If these signatures are simulated, then the location of the mine can be determined by comparing these thermal signatures with one that has actually been retrieved, using a thermal-imaging camera. The details of formulation of the problem, solution procedure using the finite element method and simulation results will be discussed in the second part of this chapter.

5.1 HEAT TRANSFER PROBLEM – ONE-DIMENSIONAL CONDUCTION WITH HEAT GENERATION

Conduction heat transfer with heat generation finds very interesting applications in the field of conversion of nuclear energy to thermal energy. A nuclear fuel rod generates heat by nuclear fission reaction and the heat released during the reaction is used to heat water to generate high-pressure steam which is used to run a turbine coupled with an alternator. During heat transport from the fuel rod assembly to the coolant, that is water, there should be minimum heat loss. Using insulating materials, the heat loss from a surface can be controlled, however, the best natural way of achieving heat loss is to maintain adiabatic boundary condition at the surface from where the heat loss is expected to take place. The adiabatic boundary condition at a surface can be maintained by suitably adjusting the controlling parameters which have to be determined by solving the governing equation for heat conduction with heat generation as discussed below.

5.1.1 Derivation of Energy Conservation Equation

The generalized heat conduction equation in different coordinate systems has been detailed in Chapter 2. Under steady state conditions, the energy balance for heat transfer by conduction through one-dimensional solid medium with heat generation can be derived by the application of the energy conservation principle to the medium shown in Figure 5.1(a) as follows.

$$\begin{aligned}&(\text{Rate at which heat enters the system at } x \text{ by conduction})\\&\quad+(\text{Amount of heat generated within the medium})\\&=(\text{Rate at which heat leaves the system at } x+\Delta x \text{ by conduction})\end{aligned} \tag{5.1}$$

Expressing the above statement in mathematical form, we get

$$q_x + Q_{gen} = q_{x+\Delta x}$$

After simplification, we can write

$$(q_x - q_{x+\Delta x}) + Q_{gen} = 0$$

Using Taylor's series expansion, the first term on the left-hand side of the above equation can be expanded across Δx as

$$-\frac{dq_x}{dx} A\Delta x + Q_{gen} = 0$$

In the above equation, Q_{gen} indicates the heat generated within the medium in W and A is the area of cross-section of the medium. It is assumed that the area of cross-section A remains constant along the length of the medium.

The conduction heat transfer can be expressed in terms of the Fourier law of heat conduction in the x-direction as

$$q_x = -k_x A \frac{dT}{dx}$$

$$\frac{d}{dx}(k_x \frac{dT}{dx})(\Delta x)A + Q_{gen} = 0$$

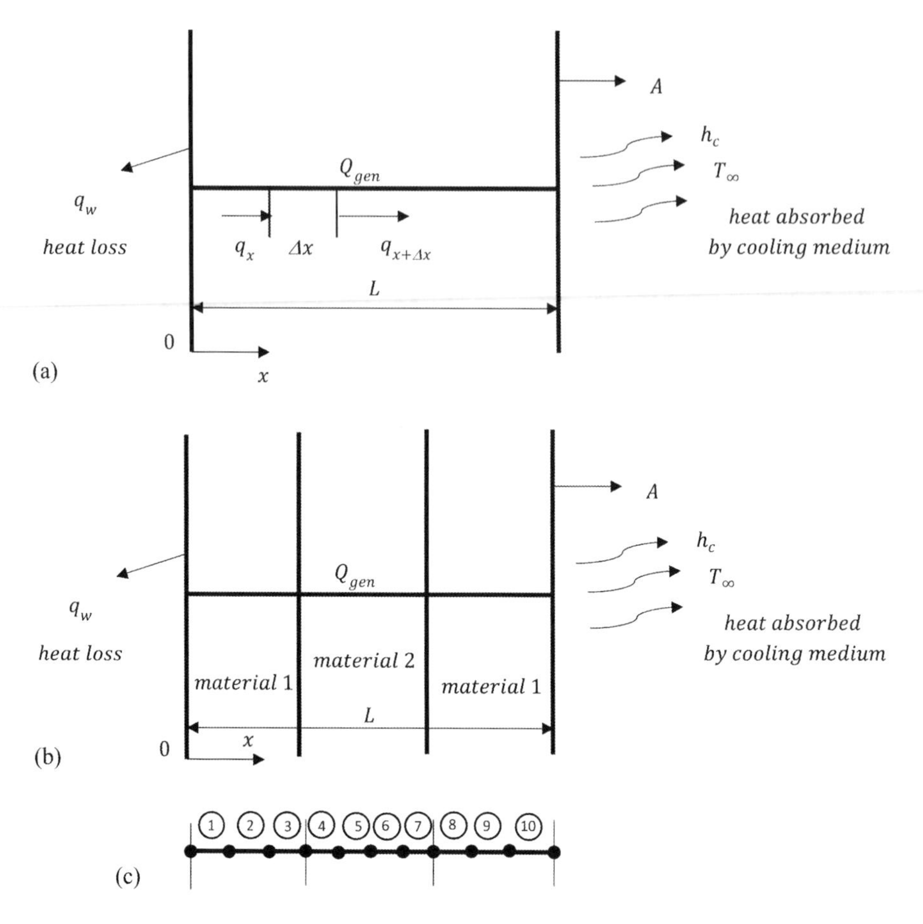

FIGURE 5.1 (a) Schematic diagram for heat conduction with generation, (b) composite wall of two materials and (c) mesh details for composite wall.

The governing equation is always written per unit volume of the control volume, $A\Delta x$, and the total heat generation can be expressed in terms of heat generation per unit volume as

$$Q_{gen} = \dot{Q}_{gen}\Delta xA.$$

Now the conduction equation takes the form

$$\frac{d}{dx}(k_x \frac{dT}{dx})(dx)A + \dot{Q}_{gen}(\Delta xA) = 0$$

and for constant thermal conductivity of the medium, the above equation can be expressed in differential form as

$$k\frac{d^2T}{dx^2} + \dot{Q}_{gen} = 0 \tag{5.2}$$

The boundary conditions for Equation (5.2) have to be identified depending on the application to a given situation. In the present exercise, it is assumed that the heat generated within the medium is absorbed at the right end of the domain at $x=L$ by means of convective heat transfer, whereas some amount of heat loss is assumed to take place through the left end of the domain at $x=0$. Now, the equations for the boundary conditions can be derived in the following section.

5.1.2 Identification of Boundary Conditions

Inside the one-dimensional medium, the source of heat is the heat generated due to some chemical reaction inside the medium with heat loss assumed to take place at x=0 end and heat extraction at $x=L$.

@ $x=0$ - Heat flux boundary condition

$$-k\frac{dT}{dx} = q_{x0} \tag{5.3a}$$

@ $x=L$ – Convective boundary condition

$$-k\frac{dT}{dx} = h_c\left(T_w - T_\infty\right) \tag{5.3b}$$

Equation (5.2) can be solved using the boundary conditions expressed by Equation (5.3).

5.1.3 Solution Using Finite Element Method

Galerkin's weighted residual finite element method will be employed to solve Equation (5.2) as follows.

$$\int_\Omega [N]^T \left(k\frac{d^2T}{dx^2} + \dot{Q}_{gen}\right)dx = 0 \tag{5.4}$$

Consider the term

$$k[N]^T\frac{d^2T}{dx^2} = k\frac{d}{dx}\left([N]^T\frac{dT}{dx}\right) - k\frac{dT}{dx}\frac{d[N]^T}{dx}$$

Application of Green's theorem converts the first term on the right-hand side of the equation from line integral to an integral to a point, that is to a dimension one order less than the dimension of the domain. Hence, the above equation is written as

$$k\int_{\Omega}[N]^T \frac{d^2T}{dx^2}dx = k\int_{\Gamma}\left([N]^T \frac{dT}{dx}\right)d\Gamma - k\int_{\Omega}\frac{dT}{dx}\frac{d[N]^T}{dx}dx$$

$$k\int_{\Omega}[N]^T \frac{d^2T}{dx^2}dx = k\int_{\Gamma}[N]^T \frac{dT}{dn}d\Gamma - k\int_{\Omega}\frac{d[N]^T}{dx}\frac{d[N]}{dx}dx\{T\} \tag{5.5}$$

In the above equation, the first term on the right-hand side is considered only while applying the boundary conditions after the formation of the element level stiffness and load vector matrices. Now, consider the heat generation term,

$$\int_{\Omega}[N]^T \dot{Q}_{gen}dx = 0$$

After applying Galerkin's method for every term in Equation (5.4), we get the final equation as

$$-kA\int_{\Omega}\frac{d[N]^T}{dx}\frac{d[N]}{dx}dx\{T\} + \int_{\Omega}[N]^T \dot{Q}_{gen}dx = 0 \tag{5.6}$$

After multiplying the above equation by negative sign throughout, we get

$$k\int_{\Omega}\frac{d[N]^T}{dx}\frac{d[N]}{dx}dx\{T\} = \dot{Q}_{gen}\int_{\Omega}[N]^T dx$$

The element level stiffness matrices can be evaluated using the method discussed in Chapter 4. After multiplying by the area, the final matrix form of the equation is obtained as
$\frac{kA}{l_e}\begin{bmatrix} 1 & -1 \\ -1 & 1 \end{bmatrix}\begin{Bmatrix} T_i \\ T_j \end{Bmatrix} = \frac{\dot{Q}_{gen}Al_e}{2}\begin{Bmatrix} 1 \\ 1 \end{Bmatrix}$. The stiffness and load vector matrices can be identified as
$[K]^{(e)} = \frac{kA}{l_e}\begin{bmatrix} 1 & -1 \\ -1 & 1 \end{bmatrix}$ and $\{f\}^{(e)} = \frac{\dot{Q}_{gen}Al_e}{2}\begin{Bmatrix} 1 \\ 1 \end{Bmatrix}$. Equation (5.4) for an element can be expressed as

$$[K]^{(e)}\{T\}^{(e)} = \{f\}^{(e)} \tag{5.7}$$

For elements with constant length and thermal conductivity, we can write

$$[K]^{(1)} = [K]^{(2)} = [K]^{(3)} = \ldots = [K]^{(M-1)}, \{f\}^{(2)} = \{f\}^{(3)} = \{f\}^{(4)}$$
$$= \{f\}^{(5)} = \ldots = \{f\}^{(M-1)}$$

These elements are generated in accordance with the mesh generation of the domain. Here, it is worth noting one important capability of the finite element method to handle different materials. In the case of nuclear fuel rod problems, let us assume it is a cylindrical enclosure consisting of two materials, material 1 and material 2. If heat transfer is significant only in the radial direction and longitudinal conduction is ignored, then the system can be modeled as a one-dimensional domain as shown in Figure 5.1(b). Let us assume the computational domain consisting of two materials, is discretized into 10 elements with 11 nodes as depicted in Figure 5.1(c). Then, referring to the above figure, one can observe that elements 1, 2, 3 and 8, 9, 10 represent material 1, whereas elements, 4, 5, 6, 7 illustrate material 2. For forming the element level matrices, the data for thermal conductivity, k and length of element, l_e, area of cross section, A and heat generation, $\dot{Q}_{gen}$ are required. Hence, if the properties of both the material are known, then the element level matrices for different materials as distinguished by the element numbers can be generated. Generally, the area of cross-section remains constant, and only the length of the element may vary for material 1 and material 2 and this will result in two values of length of element. As the problem has to be simulated for heat flux and convective boundary conditions, that involves temperature gradients at both ends of the domain, the length of element of material 1 will be smaller than that of material 2. It is also assumed that continuity exists between the interface of the materials for both heat flux and temperature. In the present example, initially simulation results will be obtained for single material and then for two materials.

5.1.4 Incorporation of Boundary Condition

The boundary conditions for the given problem have to be incorporated into the respective element stiffness and load vector matrices at the boundary. Referring to Figure 5.1(c), it is clear that element number 1 will be altered for the heat flux boundary condition and element number 10 will be modified for the convective boundary condition.

$$@\,x = 0, -k\frac{dT}{dx} = q_{x0} \tag{5.7a}$$

The first term on the right-hand side of Equation (5.4) should be made equal to the above equation for element number 1 as follows. That means, at node number 1, the temperature gradient multiplied by thermal conductivity becomes a constant equal to q_{x0}.

$$k\int_{\Omega}[N]^T\frac{d^2T}{dx^2}dx = q_{w0} - k\int_{\Omega}\frac{d[N]^T}{dx}\frac{d[N]}{dx}dx\{T\} \tag{5.8}$$

In Equation (5.5), the term $k\int_{\Gamma}[N]^T\frac{dT}{dn}d\Gamma$ is replaced with q_{x0}. In Equation (5.3a), there is a negative sign with the temperature gradient and this is a general sign written along with the temperature gradient to indicate that heat transfer takes place from a higher temperature to a lower temperature. However, in the present case due to heat generation the heat will flow from the center of the domain to the left end and it is expected that the maximum temperature of the medium is located only at the center of the medium. Hence, the temperature gradient generated by the term $\frac{(T_2 - T_1)}{l_e}$ will cause the heat conduction at the boundary node 1 and this temperature gradient is positive in mathematical notation. After the application of Galerkin's method, Equation (5.7) has been multiplied by a negative sign throughout the equation, and this makes q_{x0} a negative quantity on the left side, and by taking it to the right-hand side it contributes to the load vector. It is clear from the above discussion that the heat flux boundary condition makes changes only in the load vector of the respective node of the boundary element. Thus the matrices for element 1 can be written as

$$\left[K\right]^{(1)} = \frac{kA}{l_e}\begin{bmatrix} 1 & -1 \\ -1 & 1 \end{bmatrix} \quad \text{and} \quad \left\{f\right\}^{(1)} = \frac{\dot{Q}_{gen} A l_e}{2}\begin{Bmatrix} 1 \\ 1 \end{Bmatrix} + q_{w0} A \begin{Bmatrix} 1 \\ 0 \end{Bmatrix} \tag{5.9}$$

Convective boundary condition at x=L

$$-k\frac{dT}{dx} = h_c\left(T_w - T_\infty\right) \tag{5.3b}$$

The convective boundary condition is applied for the last element, M at node M. Assume that the temperature derivative term on the right-hand side of Equation (5.5) is already multiplied by negative sign to produce the following equation.

$$-\int \left[N\right]^T kA\frac{dT}{dn}d\Gamma = \int \left[N\right]^T h_c A(T - T_\infty)d\Gamma \tag{5.10}$$

The right-hand side of Equation (5.10) produces two terms, one for the stiffness matrix because of the unknown variable temperature, T and the other due to the constant ambient temperature as load vector. The details of derivation have already been discussed in Chapter 4. Now, the contribution due to convective boundary condition at $x=L$ produces the following terms.

$$h_c A\int \begin{bmatrix} 0 \\ N_j \end{bmatrix}\begin{bmatrix} 0 & N_j \end{bmatrix} d\Gamma \left\{T\right\} - h_c A T_\infty \int \begin{bmatrix} 0 \\ N_j \end{bmatrix} d\Gamma$$

$$h_c A\begin{bmatrix} 0 & 0 \\ 0 & 1 \end{bmatrix}\left\{T\right\} - h_c A T_\infty \begin{Bmatrix} 0 \\ 1 \end{Bmatrix}$$

Finally, the stiffness matrix and load vector matrices for the last element M can be expressed as

$$\left[K\right]^{(M)} = \frac{kA}{l_e}\begin{bmatrix} 1 & -1 \\ -1 & 1 \end{bmatrix} + h_c A \begin{bmatrix} 0 & 0 \\ 0 & 1 \end{bmatrix} \text{ and}$$
$$\left\{f\right\}^{(M)} = \frac{\dot{Q}_{gen} A l_e}{2}\begin{Bmatrix} 1 \\ 1 \end{Bmatrix} + h_c A T_\infty \begin{Bmatrix} 0 \\ 1 \end{Bmatrix} \tag{5.11}$$

Equations (5.9) and (5.11) represent the element matrices for the boundary elements after incorporating the boundary conditions.

5.1.5 Computational Algorithm

A computational algorithm is developed to solve the final set of simultaneous equations obtained as a result of assembly of element matrices shown in Equation (5.5). Following is the computational algorithm for any number of materials considered in the medium for conduction heat transfer in the presence of heat generation within the medium.

- (i) Read input data such as number of materials considered, number of elements selected for each material, properties of materials, heat generation per unit volume and length of each material that becomes part of the whole computational domain.
- (ii) Compute the length of each material
  - (a) Start a do loop for the given number of materials
  - (b) Compute the length of each element for current material
  - (c) End the loop
- (iii) Compute constants for the formation of stiffness and load vector matrices
- (iv) Identify the element number for each material and store the connectivity of nodes for each element for the whole computational domain.
- (v) Form element level matrices for stiffness matrix and load vector matrix using different constants which take care of different materials.
- (vi) Assemble all the stiffness matrices to get the global stiffness matrix.
- (vii) Assemble all the load vector matrices to form the global load vector.
- (viii) Incorporate the boundary conditions for both the ends of the computational domain.
- (ix) Form the simultaneous equations of the form $Ax=b$.
- (x) Call the solver to get solutions for temperatures at all the nodes in the domain.
- (xi) Output the results in the form required for plotting the results in the graphing software.
- (xii) End the program.

5.1.6 Computer Programming

Using the computational algorithm discussed in the previous section, a computer code can be developed to solve for temperature distribution along the length of the medium with heat generation. Figure 5.2 shows the details of formation of element level stiffness matrices and load vector for '*nelem*' elements.

The procedure to assemble all these elements is also depicted in the same figure. The inclusion of boundary conditions at $x=0$ and $x=L$ is explained in the form of computer program in Figure 5.3. There is no contribution for stiffness matrix due to heat flux boundary condition, because the heat conducted at $x=0$ is just equated to a constant, called the heat flux. However, this constant heat flux value will contribute to the load vector at node 1. Hence, the global load vector at node 1 is modified as depicted in Figure 5.3. It has to be noted that modification of boundary nodes is carried out only on the final global level stiffness and load vector matrices.

As far as the convective boundary condition at $x=L$ is concerned, the conduction heat flux at $x=L$ is equated to convective heat transfer through unknown temperature, T and some constant temperature, T_∞ and h_c. From the very basic principle of application of Galerkin's weighted residual method and definition of shape function to represent the unknown temperature, the presence of unknown temperature in the boundary condition will always produce $nne \times nne$ matrix by the product of $[N]^T$ and the shape function $[N]$ associated with temperature, T. Hence, this contribution will be added to the respective element level stiffness matrix for the boundary element. For this reason, one can notice that stiffness matrix corresponding to the last element, M gets contribution, h_cA and similarly the load vector also gets the contribution from T_∞ and h_c. It is also worth observing that for the heat flux boundary condition at $x=0$, only the shape function at node, '*i*', that corresponds to node 1, provides the contribution and the contribution from the other node, '*j*' becomes zero. Similarly, for the convective boundary condition at $x=L$, the node, '*j*' contributes to the stiffness and load vector matrices and the node, '*i*' does not contribute.

Once the global stiffness and load vector matrices are modified to include the effect of boundary conditions at nodes, 1 and M, now, the matrices have been brought to the form $Ax=b$, which can be solved using a solver. From the nature of assembly of element level stiffness matrices, one can easily notice that the final global stiffness matrix becomes a sparse matrix, that means, the matrix becomes diagonally dominant. With increase in number of nodes in the computational domain, the global stiffness matrix becomes highly sparse, resulting in only a small percentage of the final $N \times N$ matrix with non-zero entries, in which N represents the total number of nodes in the computational domain. As the value of N increases, then the sparseness of the matrix also increases. It is no wonder to notice that hardly 1% of $N \times N$ coefficients produce non-zero coefficients, making the remaining equal to zero. Hence, most of the solvers employed with the solution of finite element equations are iterative solvers. The solver is coded as a separate subroutine and hence it is called to solve the equations to obtain the solution for temperature field. Once the temperature values at all the nodes of the domain are obtained, then they can be written in an output file to make the required plots as results.

Computer code 5.1 – Formation of element matrices and global matrices – Heat conduction with heat generation

Define matrices

estiff (2, 2) – element level stiffness matrix for one-dimensional linear element

estiffg (idm, idm) – global stiffness matrix

bload (2) – element level load vector

bloadg (idm) – global load vector

l (idm, 2) – element nodal connectivity array

Define constants

idm – array size, nne – number of nodes per element, nelem – number of elements

$$aa1 = \frac{kA}{l_e},\ aa2 = \frac{\dot{Q}_{gen} A l_e}{2}\begin{Bmatrix}1\\1\end{Bmatrix}$$

Define element matrices

$estiff(1,1) = aa1$, $estiff(1,2) = -aa1$, $estiff(2,1) = estiff(1,2)$ and $estiff(2,2) = estiff(1,1)$

$bload(1) = aa2$ and $bload(2) = bload(1)$.

Assembly of matrices

```
do ielem = 1, nelem                                    Do this for all elements
   do I = 1, nne                                       Do it for the row
   lnode = l (ielem, i)                                Designate the global node number in row
  do j = 1, nne                                        Do it for the column
   knode = l (ielem, j)                                Designate the global node number in column
estiffg (lnode, knode) = estiffg (lnode, knode) +      Global stiffness matrix
                         estiff (i, j)
  enddo                                                End the column loop
 bloadg (lnode) = bloadg (lnode) + bload (i)           Do it for global load vector
   enddo                                               End the row loop
enddo                                                  End the element loop
```

FIGURE 5.2 Computer code for formation of element and global matrices.

Computer code 5.2 – Heat flux and Convective boundary conditions – Heat conduction with heat generation problem

Define matrices

estiffg (idm, idm) – global stiffness matrix

bloadg (idm) – global load vector

Define constants

idm – array size, nnode – number of nodes, nelem – number of elements,

$$aa3 = \frac{\dot{Q}_{gen} A l_e}{2} + q_{w0}, aa4 = h_c A \ , \ aa5 = h_c A T_\infty$$

Include heat flux boundary condition at node 1

Global load vector

bbload (1) = aa3

Include convective boundary condition

Global stiffness matrix

estiffg (nnode, nnode) = estiffg (nnode, nnode) + aa4

Global load vector

bloadg (nnode) = bloadg (nnode) + aa5

Solve the simultaneous equations using a solver

call solver (estiffg, bloadg, t, nnode)

Write the output

Write (File number, format) x(i), t(i)

FIGURE 5.3 Computer code for inclusion of heat flux and convective boundary conditions.

5.1.7 Mesh Sensitivity and Validation Results

The finite element formulation for the solution of one-dimensional heat conduction with heat generation discussed above, is coded into a computer program, ***Oned-heat-gen_FEM.for***. Initially this computer program has to be tested for mesh sensitivity study in order to make sure that the numerical results obtained using the finite element method are free from the mesh size. For this purpose, three meshes of 50, 250 and 500 elements have been chosen and the temperature distribution along the length of the solid is considered as the result for the mesh sensitivity study. The example considered for finite element solution has heat flux boundary condition at $x=0$ and convective boundary condition at $x=L$. The solid is considered one-dimensional with constant heat generation. It is assumed that there may be some heat loss through the left face, when all the heat generated within the solid is supposed to be dissipated by convection to the fluid medium on the right side of the solid. As the amount of heat loss through the left side increases, the maximum temperature achieved within the solid will be well within the solid, however, when heat loss becomes zero, then that surface becomes adiabatic for heat flow and attains the peak temperature. Mesh sensitivity and validation results are considered with the adiabatic boundary condition at $x=0$ end. The following parameters were used as input for the computer code to get the temperature distribution: length=60 mm, $\dot{Q}_{gen} = 0.3$ MW/m^3, T_∞=93°C, k=21 W/m K and h_c=570 W/m^{2}K. Figure 5.4(a) shows the mesh sensitivity results for the three meshes considered. It can be seen that results obtained with 250 and 500 elements almost coincide with each other and hence mesh 2 with 250 elements will be used for future computations.

The results obtained by the computer code is validated with an analytical solution which can be easily obtained from any textbook on heat transfer. Figure 5.4(b) depicts the comparison of temperature distributions obtained by the finite element method and analytical solution for the same input conditions used for mesh sensitivity study. The temperature distribution along the length of the solid predicted by the finite element method is almost close to the analytical solution. However, the numerical results are very close at the convective boundary end compared to the adiabatic end. The reason is for convective boundary condition, the heat conducted by the solid at $x=L$ is just equated to the convective heat flux which is a linear function of temperature which is unknown in the present simulation. In contrast, at the adiabatic end, the heat conducted becomes zero at $x=0$, this means the temperature gradient at this point must be made equal to zero. During computation, the convergence of a solution at the adiabatic boundary has to converge after satisfying the condition that the temperature gradient is equal to zero. That means the temperature gradient becomes normal to the boundary surface at $x=0$ and this can be achieved only by satisfying the energy balance such that whatever heat generated within the solid is dissipated through the convective surface at $x=L$. In other words, satisfying the adiabatic boundary condition becomes challenging in numerical computation, however, the accuracy can be improved by using non-uniform mesh at the boundary. In the present computation only uniform mesh has been used for the computations. From energy balance point of view, it is seen from Figure 5.4 that the heat loss at the left end is zero and hence whatever heat generated within the solid must be dissipated to the ambient fluid at the right end boundary by convective heat transfer. It can be observed that the wall temperature at $x=L$ is 125°C and after multiplying by unit area of cross-section for heat transfer, temperature difference and convective heat transfer coefficient, the heat dissipated by convection arrives at 0.304 W/m^3, which is very close to the assumed heat generation.

5.1.8 Simulation Parameters and Results

Once the mesh sensitivity study and validation of the code are completed, then the computer program can be used to obtain the simulation results. The parameters for simulation have to be decided based on the objective of the analysis of the problem. Just for example, let us assume that in the present problem, our main aim

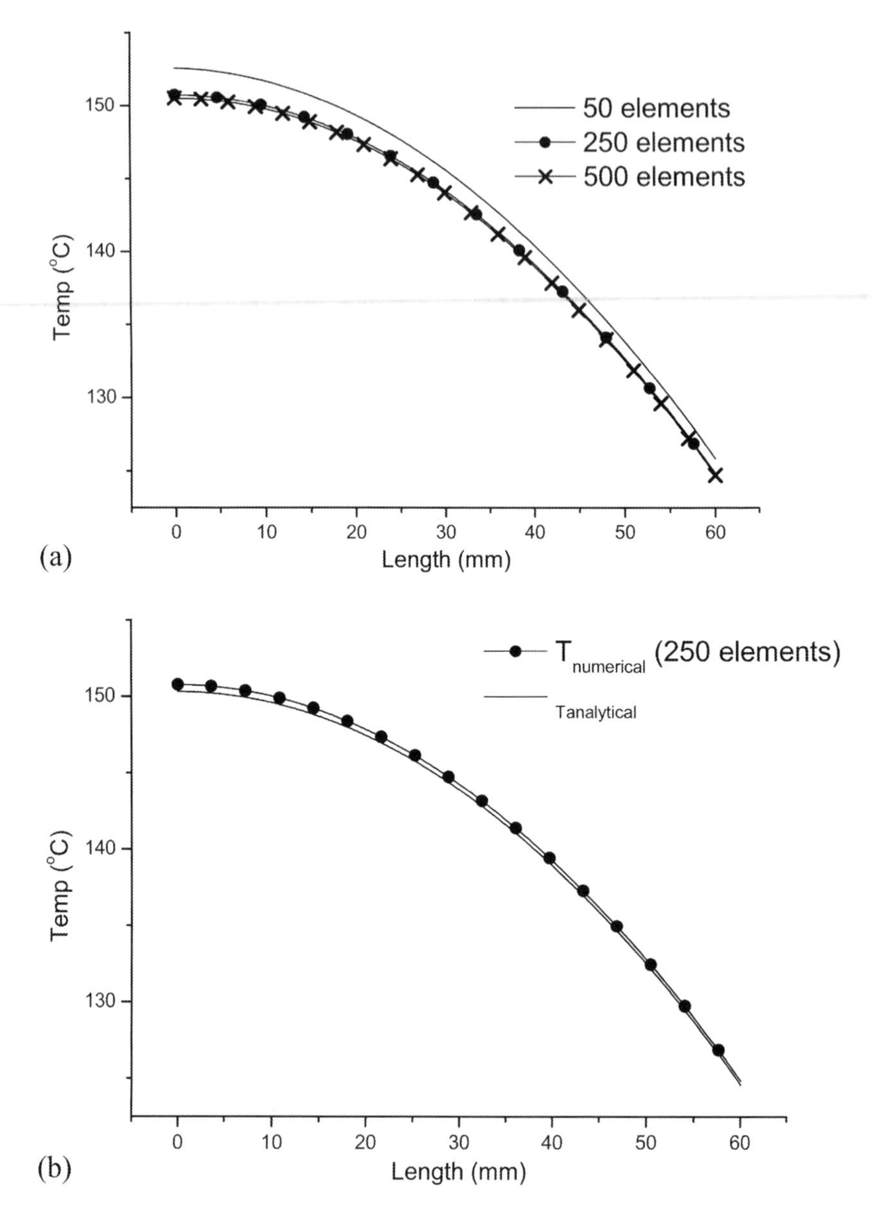

FIGURE 5.4 (a) Mesh sensitivity study and (b) validation of FEM results with analytical solution.

is to obtain all the heat generated within the solid, say by nuclear reaction, which has to be dissipated to the fluid at the convective end. For this purpose, the heat flow at $x=0$ boundary has to be insulated. In the present problem let us consider the heat loss at the left boundary, convective heat transfer coefficient and ambient fluid temperature at the right boundary are the simulation parameters which will vary within certain range, and heat generation within the solid is assumed to be constant. The following are the input parameters for simulation: length of solid – 200 mm, k=40 W/m K, $\dot{Q}_{gen} = 0.3$ MW/m^3, area=0.25 m^2, T_∞ =20°C, 25°C, 30°C, 35°C and 40°C and h_c=110, 120, 130, 140 and 150 W/m^{2}K, heat loss through the left end, q_{w0}=20, 22.5, 25.0, 27.5, 30.0 kW/m^2. Simulation results have been obtained by using the above input parameters in the computer code, and the results are plotted for temperature distribution along the length of the solid. Figure 5.5(a) shows the temperature distribution with no heat loss at the left boundary of the solid. When there is no heat transfer at *x=0*, the condition leads to adiabatic boundary condition, and hence it is expected that the temperature distribution becomes normal to the surface at *x=0*.

Figure 5.5(a) depicts the same behavior with temperature reaching the maximum value at *x=0* and the temperature gradient is normal to the boundary. The energy balance indicates that all the heat generated within the solid should be dissipated by convection to the right end of the solid. The temperature at *x=L* is noted to be 425°C, and for the assumed heat transfer coefficient and length of solid, one can easily compute the total heat dissipated per unit volume is 0.3 MW, which is equal to the heat generated within the solid. This indicates clearly that the numerical results obtained using the finite element method satisfy the energy conservation principle with which the problem was defined. Figure 5.5(b) depicts the temperature distribution for

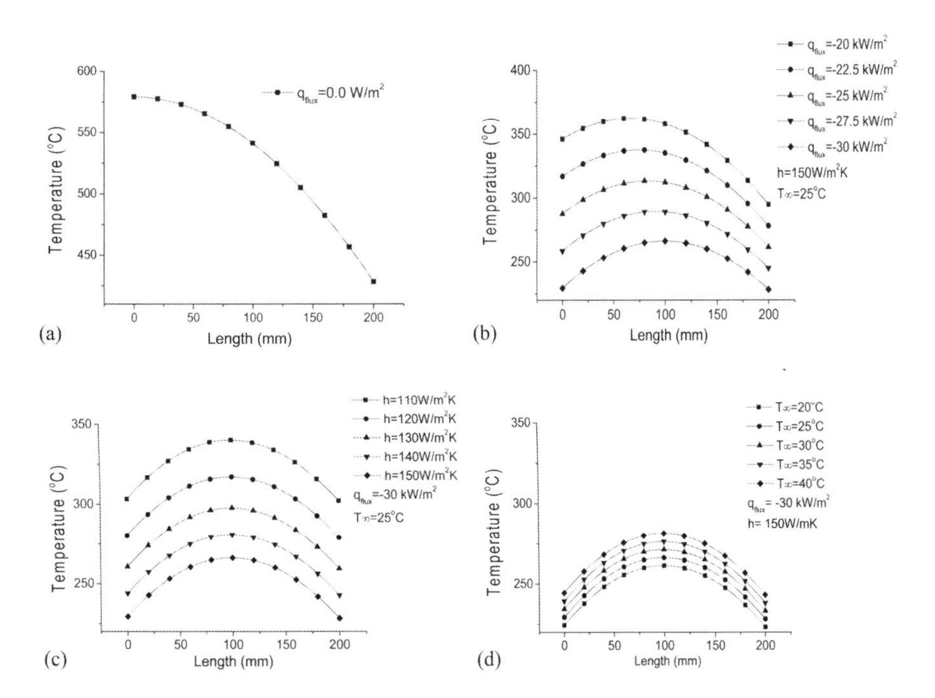

FIGURE 5.5 (a) Temperature distribution with adiabatic left end, (b) temperature distribution influenced by variation in heat loss from left end, (c) convective heat transfer coefficient and (d) ambient temperature.

variation in heat loss at the left end of the solid. As heat loss increases, the maximum temperature attained by the solid decreases and maximum value is achieved for the least heat loss. With a decrease in heat loss, the maximum temperature attained inside the solid increases and this maximum value occurs at a region closer to the left end of the solid. However, with an increase in heat loss, the maximum temperature decreases and shifts towards the center of the solid. Furthermore, increased heat loss at the left end of the solid results in reduction of wall temperature at the right end of the solid, where the heat is dissipated by convection. Figure 5.5(c) shows the effect of variation in convective heat transfer coefficient, h_c on temperature distribution. An increase in the value of the convective heat transfer coefficient ensures better convective heat transfer, thus reducing the wall temperature at the convective end. The peak temperature attained within the solid decreases with an increase in the convective heat transfer coefficient, however, the location of maximum temperature remains almost constant. Similarly, a decrease in ambient temperature assures better convective heat transfer as shown in Figure 5.5(d). With an increase in ambient temperature, the peak temperature attained by the solid increases and the location of maximum temperature is not affected by the ambient temperature. Thus, the above simulation results for parametric study indicates that (i) heat loss at one end of the solid results in a decrease in the value of peak temperature, and as the heat loss increases, the peak temperature gets shifted from the left boundary towards the center of the solid and (ii) convective heat transfer coefficient and ambient temperature do affect the peak temperature, however, the location of peak temperature is independent of these parameters.

Analysis of conduction heat transfer problem with heat generation discussed above is repeated with two materials. It is assumed the nuclear fuel rod assembly may consist of two concentric cylinders, the center one being the nuclear fuel and the surrounding material for transporting the heat released to the fluid to be heated. With the assumption of negligible longitudinal heat conduction, the domain can be treated as a composite wall made of two materials as depicted in Figure 5.1(c). From mesh sensitivity analysis, 250 elements have been chosen to obtain the simulation results. Now, referring to Figure 5.1(c), it is assumed that material 2 is the nuclear fuel rod, which is surrounded by material 1 as concentric cylinders. Hence, elements 1 to 50 and elements 201 to 250 represent material 1 and material 2 is indicated by elements 51 to 200. The finite element formulation remains the same except the formation of element level matrices for material 1 and material 2. A computer program in FORTRAN, ***onedcw-heatgen_FEM.for*** has been developed to obtain simulation results for this problem. All the input data used for the previous simulation results remain the same except the variation in thermal conductivity for materials 1 and 2 as k_1 and k_2 respectively. For q_{w0}=30 kW/m^2, two different cases were studied case I: k_1=50 W/m K and k_2=100 W/m K, case II: k_1=50 W/m K and k_2=250 W/m K. Figure 5.6 shows the temperature distribution obtained from the simulation using the above data. The temperature variation indicates two distinct behaviors for material 1 and material 2. As heat is generated within material 2, a non-linear temperature behavior is noticed, whereas material 1 is free from heat generation, and hence linear temperature distribution is expected as per the underlying physics of the problem.

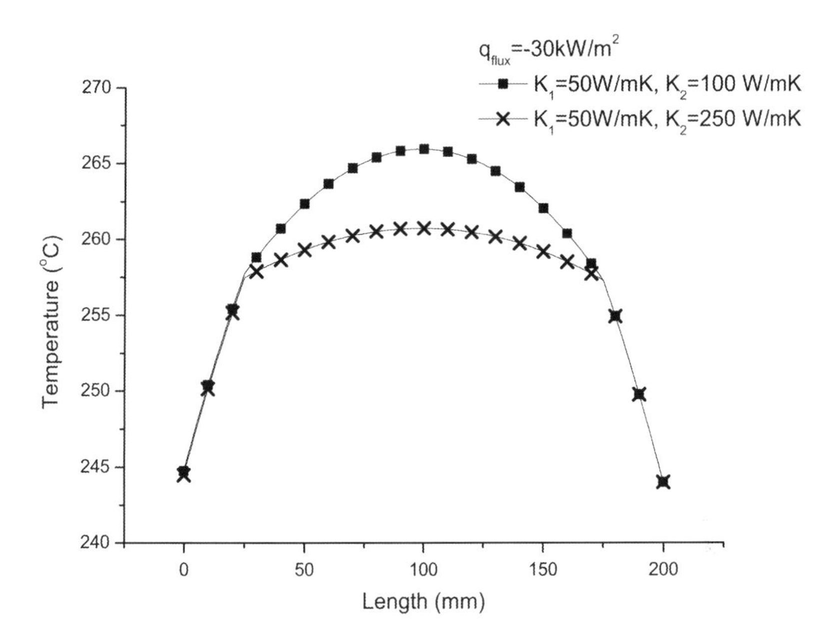

FIGURE 5.6 Conduction with heat generation in composite materials.

For the selected quantity of heat loss at the left end of the material, a symmetric temperature pattern is observed for material 2 and material 1. It is worth noticing that in the earlier case, material 2 was exposed to the boundary conditions, whereas in the present case, it is material 1 that is exposed to heat loss at the left end and convective heat transfer at the right end. When thermal conductivity of material 2 increases, conduction heat transfer within material 2 increases, and hence the rise in temperature distribution along the length of material 2 is lower than that of the case I as seen from the above figure

5.2 TWO-DIMENSIONAL PROBLEM – HEAT AND MASS TRANSFER THROUGH SOIL: LANDMINE DETECTION

Landmine detection has become an important environmental issue in many countries such as Afghanistan, Angola, Croatia, Egypt and Cambodia etc. It is a highly challenging task for the armed forces to de-mine the areas where landmines were installed either deliberately or buried by anti-social elements. More than 110 million landmines are hidden under the ground of 70 countries all over the world [2] and have caused great damage to humans. One of the deadliest legacies of the 20th century is the use of landmines in warfare. Anti-personnel landmines continue to have tragic, unintended consequences for years after a battle and even the entire war has ended. As time passes, the location of landmines is often forgotten, even by those who planted them, and they continue to be functional for many decades, causing further damage, injury and death. Landmines are basically explosive devices that are designed to explode when triggered by pressure or a tripwire. These devices are typically found on or just below the surface of the ground. The purpose of mines when used by armed forces is to disable any person or vehicle that comes into contact with it by an explosion or fragments released at high speeds.

Detecting these small objects in large areas is especially difficult where the character of the large areas is highly heterogeneous with the features that can mask the presence of the mine. It is this complex and dynamic environment that presents both the problem and the opportunity to more effectively detect minefields and mines. The conventional method of using metal detectors has become inefficient to detect landmines made of plastic or wood. Researchers have been therefore working to find better ways to detect landmines. Landmines buried more than 5 cm in depth are also difficult to detect by the conventional approach. Hence, this field has become an important research area where a number of researchers are working to find out a methodology to locate the landmine without loss of lives. Recently, the application of a thermal infrared (IR) technique to landmine detection has received a great amount of attention in the scientific community. This detection principle relies on a variation of *thermal signature* [3–5]. The presence of a landmine made of plastic or wood below the ground at a shallow depth causes variation in heat absorbed by the ground compared to the neighbouring region. The soil in the ground contains some amount of moisture and due to exposure to solar radiation during sunny hours, a certain quantity of moisture will evaporate from the surface, thus causing changes in temperature. This change in temperature depends on many factors, and this problem will be analyzed in detail in the following sections.

5.2.1 Derivation of Conservation Equations

Heat and moisture transport through soil in the presence of a mine has to be investigated in order to obtain the difference in temperature on the ground at a point just above the mine and an adjacent region. This temperature called thermal signature is the basic concept of this method to detect the mines buried under the ground. Figure 5.7 shows the schematic diagram of the heat and moisture transport through the soil under the ground in the presence of a rectangular landmine. When atmospheric air passes over the surface of the ground, there is an exchange of heat energy between the air and ground by convection and solar radiation.

The moisture present in the soil in the ground gets evaporated by taking heat from the ground itself by means of evaporative cooling as the mass transfer from the soil to the ambient takes place purely due to the difference in the vapor pressure of water vapor present in the soil in the ground and the ambient air. The very purpose of modeling this heat and moisture transport phenomenon is to capture the thermal signatures during the entire diurnal cycle. In the early morning around 6 AM, there will not be much of solar radiation on the ground, however, the vapor pressure difference will cause the moisture transport from the soil in the ground. The ground above the mine is not well connected for heat transmission to the region just below the mine because the soil domain is discontinued by the presence of the mine, thus causing some disruption in heat transfer. At the same time, the ground regions toward the right or left side of the mine are always well connected with the soil below without

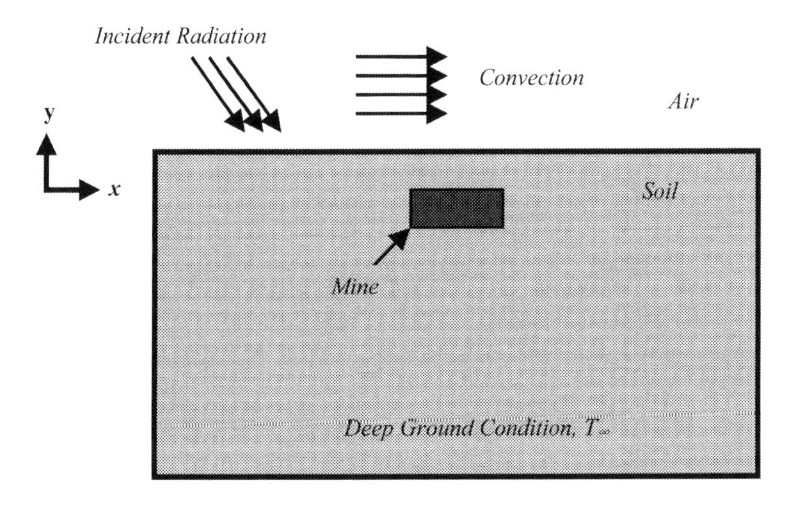

FIGURE 5.7 Schematic diagram of landmine detection using thermal signature concept.

any discontinuity of soil medium, thus the heat and mass transfers take place in a normal way. Now, as time passes to 9 AM, 12 noon, 1 PM, 2 PM, the amount of solar radiation keeps increasing. Of course, the amount depends on many factors. Until evening 6 PM and after sunset towards midnight, the soil may lose heat to the ambient, as ambient temperature decreases in the nighttime. Hence, during 24 hours, the region of soil above the mine and the region adjacent to the mine behave entirely differently, causing a definite temperature difference, called the *thermal signature*.

The type of differential equations and other data required for the analysis of the problem can be discussed in detail. Variation of temperature and moisture content within the soil/ground depends on the type of governing equations derived for the soil/ground. Similarly, a separate governing equation has to be developed for the mine. It has to be appreciated that the temperature and moisture distribution within the soil just above the mine depends on the interaction of heat and moisture transport between mine and soil. It is assumed that there is no moisture interaction between the mine and the soil except heat interaction. Hence, governing equations for heat and moisture transport within soil need to be derived with time variation, and transient heat conduction by the mine also has to be modeled. As the mine and soil interact with each other thermally, their boundary conditions have to be suitably coupled. The thermal signature is monitored during a diurnal cycle and hence the variation of ambient temperature and solar irradiation with time during a day have to be represented mathematically so that the data can be used as a boundary condition for the top side of the soil.

5.2.1.1 Conservation Equation for Heat and Moisture Transport within the Soil Medium

Let us assume the soil medium that forms the ground consists of four continuums, solid, liquid, vapor and air. In general, the effect of air presence is not significant for heat and moisture transport because the amount of moisture transport due to solar energy heat interaction at the soil surface is small. It is also assumed that the soil medium is a homogenous medium in which all the three phases, solid, liquid and vapor, are in equilibrium. The solid medium present in the soil is assumed to be non-deformable and its presence is accounted for calculating the average thermo-physical properties of the soil as a porous medium. The governing equations should be capable of predicting change in moisture content of soil and temperature with time in the presence of mine and also in the neighboring regions. In a sense, the problem is reduced to a problem of heat and moisture transport in porous medium. There are different types of approaches that can be followed to model such problems. In the present work, the approach proposed by Kallel et al. [6] and later implemented by Murugesan et al. [7] will be discussed. As far as the moisture transport is concerned, both the liquid and vapor form of moisture contribute to the governing equation. During a good sunshine period, the solar radiation received by the ground will be used to evaporate some of the moisture present at the surface of the soil and thereby sets up a diffusion path for moisture to get transported from the interior region of the soil to the surface. Moisture in liquid form gets evaporated to vapor and vice versa – condensation of water vapor present in the ambient air may take place during night time when the temperature is sufficient enough for condensation. The generalized mass conservation of the two mobile components, liquid or vapor can be expressed as

$$\frac{\partial w_j}{\partial t} = -\frac{1}{\rho_0}\frac{\partial J_j}{\partial x} \pm \frac{\dot{m}}{M_0} \tag{5.12}$$

A positive value of the source term in the above equation indicates condensation of vapor. Therefore, the plus sign appears in the equation for liquid and the minus for vapor. In case air is considered as one of the mobile components, then the source term is zero. Thus, for the two-dimensional domain, the mass conservation is expressed as:

Liquid phase

$$\frac{\partial w_l}{\partial t} = -\frac{1}{\rho_0}\left(\frac{\partial J_{lx}}{\partial x} + \frac{\partial J_{ly}}{\partial y}\right) + \frac{\dot{m}}{M_0} \tag{5.13}$$

Vapor phase

$$\frac{\partial w_v}{\partial t} = -\frac{1}{\rho_0}\left(\frac{\partial J_{vx}}{\partial x} + \frac{\partial J_{vy}}{\partial y}\right) - \frac{\dot{m}}{M_0} \tag{5.14}$$

where $\dot{m}$ denotes the rate of phase change, M_0 is the mass of dry porous solid, J denotes the diffusion flux of moisture and w represents the moisture content per kg of dry solid. Diffusion of liquid flux can be expressed using Darcy's equation [6] as

$$J_l = -\rho_l \frac{K_l}{\mu_l}\frac{\partial P_c}{\partial x} \tag{5.15}$$

Since the capillary pressure P_c is a function of total moisture content w and the function depends on temperature T, the flux due to capillary forces can be rewritten as

$$J_l = -\rho_l \frac{K_l}{\mu_l}\left(\frac{\partial P_c}{\partial w}\frac{\partial w}{\partial x} + \frac{\partial P_c}{\partial T}\frac{\partial T}{\partial x}\right) - \rho_l \frac{K_l}{\mu_l}\left(\frac{\partial P_c}{\partial w}\frac{\partial w}{\partial y} + \frac{\partial P_c}{\partial T}\frac{\partial T}{\partial y}\right)$$

After introducing the isothermal and non-isothermal mass diffusion coefficients for liquid transport, D_{ml}, D_{Tl} respectively, the above equation for mass flux can be simplified as

$$J_l = -\rho_0\left(D_{ml}\frac{\partial w}{\partial x} + D_{Tl}\frac{\partial T}{\partial x}\right) - \rho_0\left(D_{ml}\frac{\partial w}{\partial y} + D_{Tl}\frac{\partial T}{\partial y}\right) \tag{5.16}$$

Similarly, the diffusion of moisture in vapor form can be expressed as

$$J_v = -\rho_0\left(D_{mv}\frac{\partial w}{\partial x} + D_{Tv}\frac{\partial T}{\partial x}\right) - \rho_0\left(D_{mv}\frac{\partial w}{\partial y} + D_{Tv}\frac{\partial T}{\partial y}\right) \tag{5.17}$$

where D_{mv}, D_{Tv} respectively represent isothermal mass diffusion coefficient for vapor and non-isothermal mass diffusion coefficient for vapor. The dependence of these mass transfer coefficients on capillary pressure, permeability, viscosity and other porous media-related properties have been tested through experiments and empirical correlations [6]. Substituting Equations (5.16) and (5.17) in Equations (5.13) and (5.14) respectively, the conservation equation for liquid and vapor can be expressed as

$$\frac{\partial w_l}{\partial t} = \frac{\partial}{\partial x}\left(D_{ml}\frac{\partial w}{\partial x} + D_{Tl}\frac{\partial T}{\partial x}\right) + \frac{\partial}{\partial y}\left(D_{ml}\frac{\partial w}{\partial y} + D_{Tl}\frac{\partial T}{\partial y}\right) + \frac{\dot{m}}{M_0} \tag{5.18}$$

$$\frac{\partial w_v}{\partial t} = \frac{\partial}{\partial x}\left(D_{mv}\frac{\partial w}{\partial x} + D_{Tv}\frac{\partial T}{\partial x}\right) + \frac{\partial}{\partial y}\left(D_{mv}\frac{\partial w}{\partial y} + D_{Tv}\frac{\partial T}{\partial y}\right) - \frac{\dot{m}}{M_0} \tag{5.19}$$

where D_{ml}, D_{Tl}, D_{mv}, D_{Tv} represent the isothermal and non-isothermal diffusion coefficients for liquid and vapor respectively.

During moisture transport process within the soil, $w_v << w_l$ for major part of the transient period, hence, the contribution by vapor can be ignored in comparison to the liquid phase. Adding Equations (5.18) and (5.19), we get

$$\begin{aligned}\frac{\partial w}{\partial t} &= \frac{\partial}{\partial x}\left[\left(D_{mv} + D_{ml}\right)\frac{\partial w}{\partial x} + \left(D_{Tv} + D_{Tl}\right)\frac{\partial T}{\partial x}\right] \\ &+ \frac{\partial}{\partial y}\left[\left(D_{mv} + D_{ml}\right)\frac{\partial w}{\partial y} + \left(D_{Tv} + D_{Tl}\right)\frac{\partial T}{\partial y}\right]\end{aligned} \tag{5.20}$$

With the assumption that the rate of accumulation of water vapor is very small compared to the liquid phase, the expression for the rate of vapor condensation can be written from Equation (5.19) as

$$\frac{\dot{m}}{M_0} = \frac{\partial}{\partial x}\left(D_{mv}\frac{\partial w_l}{\partial x} + D_{Tv}\frac{\partial T}{\partial x}\right) + \frac{\partial}{\partial y}\left(D_{mv}\frac{\partial w_l}{\partial y} + D_{Tv}\frac{\partial T}{\partial y}\right) \tag{5.21}$$

Energy Conservation Equation for the Soil Medium

The energy conservation equation for the soil medium can be derived by performing an energy balance over a two-dimensional control volume of the medium and the energy conservation principle can be stated as

(Rate of change of total enthalpy of porous solid + liquid + water vapour)

= (Net heat conducted out of the control volume by all the three phases (5.22)

+Enthalpy transport of liquid and vapour species by convection)

This expression can be written in mathematical form as

$$\rho_0 \frac{\partial}{\partial t}\left(\sum w_j H_j\right) = \frac{\partial}{\partial x}\left(k\frac{\partial T}{\partial x}\right) + \frac{\partial}{\partial y}\left(k\frac{\partial T}{\partial y}\right) - \frac{\partial}{\partial x}\left(\sum J_j H_j\right) - \frac{\partial}{\partial y}\left(\sum J_j H_j\right) \tag{5.23}$$

The left-hand side of the above equation can be written in terms of average properties of porous solid, liquid and vapor phases and temperature difference during a given time interval. The change in enthalpy due to convective transport of liquid and vapor phases present in the soil medium as indicated by the last two terms on the right-hand side of Equation (5.23) can be neglected in comparison to the significant contribution by the phase change term. During phase change the change in enthalpy is very large and hence its contribution for energy balance cannot be ignored. Hence, Equation (5.23) can be rewritten as

$$C^* \frac{\partial T}{\partial t} = \frac{1}{\rho_0}\left[\frac{\partial}{\partial x}\left(k\frac{\partial T}{\partial x}\right) + \frac{\partial}{\partial y}\left(k\frac{\partial T}{\partial y}\right)\right] + \frac{\dot{m}}{M_0}\left(H_v - H_l\right) \tag{5.24}$$

where $C^* = C_0 + w_l C_l + w_v C_v$ and $\left(H_v - H_l\right)$ is the latent heat of vaporization during the phase change process. Equation (5.24) can be further simplified to the following form.

$$\frac{\partial T}{\partial t} = K_1\left[\frac{\partial}{\partial x}\left(\frac{\partial T}{\partial x}\right) + \frac{\partial}{\partial y}\left(\frac{\partial T}{\partial y}\right)\right] + K_2\left[\frac{\partial}{\partial x}\left(\frac{\partial w}{\partial x}\right) + \frac{\partial}{\partial y}\left(\frac{\partial w}{\partial y}\right)\right] \tag{5.25}$$

where $K_1 = \left[\frac{k}{\rho_0 C^*} + \frac{(H_v - H_l) D_{Tv}}{C^*}\right]$ and $K_2 = \frac{(H_v - H_l) D_{mv}}{C^*}$

Now, Equation (5.25) looks like a simple transient heat conduction equation for a two-dimensional domain. Similarly, the moisture conservation equation expressed by Equation (5.20) can be written as

$$\frac{\partial w}{\partial t} = K_3\left[\frac{\partial}{\partial x}\left(\frac{\partial w}{\partial x}\right) + \frac{\partial}{\partial y}\left(\frac{\partial w}{\partial y}\right)\right] + K_4\left[\frac{\partial}{\partial x}\left(\frac{\partial T}{\partial x}\right) + \frac{\partial}{\partial y}\left(\frac{\partial T}{\partial y}\right)\right] \tag{5.26}$$

where $K_3 = (D_{mv} + D_{ml})$ and $K_4 = (D_{Tv} + D_{Tl})$

Equations (5.25) and (5.26) are the final energy conservation and moisture conservation equations to be solved for time evolution of temperature and moisture content within the soil medium, and these equations are coupled with each other because both temperature and moisture evolution with time are functions of temperature and moisture content. Hence, a suitable computational algorithm has to be established to solve these equations.

Energy Conservation Equation for Landmines

The landmine is assumed to be made of non-metallic material, and it interacts with the soil only through thermal energy interaction, and it is adiabatic for moisture transport. For an isotropic material, the energy conservation equation for mine can be expressed as

$$\rho C \frac{\partial T_m}{\partial t} = \nabla(k \nabla T)_m \tag{5.27}$$

where, ρ, c, k and T are density, specific heat, thermal conductivity and temperature of mine respectively, and the subscript m refers to mine. The conservation equations for temperature and moisture content of soil medium and energy equation for mine have been established. In order to solve these equations, the boundary conditions have to be identified.

5.2.2 Initial and Boundary Conditions

5.2.2.1 Initial Conditions

The problem under analysis is a transient problem, in which all the field variables have to be computed with time starting from a specified initial time up to an end time, covering the total transients. Hence, the values of the field variables at all the nodes of the computational domain should be known before the start of computation for the next time level. For the present problem, the initial conditions are specified as

$$@\ t = 0,$$

Soil medium: $T(j) = T_{inis}$ and $w(j) = w_{inis}$ for $0<x<L_s$, $y=0$ and H_s; for $0<y<H_s$, $x=0$ and L_s

Landmine: $T(j)=T_{inim}$ for $0<x<L_m$, $y=0$ and H_m; for $0<y<H_m$, $x=0$ and L_m

where T_{inis} – initial temperature of soil, w_{inis} – initial moisture content of soil, T_{inim} – initial temperature of landmine, L_s, H_s – length and height of soil medium, L_m, H_m – length and height of landmine.

5.2.2.2 Boundary Conditions – Soil Medium

The boundary conditions for soil medium and the landmine have to be specified so that the conservation equations given by Equations (5.25) to (5.27) can be solved to obtain temperature and moisture distributions within the soil medium. Let us consider the boundary conditions of the soil medium by defining the four sides. Referring to Figure 5.7 it can be noticed that the rectangular soil domain shown in this figure is part of the soil medium and the boundary conditions for the four sides of this rectangular domain have to be specified.

5.2.2.3 Top Side

The net heat flux from the solar radiation interacting with the top side of the rectangular domain of the soil medium can be expressed as

$$q_{net} = q_{conv} + q_{rhs} - q_{evap} \tag{5.28}$$

where q_{conv} is the heat transfer due to convective heat and mass transfer between the surface of the soil and the ambient air, q_{rhs} is the incident radiation heat flux on the soil surface, and q_{evap} is the evaporative heat loss due to evaporative cooling of the soil. The emission loss from soil to the atmosphere is negligible.

The convective boundary condition for the energy equation is written as

$$\begin{aligned}\left(\frac{k}{\rho_0 C^*} + \frac{(H_v - H_l)D_{Tv}}{C^*}\right)\frac{\partial T}{\partial n} + \frac{(H_v - H_l)D_{mv}}{C^*}\frac{\partial w}{\partial n} \\ = \frac{h_c}{\rho_0 C^*}(T_s - T_\infty) + \frac{(H_v - H_l)h_m}{\rho_0 C^*}(\rho_v - \rho_\infty) + \frac{q_{rhs}}{\rho_0 C^*}\end{aligned} \tag{5.29}$$

and the boundary condition for mass transfer or evaporative mass transfer is

$$\left(D_{mv} + D_{ml}\right)\frac{\partial w}{\partial n} + \left(D_{Tv} + D_{Tl}\right)\frac{\partial T}{\partial n} = h_m\left(\rho_v - \rho_\infty\right) \tag{5.30}$$

In the above equations, D_{mv}, D_{ml}, D_{Tv}, D_{Tl}, are the isothermal and non-isothermal diffusion coefficients for liquid and vapor respectively. Here, the ambient temperature T_∞ varies continuously with time throughout the diurnal cycle and can be represented as [8]:

$$T_\infty = T_{avg} - A_m \cos\left[\frac{2\pi\left(t_h - 2\right)}{24}\right] \tag{5.31}$$

where T_{avg} is the average temperature of the day, A_m is the amplitude of daily temperature variation from the average, and t_h is the day time in hours on a scale of 1 to 24 starting from midnight.

The solar radiation incident on the top surface of the soil is expressed as [1]

$$q_{rhs} = (1 - C_L) S_0 (1 - C_G) M(\phi) \cos(\phi) \tag{5.32}$$

where C_L (=0.2) is the cloud cover, C_G (=0.3) is the ground albedo, and S_0 (= 1385 W/m^2) is the solar constant. The atmospheric transmissivity $M(\phi)$ can be expressed as

$$M(\phi) = 1 - 0.2\left[\cos(\phi)\right]^{-0.5} \tag{5.33}$$

where, ϕ is the zenith angle calculated using the following expression:

$$\cos(\phi) = \max\left[0, -\cos(\lambda)\cos(\delta)\cos\left(\frac{2\pi t_h}{24}\right) + \sin(\lambda)\sin(\delta)\right] \tag{5.34}$$

where λ is the local latitude and δ is the declination angle calculated by:

$$\delta = -23.43^\circ \cos\left(\frac{2\pi m}{12}\right) \tag{5.35}$$

where m is the month of the year on a scale of 1 to 12 starting from January.

5.2.2.4 Bottom Side

The bottom side of the soil is considered deep enough to influence the change in temperature and moisture content of the domain. Hence, the bottom side is assumed to be subjected to Dirichlet boundary condition for both temperature, T and moisture content, w.

At the bottom side of soil deep below mine:

$$\text{at } y = 0,\ 0 \le x \le L, T_{soil} = T_b \text{ and } w_{soil} = w_b \tag{5.36}$$

where T_b and w_b are the deep soil temperature and moisture content below the buried mine.

5.2.2.5 Vertical Sides

$$\vec{n} \cdot \nabla T = \vec{n} \cdot \nabla w = 0 \tag{5.37}$$

where, $\vec{n}$ represents the normal unit vector.

5.2.2.6 Mine-Soil Interface Boundary Condition

The landmine is treated as a homogeneous solid object with zero moisture content. In the present study, conducting mine is incorporated instead of heat source and the mine-soil interface is treated as impermeable for moisture transfer. Therefore, the boundary condition for moisture for mine-soil interface is given as

$$\frac{\partial w}{\partial \vec{n}} = 0 \tag{5.38}$$

According to energy conservation principle, there is continuous heat flux boundary condition across the mine-soil interface and hence, the following equation has to be satisfied.

$$\left(k\frac{\partial T}{\partial \vec{n}}\right)_m = \left(k\frac{\partial T}{\partial \vec{n}}\right)_s \tag{5.39}$$

where, the subscripts m and s refer to mine and soil respectively.

5.2.3 Solution Using Finite Element Method

The conservation equations derived for the soil medium and landmine represented by Equations (5.25) to (5.27) along with the boundary conditions given by Equations (5.29), (5.30) and Equations (5.36) to (5.39) have to be solved using Galerkin's weighted residual finite element method. Application of the finite element method for the energy equation of the soil medium gives the following equation.

$$\int_{\Omega} N^T\left[\frac{\partial T}{\partial t} - K_1\left(\frac{\partial^2 T}{\partial x^2}+\frac{\partial^2 T}{\partial y^2}\right) - K_2\left(\frac{\partial^2 w}{\partial x^2}+\frac{\partial^2 w}{\partial y^2}\right)\right]dxdy = 0 \tag{5.40}$$

Integration of the first term, the time derivative can be expressed as

$$\int_{\Omega} N^T \frac{\partial T}{\partial t} dxdy$$

$$\int_{\Omega} [N]^T [N] dxdy \frac{d\{T\}}{\partial t} = [C]^{(e)} \frac{d\{T\}}{\partial t} \tag{5.41}$$

After integrating the remaining two terms, which will give rise to the respective stiffness matrices, the above term is included to get the final form of the transient equation. Now, let us integrate the other two terms as follows.

Consider $\int_{\Omega} N^T \frac{\partial^2 T}{\partial x^2} d\Omega = \int_{\Omega} \frac{\partial}{\partial x}\left(N^T \frac{\partial T}{\partial x}\right) d\Omega$

Now, $\int_{\Omega} \frac{\partial}{\partial x}\left(N^T \frac{\partial T}{\partial x}\right) d\Omega = \int_{\Omega} N^T \frac{\partial^2 T}{\partial x^2} d\Omega + \int_{\Omega} \frac{\partial N^T}{\partial x}\frac{\partial T}{\partial x} d\Omega$

$$\text{or, } \int_\Omega N^T \frac{\partial^2 T}{\partial x^2} d\Omega = \int_\Omega \frac{\partial}{\partial x}\left(N^T \frac{\partial T}{\partial x}\right) d\Omega - \int_\Omega \frac{\partial N^T}{\partial x}\frac{\partial T}{\partial x} d\Omega$$

Using Green's theorem,

$$\int_\Omega \frac{\partial}{\partial x}\left(N^T \frac{\partial T}{\partial x}\right) d\Omega = \int_\Gamma N^T \frac{\partial T}{\partial n} d\Gamma$$

$$\text{Therefore, } \int_\Omega N^T \frac{\partial^2 T}{\partial x^2} d\Omega = \int_\Gamma N^T \frac{\partial T}{\partial n} d\Gamma - \int_\Omega \frac{\partial N^T}{\partial x}\frac{\partial T}{\partial x} d\Omega$$

$$= -\int_\Omega \frac{\partial N}{\partial x}\frac{\partial N^T}{\partial x} d\Omega \{T\} + \int_\Gamma N^T \frac{\partial T}{\partial n} d\Gamma \tag{5.42}$$

Similarly, we can write the other terms in Equation (5.40) as

$$\int_\Omega N^T \frac{\partial^2 T}{\partial y^2} d\Omega = -\int_\Omega \frac{\partial N}{\partial y}\frac{\partial N^T}{\partial y} d\Omega \{T\} + \int_\Gamma N^T \frac{\partial T}{\partial n} d\Gamma \tag{5.43}$$

$$\int_\Omega N^T \frac{\partial^2 w}{\partial x^2} d\Omega = -\int_\Omega \frac{\partial N}{\partial x}\frac{\partial N^T}{\partial x} d\Omega \{w\} + \int_\Gamma N^T \frac{\partial w}{\partial n} d\Gamma \tag{5.44}$$

$$\int_\Omega N^T \frac{\partial^2 w}{\partial y^2} d\Omega = -\int_\Omega \frac{\partial N}{\partial y}\frac{\partial N^T}{\partial y} d\Omega \{w\} + \int_\Gamma N^T \frac{\partial w}{\partial n} d\Gamma \tag{5.45}$$

It should be observed that the second term on the right-hand side of Equations (5.42) to (5.45) will contribute only to the boundary nodes and hence they can be considered at the time of incorporating the boundary conditions. Now substituting the terms corresponding to the two-dimensional spatial derivatives and time derivative term in Equation (5.40), we get

$$\int_\Omega [N]^T [N] \left\{\frac{T^{n+1} - T^n}{\Delta t}\right\} d\Omega + K_1 \int_\Omega \frac{\partial N}{\partial x}\frac{\partial N^T}{\partial x}\{T\} d\Omega + K_1 \int_\Omega \frac{\partial N}{\partial y}\frac{\partial N^T}{\partial y}\{T\} d\Omega$$

$$+ K_2 \int_\Omega \frac{\partial N}{\partial x}\frac{\partial N^T}{\partial x}\{w\} d\Omega + K_2 \int_\Omega \frac{\partial N}{\partial y}\frac{\partial N^T}{\partial y}\{w\} d\Omega = 0$$

5.2.4 Inclusion of Convective Boundary Condition on the Top Surface of the Soil Medium

For the solution of energy equation, the convective boundary condition on the top surface of the soil medium has to be included in the respective stiffness and load vector matrices. From Equation (5.28) it is understood that the contribution for net heat transfer on the top surface of the soil medium, q_{net} comes from convective heat transfer, q_{conv}, solar radiation falling on the surface, q_{rhs} and evaporative heat loss, q_{evap}. Among the three quantities contributing for q_{net}, q_{conv} and q_{rhs} are directly related to the temperature gradient and both these quantities are expressed in finite element form as below.

$\frac{h_c}{\rho_0 C^*}\int\limits_{\Gamma}[N]^T[N]d\Gamma\{T\}-\frac{h_c T_\infty}{\rho_0 C^*}\int\limits_{\Gamma}[N]^T d\Gamma+\frac{q_{rhs}}{\rho_0 C^*}\int\limits_{\Gamma}[N]^T d\Gamma$. In this equation the first term contributes to the stiffness matrix corresponding to the temperature, and the second and third terms contribute to the load vector. The line integral has to be evaluated only on the side of the element corresponding to the top surface of the soil. Similarly, the convective term related to moisture transport term in the energy equation can be written as

$\frac{(H_v-H_l)h_m}{C^*}\int\limits_{\Gamma}[N]^T[N]d\Gamma\{w\}-\frac{h_m\rho_\infty}{\rho_0 C^*}\int\limits_{\Gamma}[N]^T d\Gamma$. Now, consider an element on the surface of the soil medium represented by '*ijkm*' with side '*ij*' on the surface. The effect of convective boundary condition is contributed only by the line '*ij*' and hence for the respective boundary element, the above terms related to temperature and moisture will contribute only at these two nodes, '*i*' and '*j*'. Similar contributions are obtained for the load vector at these nodes only as explained below. For the energy equation of the soil medium, let us represent the stiffness matrix contribution of convective boundary condition as $[K_{TCB}]$ and load vector $\{f_{TCB}\}$ which are finally written as

$[K_{TCB}]_{ij}=\frac{h_c}{\rho_0 C^*}\int\limits_{\Gamma_{ij}}[N]^T[N]d\Gamma_{ij}\{T\}$ and $\{f_{TCB}\}_{ij}=\frac{h_c T_\infty}{\rho_0 C^*}\int\limits_{\Gamma_{ij}}[N]^T d\Gamma_{ij}-\frac{q_{rhs}}{\rho_0 C^*}\int\limits_{\Gamma_{ij}}[N]^T d\Gamma_{ij}$.

The contribution of $[K_{TCB}]$ will be added to the stiffness matrix, $[K]_{ET}^{(e)}$ of the energy equation and contribution of $\{f_{TCB}\}$ will be added to the load vector, $\{f\}_{ET}^{(e)}$. Similarly, the moisture component of the energy equation will get the following contributions.

$[K_{WCB}]_{ij}=\frac{(H_v-H_l)h_m}{C^*}\int\limits_{\Gamma_{ij}}[N]^T[N]d\Gamma_{ij}\{w\}$ and $\{f_{WCB}\}_{ij}=\frac{h_m\rho_\infty}{\rho_0 C^*}\int\limits_{\Gamma_{ij}}[N]^T d\Gamma_{ij}$.

The stiffness matrix contribution, $[K_{WCB}]_{ij}$ will be added to the stiffness matrix, $[K]_{EW}^{(e)}$ of the energy equation and $\{f_{WCB}\}_{ij}$ will contribute to the load vector, $\{f\}_{ET}^{(e)}$ of the energy equation. The following matrices illustrate the contributions from the line '*ij*' of the surface element to the stiffness and load vector matrices.

The effect of convective mass transfer on the top surface of the soil medium will give the following contribution from the line '*ij*' of the surface element.

$$[K_{WCB}]_{ij}=h_m\int\limits_{\Gamma_{ij}}[N]^T[N]d\Gamma_{ij}\{w\}\text{ and }\{f_{WCB}\}_{ij}=h_m\rho_\infty\int\limits_{\Gamma_{ij}}[N]^T d\Gamma_{ij}$$

The stiffness matrix contribution $\left[K_{WCB}\right]_{ij}$ will be added to the stiffness matrix, $\left[K\right]_{WW}^{(e)}$ and the load vector contribution $\left\{f_{WCB}\right\}_{ij}$ will be added to the load vector, $\left\{f\right\}_{W}^{(e)}$.

$$\left[K_{CB}\right]_{ij}=\begin{bmatrix} x & x & 0 & 0 \\ x & x & 0 & 0 \\ 0 & 0 & 0 & 0 \\ 0 & 0 & 0 & 0 \end{bmatrix}\begin{matrix} i \\ j \\ k \\ m \end{matrix} \quad \text{and} \quad \left\{f_{CB}\right\}_{ij}=\begin{Bmatrix} x \\ x \\ 0 \\ 0 \end{Bmatrix}\begin{matrix} i \\ j \\ k \\ m \end{matrix}.$$

where 'x' indicates the contribution, which is non-zero. Now, the final form of the energy equation can be written as

$$\left[C\right]_{E}^{(e)}\left\{\frac{T^{n+1}-T^{n}}{\Delta t}\right\}+\left[K\right]_{ET}^{(e)}\left\{T\right\}+\left[K\right]_{EW}^{(e)}\left\{w\right\}+\left\{f\right\}_{E}^{(e)}=0 \tag{5.46}$$

where $\left[C\right]_{E}^{(e)}=\int_{\Omega}\left[N\right]^{T}\left[N\right]dxdy$

$$\left[K\right]_{ET}^{(e)}=K_{1}\int_{\Omega}\frac{\partial N}{\partial x}\frac{\partial N^{T}}{\partial x}\left\{T\right\}d\Omega+K_{1}\int_{\Omega}\frac{\partial N}{\partial y}\frac{\partial N^{T}}{\partial y}\left\{T\right\}d\Omega$$

$$\left[K\right]_{EW}^{(e)}=K_{2}\int_{\Omega}\frac{\partial N}{\partial x}\frac{\partial N^{T}}{\partial x}\left\{w\right\}d\Omega+K_{2}\int_{\Omega}\frac{\partial N}{\partial y}\frac{\partial N^{T}}{\partial y}\left\{w\right\}d\Omega$$

In Equation (5.46), the time discretization can be carried out as discussed in Chapter 4 using finite difference method and mean value theorem. Crank-Nicholson time marching scheme will yield a stable solution for temperature distribution in the soil medium. However, instead of solving simultaneous equations using the semi-implicit method, a fully explicit scheme called mass lumping technique can be implemented. This method will give a solution for temperature at time level *(n+1)* directly using the values at time level *(n)*. The mass lumping technique is explained as below [9]. Equation (5.46) can be rewritten after assembling all the elements as

$$\left[C_{ij}\right]_{E}\left\{\frac{T_{j}^{n+1}-T_{j}^{n}}{\Delta t}\right\}+\left[K_{ij}\right]_{ET}\left\{T_{j}\right\}+\left[K_{ij}\right]_{EW}\left\{w_{j}\right\}+\left\{f_{j}\right\}_{E}=0 \text{ with indices, '}i, j\text{'}$$

for all the matrices, where 'i' indicates rows and 'j' indicates column. The above equation can be further simplified as

$$\left[C_{ij}\right]_{E}\left\{T_{j}^{n+1}\right\}=\left[C_{ij}\right]_{E}\left\{T_{j}^{n}\right\}-\Delta t\left[K_{ij}\right]_{ET}\left\{T_{j}\right\}-\Delta t\left[K_{ij}\right]_{EW}\left\{w_{j}\right\}-\Delta t\left\{f_{j}\right\}_{E}=0 \tag{5.47}$$

Let us expand the capacitance matrix $\left[C_{ij}\right]_{E}$ for a triangular element as

$\left[C_{ij}\right]_E = \begin{bmatrix} 2 & 1 & 1 \\ 1 & 2 & 1 \\ 1 & 1 & 2 \end{bmatrix}$. Mass lumping involves the procedure of adding up all the entries on a given row and assign the sum at the diagonal entry of the respective row. That means mass lumping of the above capacitance matrix takes the following final form.

$$\left[C_{ij}\right]_E = \begin{bmatrix} 2 & 1 & 1 \\ 1 & 2 & 1 \\ 1 & 1 & 2 \end{bmatrix} = \begin{bmatrix} 4 & 0 & 0 \\ 0 & 4 & 0 \\ 0 & 0 & 4 \end{bmatrix}$$

That is,

$$\sum_{j=1}^{nne}\left[C_{ij}\right] = \left[CL_{ii}\right] \tag{5.48}$$

where '*nne*' is the number of nodes per element.

After substituting Equation (5.48) in Equation (5.47), we get the expression for temperature at time level *(n+1)* as

$$\left\{T_j^{n+1}\right\} = \left\{T_j^{n}\right\} - \frac{\Delta t\left[K_{ij}\right]_{ET}}{\left[CL_{ii}\right]_E}\left\{T_j^n\right\} - \frac{\Delta t\left[K_{ij}\right]_{EW}}{\left[CL_{ii}\right]_E}\left\{w_j^n\right\} - \frac{\Delta t\left\{f_j\right\}_E}{\left[CL_{ii}\right]_E} \tag{5.49}$$

Similarly, the application of Galerkin's method and mass lumping technique on the moisture conservation equation given by Equation (5.26) will result in the following equation.

$\left[C\right]_W^{(e)}\left\{\dfrac{w^{n+1} - w^n}{\Delta t}\right\} + \left[K\right]_{WW}^{(e)}\left\{w\right\} + \left[K\right]_{WT}^{(e)}\left\{T\right\} + \left\{f\right\}_W^{(e)} = 0$. After application of the mass lumping technique, the above equation is simplified as

$$\left\{w_j^{n+1}\right\} = \left\{w_j^{n}\right\} - \frac{\Delta t\left[K_{ij}\right]_{WW}}{\left[CL_{ii}\right]_W}\left\{w_j^n\right\} - \frac{\Delta t\left[K_{ij}\right]_{WT}}{\left[CL_{ii}\right]_W}\left\{T_j^n\right\} - \frac{\Delta t\left\{f_j\right\}_W}{\left[CL_{ii}\right]_W} \tag{5.50}$$

where, $\left[C\right]_W^{(e)} = \int_\Omega \left[N\right]^T\left[N\right]dxdy$

$$\left[K\right]_{WW}^{(e)} = K_1\int_\Omega \frac{\partial N}{\partial x}\frac{\partial N^T}{\partial x}\left\{w\right\}d\Omega + K_1\int_\Omega \frac{\partial N}{\partial y}\frac{\partial N^T}{\partial y}\left\{w\right\}d\Omega$$

$$\left[K\right]_{WT}^{(e)} = K_2\int_\Omega \frac{\partial N}{\partial x}\frac{\partial N^T}{\partial x}\left\{T\right\}d\Omega + K_2\int_\Omega \frac{\partial N}{\partial y}\frac{\partial N^T}{\partial y}\left\{T\right\}d\Omega$$

Looking at Equations (5.49) and (5.50), one can easily notice that temperature and moisture content at any given node can be obtained as simple algebraic equations without the need to solve the simultaneous equations which is the great advantage of the mass lumping technique. However, this method has its own limitations, such as it cannot take large time steps. It is always preferable to use smaller time steps for the mass lumping technique, and the time step has to be determined by trial and error for a given problem. As the mass lumping method avoids the use of an equations solver, it is worth taking the effort to carry out a few trials in order to determine the time step which will give a stable solution.

5.2.5 Solution of Energy Equation for the Landmine

The energy equation for the landmine expressed by Equation (5.27) can be expressed as

$$\frac{\partial T_m}{\partial t} = \alpha \nabla^2 T_m \tag{5.51}$$

where the thermo-physical properties of the material of the landmine are assumed to be isotropic and do not depend on the temperature and moisture content of the soil. This assumption is valid because the variation in temperature and moisture content during the process of determining the thermal signature is small enough to cause changes in the properties of the material of the landmine and soil. Application of Galerkin's finite element method will result in the following equation.

$\int_\Omega [N]^T \left(\frac{\partial T_m}{\partial t} - \alpha \nabla^2 T_m \right) d\Omega = 0$. Expanding this equation, we get

$$\int_\Omega [N]^T [N] \left\{ \frac{T_m^{n+1} - T_m^n}{\Delta t} \right\} d\Omega - \left(\frac{k}{\rho c} \right)_m \int_\Omega \frac{\partial N}{\partial x} \frac{\partial N^T}{\partial x} \{T_m\} d\Omega ,$$

$$- \left(\frac{k}{\rho c} \right)_m \int_\Omega \frac{\partial N}{\partial y} \frac{\partial N^T}{\partial y} \{T_m\} d\Omega = 0$$

which can be written in matrix form as

$\left[C_{ij}\right]_m \left\{ \frac{T_j^{n+1} - T_j^n}{\Delta t} \right\}_m + \left[K_{ij}\right]_m \{T_j\}_m + \{f_j\}_m = 0$. After application of the mass lumping technique, the following final form of equation can be obtained for temperature variation of mine at time level, *(n+1)*.

$$\left\{T_j^{n+1}\right\}_m = \left\{T_j^n\right\}_m - \frac{\Delta t \left[K_{ij}\right]_m}{\left[CL_{ii}\right]_m} \left\{T_j^n\right\}_m - \frac{\Delta t \{f_j\}_m}{\left[CL_{ii}\right]_m} \tag{5.52}$$

where $\left[C\right]_m^{(e)} = \int_\Omega [N]^T [N] dxdy$

$$\left[K\right]_m^{(e)} = \alpha \int_{\Omega} \frac{\partial N}{\partial x} \frac{\partial N^T}{\partial x} \{T_m\} d\Omega + \alpha \int_{\Omega} \frac{\partial N}{\partial y} \frac{\partial N^T}{\partial y} \{T_m\} d\Omega$$

As all the four sides of the landmine are assumed to be subjected to heat flux boundary condition with the soil medium, there will not be any contribution to the stiffness matrix. However, the load vector $\{f_j\}_m$ will have the following contribution.

$$\{f_{qs}\} = \frac{q_s}{(\rho C)_m} \int_{\Gamma} [N]^T d\Gamma.$$

The energy and moisture conservation equations for soil medium and the energy equation for the landmine are finally expressed for the solution of temperature and moisture content of soil and temperature of mine by Equations (5.49), (5.50) and (5.52) respectively. The conservation equations for soil medium expressed by Equations (5.49) and (5.50) are coupled with each other, hence, the computational algorithm has to take care of this coupling while solving these two equations. Following is the computational algorithm for the solution of the above equations to obtain temperature and moisture distributions within the soil medium.

5.2.6 Computational Algorithm

The final form of the finite element solution for equations expressed by Equations (5.49) and (5.50) for energy and moisture conservation of soil and Equation (5.52) for energy conservation of landmine have to be solved to obtain temperature and moisture content variation within soil and temperature distribution within landmine respectively. It has to be observed that both the conservation equations for soil medium are coupled with each other through temperature and moisture content and hence a separate iterative procedure has to be followed to resolve the coupling. Figure 5.8 shows the flow chart of the computational algorithm for the solution of the field variables for both the soil medium and the landmine.

In this figure, the abbreviations are expanded as: EES – energy equation of soil, MES – moisture conservation equation of soil and EEM – energy equation of landmine, T_s, w_s, T_m – temperature, moisture content of soil, and temperature of landmine respectively. After the formation of the global matrices, the boundary conditions are included in the respective stiffness and load vector matrices. Then the outermost loop called the time loop starts for time iteration. All the field variables are initialized before the start of the time loop for the computations. In the landmine detection problem, the top surface of the soil medium is exposed to convective boundary conditions in which the ambient temperature and solar irradiation continuously vary with time over a diurnal cycle. The main purpose of the present simulation is to determine the thermal signature of the soil medium and landmine combination. As was discussed in section 5.2.4 for incorporating the boundary conditions on the top

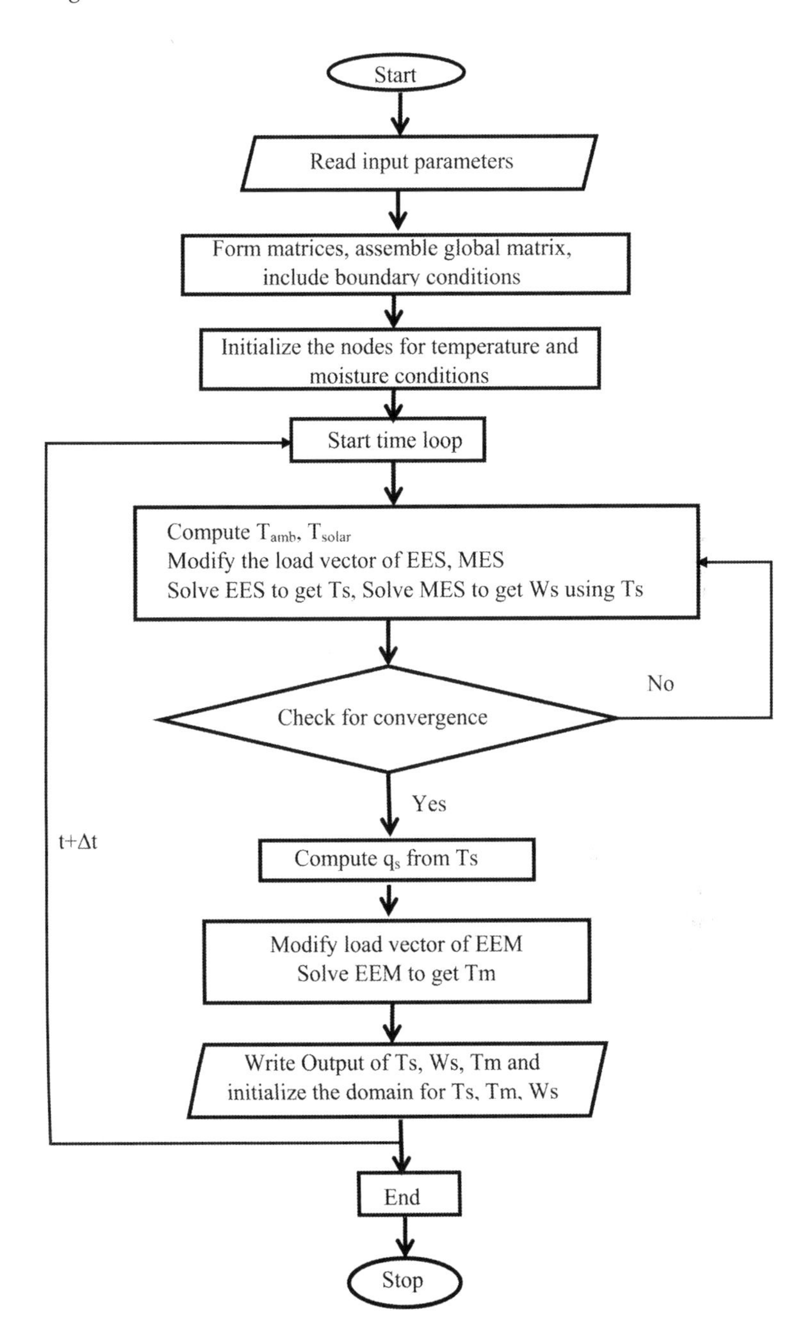

FIGURE 5.8 Flow chart for landmine detection simulation.

surface of the soil, the stiffness matrix of energy and moisture conservation equations remain constant with time with the assumption of a constant convective heat transfer coefficient. However, the load vector for both energy conservation and moisture conservation equations will change with time according to the ambient temperature, which is computed using Equation (5.31) and solar heat flux is computed using Equation (5.32). Hence, within the time loop, the load vectors of conservation equations of soil medium have to be updated based on the ambient temperature, density of water vapor in the ambient and solar incident heat flux on the top surface of the soil medium. Once the temperature and moisture content of the soil medium are converged, then the heat flux required for the boundary condition of the landmine equation can be computed and this quantity is indicated by q_s which is represented in the load vector of the energy equation of the landmine. With the heat flux boundary conditions, the energy equation for the landmine will be solved to get the temperature distribution within the landmine. It is to be noted that the temperature distribution within the mine is of not much importance because the main objective of the present landmine detection analysis is to predict the thermal signatures of the soil medium in the presence of the mine. Sometimes, conduction within the mine also can be neglected; in such cases there is no need to solve the energy equation for the mine. The present formulation is focused on conducting the mine case, however, in the present analysis, landmines with very small thermal conductivity are considered. The simulations are carried out for 24 hours duration with time steps of 30 s, because when the mass lumping technique is employed, larger time steps will result in numerical instability issues.

5.2.7 Computer Programming

The very purpose of this textbook is to motivate the readers to develop their own programs wherever possible. When someone develops his/her own computer program, he/she should be thorough enough to have good insight into the physics of the problem, and writing a program definitely gives this confidence to the individual. After completing a computer code, when it is executed to get the results, the real learning starts because many times there will be some initial issues related to syntax, incorrect typing, mistakes in logic etc. However, by patiently debugging the code, the individual will taste the final fruit of the efforts they put in developing the code. Anyone can easily appreciate, after developing a few codes and successfully getting the results, that it is not impossible to write programs even for the solution of Navier-Stokes equations. With these points in mind, the basic steps in programming the present problem are discussed in this section. It is always a good practice to split the total computational procedure into a number of logical steps; here logical steps mean the sequence of steps of computations by which the program is expected to execute. Any computer program for the solution of a given problem will consist of main three sections: pre-processing, main body and post-processing, which will be explained in detail below. In each section of the above divisions of programs, there may be some more steps required to be completed. Generally, the concept of a subroutine is employed to carry out computations that may be required or called upon a number of times during the entire simulation program. Subroutines can also be used even when they are not required to repeat the computations within the main program, with the sole aim of developing a well-structured program. The following are the various subroutines that will be used for the simulation of landmine detection problems.

I Pre-processing stage – Read input. *Subroutine input:* dimension of domain, mesh details, thermo-physical properties of soil, mine, initial values of T_s, w_s and T_m, constants required for computing T_∞, solar flux.

Subroutine inshiat: initialize the field variables, T_s, w_s and T_m

II Main body – *Subroutine matrices.* Formation of all the matrices in the solution of EES, MES, EEM, assembly of stiffness matrices

This subroutine uses subroutines shapeld, shape2d, shapeder and invdet

Subroutines shaeld, shape2d store the shape functions for the quadrilateral element in natural coordinates

Subroutines shapeder and invdet – perform coordinate transformation

Subroutine loads – form the load vectors for EES and MES

Subroutine loadm – form the load vector for EEM

Subroutine iterat – starts the time loop, computes solar heat flux, ambient temperature, updates the load vectors for EES and MES and solves for T_s and w_s by iterations.

Computes load vector for EEM and solves EEM for T_m. Initialize the field variables, then go to the next time step and end the time loop.

This subroutine uses the *subroutine solver* for the solution of simultaneous equations.

III Post-processing – *Subroutine output.* Writes the output of all the field variables in different output files for making plots. With the help of many kinds of graphing software, various plots such as *x-y* plots and contour plots of temperature and moisture can be drawn at different time levels.

5.2.8 Simulation Parameters and Results

The basic objective of the present simulation program has to be clearly defined before proceeding to run the computer code for results. During the formulation of the problem, it has been observed that there are many geometrical, physical and ambient parameters that decide the nature of variation of thermal signature. For example, the size of the landmine influences the temperature distribution on the ground surface and hence the thermal signatures. Thermo-physical properties of soil medium and landmine material play an important role in temperature and moisture distributions. As this problem involves the effect of ambient temperature and solar radiation falling on the soil, these uncontrolled parameters are crucial for understanding the behavior of thermal signatures. The depth at which the mine is buried is a very significant parameter that affects the nature of thermal signature. The temperature variation of soil also depends on the amount of moisture content present in the soil. In the present method of detecting the landmine buried under the ground, an infrared camera will be used to scan the temperature distribution over the surface of the ground where landmines are suspected to have been buried. Hence, with the variation of many influencing parameters, the thermal signatures can be obtained, and by comparing these with the thermal images obtained using the infrared camera and by trial and error method, the depth can be estimated. After deciding the simulation parameters, their range of variation has to be decided based on results available in the literature.

A systematic simulation program has to be evolved to obtain results for the variation of different simulation parameters. From computations, the field variables at different node points will be obtained for both the soil medium and landmine. Now, the type of presentation of results has to be identified for better clarity while interpreting the results. The output of the simulation program should be able to highlight the physics underlying the problem and also it should give direction on how to improve the detection methodology using both infrared images and thermal signature results. It is always a good idea to have a technical debate on how one is going to interpret the final results and accordingly the presentation of results in different modes can be decided. Presentation of results in tabular form, *x-y* plot variation, contour plots etc. are some of the modes of illustrating the output results from the simulation program. In the present problem, the depth of landmine buried below the ground, size of mine and moisture content are considered as the simulation parameters.

5.2.9 Discussion of Simulation Results

The very purpose of the present simulation program is to determine the thermal signatures of ground under which a landmine is buried and this thermal signature has to be computed for a diurnal cycle. Thermal signature is defined as the temperature difference observed between the temperature noted on the surface of ground just above the mine and an adjacent point on the surface of ground away from the mine. As the presence of a mine will affect the heat transport from the ground surface towards its bottom, these two temperatures will not be the same. Depending on the size of mine, depth of mine from the ground surface, moisture content of soil medium and solar irradiation falling on the ground, this temperature variation will differ. Initially mesh sensitivity and validation results are obtained for the geometric parameters considered in the present simulation program.

5.2.9.1 Mesh Sensitivity and Validation Results [10]

Numerical results are obtained for a rectangular domain of 1 m × 0.5 m (1 m length and 0.5 m height) for the soil medium and 10 cm × 5 cm size landmine buried at a depth of 4 cm below the ground. For the purpose of mesh sensitivity analysis, three meshes of size, 64×40 (Mesh 1), 74×48 (Mesh 2), and 80×52 (Mesh 3) were considered. Figure 5.9 shows a typical mesh obtained using a transfinite interpolation technique (TFI).

The time variation of temperature at a surface node on the soil above the mine and temperature distribution along the surface of the soil medium (along the length) at midnight are recognized as results for comparison for different meshes. Figures 5.10(a) and 5.10(b) show the time variation of temperature and temperature distribution respectively. It can be noticed that the simulation results obtained for

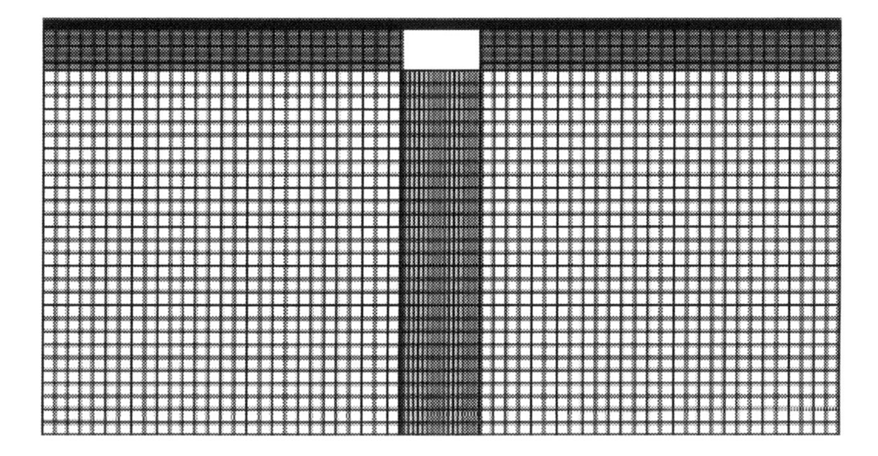

FIGURE 5.9 Computational mesh.

temperature converge as the mesh size is refined from coarse mesh (Mesh 1) towards fine mesh (Mesh 3) in both the figures.

As the results obtained with Mesh 2 and Mesh 3 almost coincide with each other, Mesh 2 is selected for further computations. The convergence of results for temperature with refinement of mesh is the proof that beyond Mesh 2, the numerical results become independent of size of discretization of the spatial domain. Mesh 2 consists of 3552 elements and 3675 nodes.

For the validation of the computer code, the wet bulb temperature attained during evaporative drying of the soil surface is considered for different relative humidity values. In this case the temperature of the soil and the drying medium (air) temperature are assumed to be equal. The bottom surface is assumed to be adiabatic for heat and moisture transfer and the rest three sides are subjected to convective boundary conditions. Simulation results were obtained for 40 hours of drying. Initially the temperature of the soil decreases because the latent heat of vaporization is taken from the soil itself and it continues to decrease until the soil reaches the wet bulb temperature corresponding to the temperature and relative humidity of the air. During initial drop in temperature of the soil, the temperature difference between the air and the soil increases, thus heat starts flowing from the drying medium (air) to the soil. When the soil reaches the wet bulb temperature, maximum heat transfer from the air takes place and hence after this point, the soil starts heating up to reach the temperature of air. At 30% relative humidity and 303 K temperature, the minimum temperature attained during the evaporative cooling, as predicted by the computer code, is 290.4 K which is very close to the wet bulb temperature, 291 K obtained from psychrometric chart. Since the evaporation takes place only at the exposed surfaces, the surface temperature decreases faster, and it attains the wet bulb temperature. Table 5.1 shows the comparison of results for wet bulb temperature obtained from numerical simulation with the values obtained from the psychrometric chart.

As the relative humidity increases, the evaporative cooling decreases, resulting in the increase in the wet bulb depression as seen from Table 5.1.

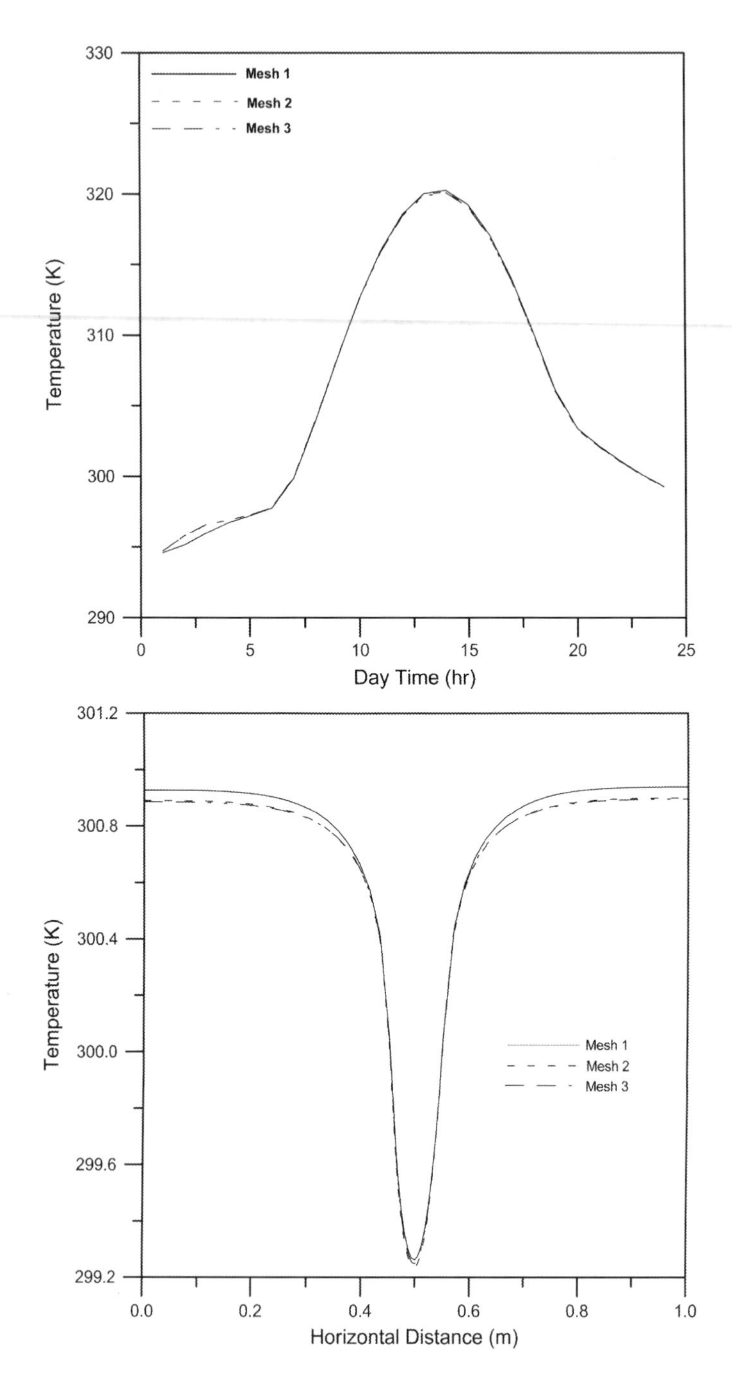

FIGURE 5.10 (a) Time variation of temperature at a surface node on the soil and (b) temperature distribution along the surface of soil medium at midnight.

TABLE 5.1
Wet Bulb Temperature Comparisons

Sl. No.	Dry Bulb Temperature (DBT) (K)	Relative Humidity ϕ(%)	Wet Bulb Temperature (WBT) (K)		% of Error
			Simulation	Psychrometric Chart	
1	303	30%	290.4	291	0.2
2	303	50%	294	295	0.3
3	303	70%	297.2	298.5	0.4

TABLE 5.2
Properties of Soil Medium Used in Simulation

Properties	Soil
ρ(kg/m^3)	2000
C (J/kg K)	837.2
k (W/m K)	2.5116
D_{ml}(m^2/s)	1.0e-12
D_{mv}(m^2/s)	1.0e-12
D_{Tl}(m^2/K s)	1.0e-12
D_{Tv}(m^2/K s)	1.0e-12

5.2.9.2 Simulation Results [10]

The effect of depth of mine buried under the ground, moisture content of the soil medium and size of mine are considered for the simulation. Thermal signatures were computed for the variation of the above parameters and were plotted against time during a diurnal cycle. The data related to soil medium and landmine are depicted in Tables 5.2 and 5.3 respectively. The other input data required for computations are shown in Table 5.4.

Effect of Depth and Size of Mine

One can visualize that as the depth of the mine increases, the strength of thermal signature decreases and similarly, the images taken by the infrared camera also will become weak, because at deeper depths, the presence of a mine will not produce any effect on heat transfer in the soil. The strength of the thermal signature creates the difference between the two temperatures at the soil surface. The higher this temperature difference, the stronger the thermal signature is. Therefore, it becomes important to investigate the effect of depth on thermal signature at the soil surface and to identify the maximum depth after which no thermal signature appears at the

TABLE 5.3
Properties of Mine Used in Simulation

Properties	Mine
ρ(kg/m^3)	1560
C (J/kg K)	1370
k (W/m K)	0.2234

TABLE 5.4
Input Parameters Used in Simulation

Parameters	Value
Number of nodes	3675
Number of elements	3552
Time step (sec)	0.5
Convective heat transfer coefficient, h_c (W/m^2 K)	15
Mass transfer coefficient, h_m (m/s)	0.015
Average ambient temperature, T_{avg} (K)	302
Amplitude of temperature variation from the average, A_m (K)	8
Relative humidity (%)	40

soil surface. In the present simulation program, computations were performed using three different sizes of landmines, 10 cm×5 cm, 16 cm×8 cm and 20 cm×10 cm buried at various depths of 1, 2, 4, 6, and 8 cm below the ground at 13% moisture content of the soil medium. The variations in thermal signatures for different depths of mines are displayed in Figures 5.11(a), 5.11(b) and 5.11(c) respectively. It can be noticed that as the depth of mine from the soil surface increases, the strength of thermal signature decreases; this means the difference in temperatures between two adjacent points decreases, and it becomes very weak at a depth of 8 cm. In the present analysis a TNT (trinitrotoluene) mine is taken as mine material for which the thermal conductivity is very low (k_{TNT}=0.2234 W/m K).

Since the thermal conductivity of the mine material is lower than that of soil the thermal resistance to heat transfer increases along the length of the path containing the mine. Therefore, heat transfer through the soil above the mine faces higher thermal resistance compared to the surrounding soil without the effect of mine. For the above reason, the upper surface temperature of soil above the mine increases during the heating period, thus providing a visible thermal contrast at the soil surface. It is also clear from the plots that the depth of mine has an influence on the time of occurrence of the maximum thermal signature at the surface and on the maximum amplitude of difference, and this maximum amplitude decreases as the depth of mine from the surface increases, and its peak value shifts during the heating phase towards later hours of the day. This time shift is mainly due to the higher storage capacity of the

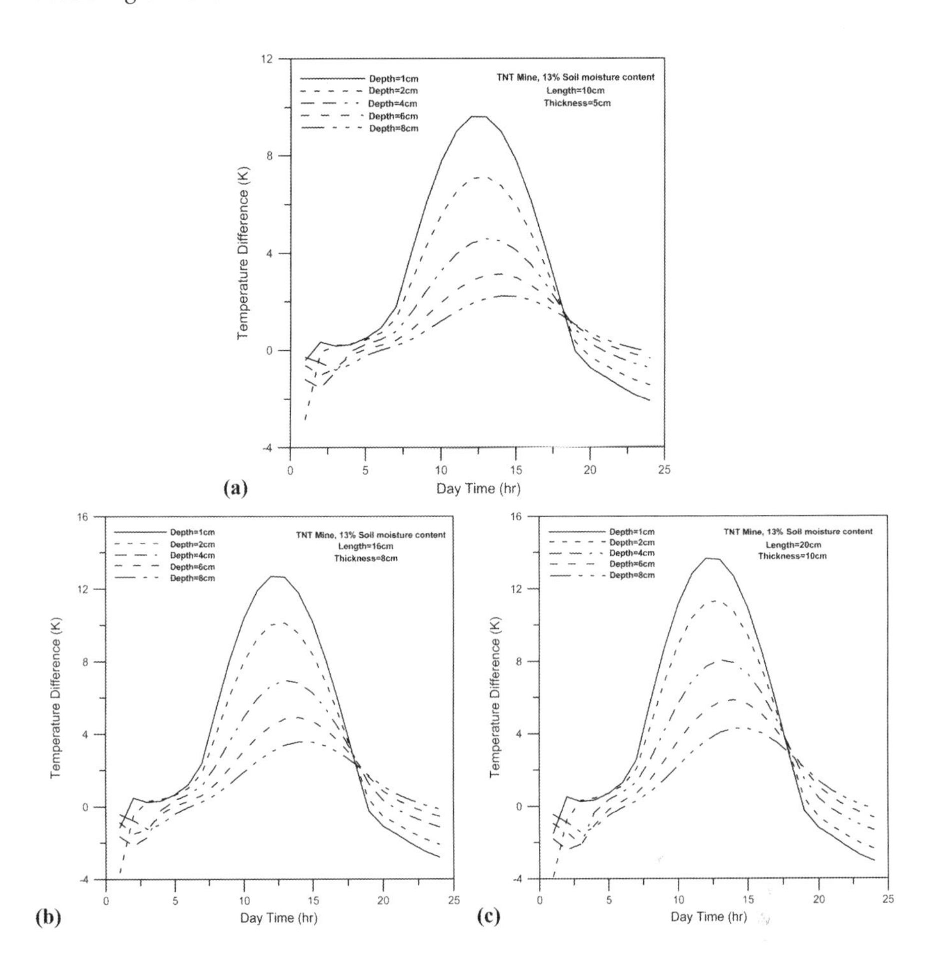

FIGURE 5.11 Effect of depth of mine for mines of (a) 10 cm × 5 cm, (b) 16 cm × 8 cm and (c) 20 cm × 10 cm.

soil above the mine, causing a slower thermal response to heating. It is interesting to observe from Figure 5.11 that with an increase in the size of the landmine the thermal signatures become stronger for any given depth of mine. A larger surface area as thermal resistance for the flow of heat through the soil, so thermal contrast increases. Due to the increase of mine width from 10 cm to 16 cm, a maximum difference in the thermal signature increases by around 3°C.

Effect of Moisture Content of Soil Medium

The variation of surface temperature of soil for different moisture content of the soil is shown in Figures 5.12(a), 5.12(b) and 5.12(c). Simulation results were obtained for three different moisture contents of soil: 5%, 10% and 13% for mines at 1 cm depth. Figure 5.12 clearly indicates that the surface temperature distribution of the soil at different times of the day varies with different moisture content of the soil.

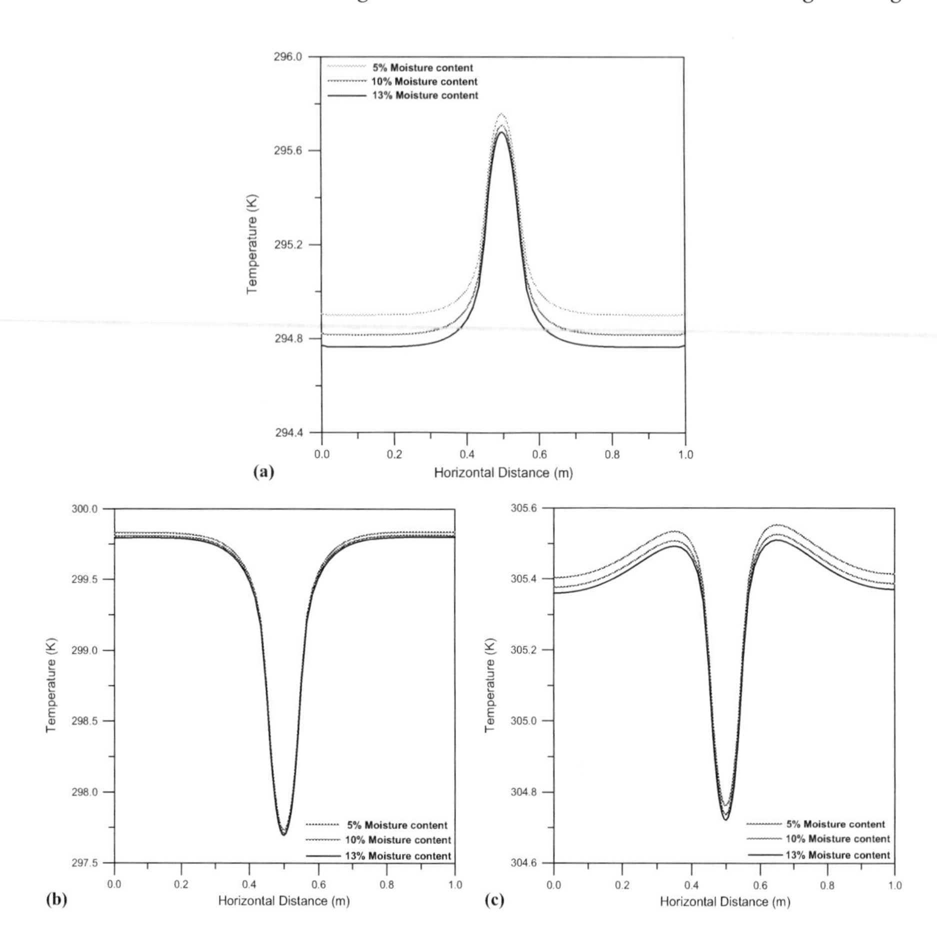

FIGURE 5.12 Temperature distribution along the length of soil medium at the surface at (a) 6 AM, (b) 8 PM and (c) 0.0 AM (midnight).

At early morning, soil with less moisture content shows higher surface temperature value compared to the soil with higher moisture content. This is because for lower moisture content of soil, the evaporative effect is not predominant and heating of the soil surface starts faster with solar radiation. Similarly, at night as the mine offers thermal resistance to the upward conductive heat transfer, the surface temperature of the soil above the mine is less compared to the surrounding, and soil with 5% moisture content shows a higher value of temperature. At the heating period, the surface temperature increases with decreasing moisture content. This phenomenon is expected because the higher the wetness of the soil, the more heat will be required for evaporation. Consequently, less heat will be available to heat up the soil and increase its temperature. This behavior is also attributed to the competing effects of soil thermal conductivity and specific heat, as both increase with an increase in the moisture content as the trapped air in the pores is replaced by water. In addition to this, the thermal diffusivity of soil which is the ratio of thermal conductivity to volumetric heat capacity depends heavily on the moisture content of the soil.

Temperature Contours of Soil Medium with Mine at Different Depths

Figures 5.13 and 5.14 show the temperature contours of the soil due to the presence of 10 cm width and 5 cm height mine at 2 cm and 4 cm depth respectively of the soil domain at early morning 6 AM and at midnight. From these figures, it is clear that the temperature near the heat-conducting mine is higher at daytime (6 AM) compared to other regions of the domain. The temperature contours become perpendicular to the two vertical side walls which are assumed to be adiabatic. Thus the present predictions satisfy the expected physics underlying the problem. The bottom surface remains at a constant temperature due to the enforced Dirichlet boundary condition.

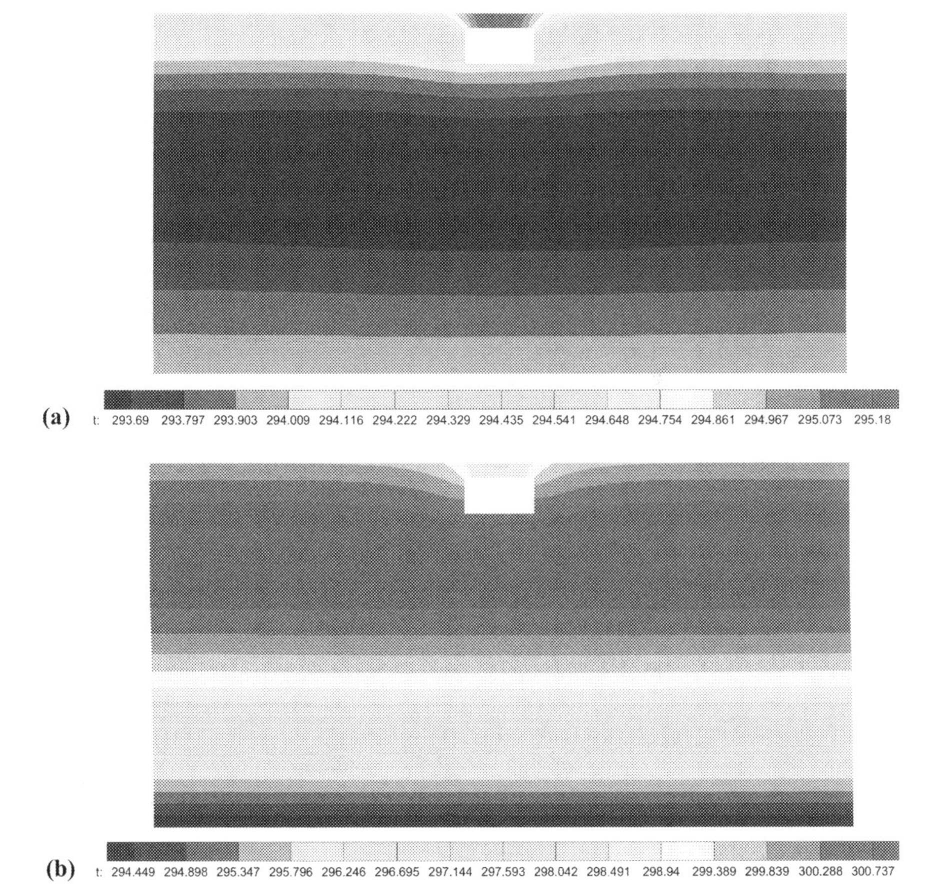

FIGURE 5.13 Temperature contours of soil medium with mine at a depth of 2 cm at (a) 6 am and (b) midnight.

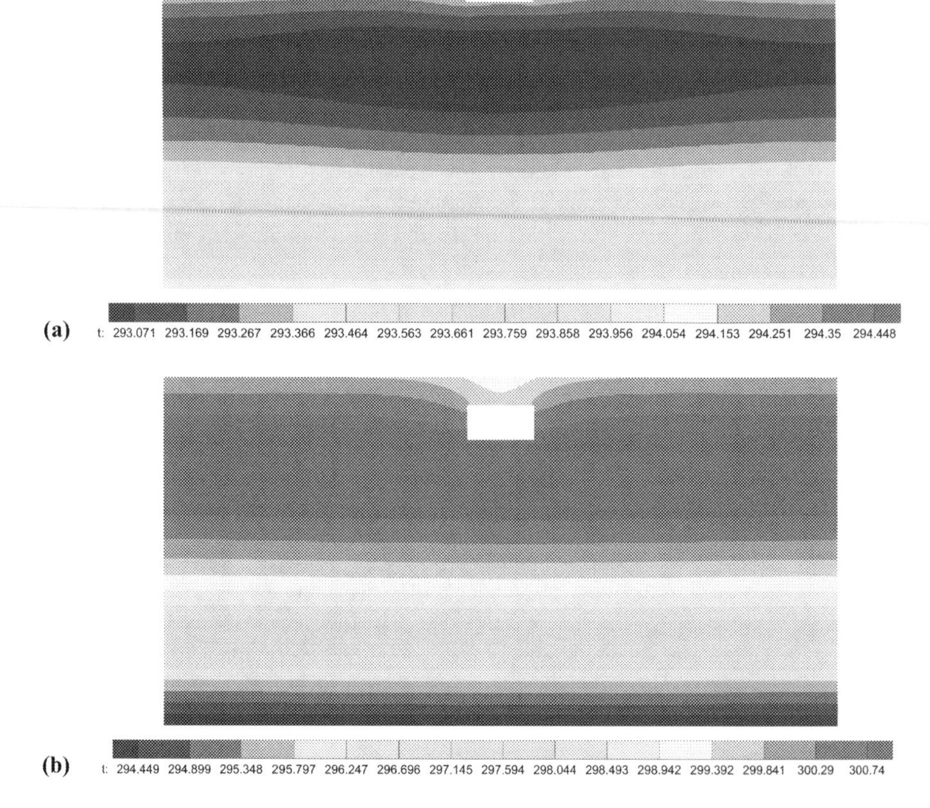

FIGURE 5.14 Temperature contours of soil medium with mine at a depth of 4 cm at (a) 6 am and (b) midnight.

Comparing Figure 5.13 with Figure 5.14, it can be noticed that for the mine buried at a lower depth of soil, the surrounding temperature around the mine is higher compared to the higher depth of mine as expected. The surface temperature of the soil above the mine is higher than the surface temperature away from the heat source. This variation of temperature provides the idea about the location of heat-conducting mine. As described previously, during the heating period the temperature above the mine surface remained higher compared to the surrounding soil due to its low thermal conductivity value. Similarly, at midnight compared to the top surface, the bottom surface of the mine remained at a higher temperature as clearly shown in the above figures.

In the present simulation problem, the thermal signatures obtained with numerical computations will be used along with infrared thermal images in order to detect the exact location and depth of mine. In a real situation, in order to locate the landmine buried under the ground, all the three coordinates, that is x, y and z of the landmine, have to be estimated so that the landmine can be safely removed. Let us assume the *x-y* coordinate plane represents the ground surface and *z*-coordinate indicates the depth of ground from the surface. Initially, the position of the landmine in the *x-y* plane has to be determined and then the depth also. In the present simulation work, two-dimensional computational domain in the *z*-direction only is considered. Using the proposed mathematical model, the thermal signatures of the ground with landmine can be computed. In a given location, the soil and mine properties should be available so that these data can be used for simulation. Assuming different depths of mine, thermal signatures can be simulated, and using a thermal imaging technique where the mines are suspected to be buried, thermal images of the ground on the *x-y* plane will be obtained. Analysis of these images will provide some data on thermal signatures, which can be compared with the simulated signatures. During the initial trials, these two thermal signatures may not match with each other, and the numerical computations will be repeated with varying depths of mine to obtain different sets of thermal signatures. These trials can be repeated until the simulated thermal signatures match with those obtained from thermal imaging techniques. Now, the *x-y* plane on which the thermal images are obtained is known from the experimental trials, whereas the depth is estimated using the numerical simulation. Thus, by making use of these two techniques in tandem, the exact location of the landmine can be detected without causing any harm to the personnel involved in de-mining process.

REFERENCES

1. Moukalled, F. and Saleh, Y. Heat and mass transfer in moist soil, Part I. Formulation and testing, *Numerical Heat Transfer, Part B*, 2006; 49: 467–486.
2. Muscio, A. and Corticelli, M. A. Land mine detection by infrared thermography: reduction of size and duration of the experiments, *IEEE Transactions on Geoscience and Remote Sensing*, 2004: 1–10.
3. Moukalled, F. and Saleh, Y. Heat and mass transfer in moist soil, part I. Formulation and testing, *Numerical Heat Transfer, Part B*, 2006; 49: 467–486.
4. Moukalled, F., Ghaddar, N., Kabbani, H., Khalid, N. and Fawaz, Z. Numerical and experimental investigation of thermal signatures of buried landmines in dry soil, *ASME*, 2006; 128: 484–494.
5. Moukalled, F., Ghaddar, N., Saleh, Y. and Fawaz, Z. Heat and mass transfer in moist soil, part II. Application to predicting thermal signatures of buried landmines, *Numerical Heat Transfer, Part B*, 2006; 49: 487–512.
6. Kallel, F., Galanis, N., Perrin, B. and Javelas, R. Effects of moisture on temperature during drying of consolidated porous materials, *ASME*, 1993; 115: 724–733.
7. Murugesan, K., Seetharamu, K. N. and Aswatha Narayana, P. A. A one-dimensional analysis of convective drying of porous materials, *Heat and Mass Transfer*, 1996; 32: 81–88.
8. Garcia-Padron, R., Loyd, D. and Sjökvist, S. Heat and moisture transfer in wet sand exposed to solar radiation—models and experiments concerning buried objects, *Subsurface Sensing Technologies and Applications*, 2002; 3(2): 125–150.
9. Murugesan, K., Lo, D. C., Young, D. L., Chen, C. W. and Fan, C. M., Convective drying analysis of three-dimensional porous solid by mass lumping finite element technique, *Heat and Mass Transfer*, 2008; 44: 401–412.

10. Nag, P. K. *Numerical prediction of thermal signatures of buried heat conducting landmines*, MTech. Thesis, Department of Mechanical & Industrial Engineering, Indian Institute of Technology Roorkee, Roorkee, India.

EXERCISE PROBLEMS

Qn: 1 Consider a wall of 0.5 m thickness having uniform internal heat generation of 2500 W/m^3. The wall has variable thermal conductivity k=25+0.015T where T represents temperature in degree Celsius. The density of the wall is 2000 kg/m^3 and specific heat is 10J/kg K. Given that temperature distribution across the wall is $T(x)=200+50x+90x^2$, determine the net rate of heat transfer.

Qn: 2 Water having an initial temperature of 20°C is used to cool a hot fluid at a temperature of 120°C. Water is flowing at the rate 10 kg/s while the hot fluid has a flow rate of 5 kg/s and specific heat 4000 J/kg K. For overall heat transfer coefficient of 2500 W/m^2K and heat transfer area 10 m^2, determine outlet temperatures for both the fluids.

Qn: 3 A square block of cross-sectional area 1 m by 1 m having thermal conductivity 2 W/m-K is exposed to a temperature of 600 K on all its sides, except the bottom side exposed to 500 K. Divide the domain into equal grid space of 0.25 m and determine the temperature distribution across the block using the finite difference method.

Qn: 4 For the above problem replace the Dirichlet boundary condition for the bottom face with convective boundary condition with a heat transfer coefficient of 15 W/m-K and temperature 25°C to determine the temperature distribution across the square block.

Qn: 5 A 5 mm wire has an outside temperature of 250°C and thermal conductivity of 25W/m-K. Given that internal heat generation in the wire is 400 W/m^3, determine the temperature distribution across the wire in the radial direction. Take a grid size of 1.25 mm to obtain the solution numerically.

Qn: 6 A nuclear rod having a length of 20 mm and thermal conductivity of 80 W/m-K is covered with steel cladding on both sides having a thickness of 5 mm and thermal conductivity of 20 W/m-K. The rod is insulated at one end and subjected to convective cooling at the other end with a fluid having a temperature of 30°C and a heat transfer coefficient 500 W/m^2K. If the heat generated by the nuclear fuel is 2MW/m^3 determine the temperature distribution within the rod.

Qn: 7 A copper rod is exposed to ambient with a heat transfer coefficient 10 W/m^2K and ambient temperature 35°C. The thermal conductivity of the rod is 380 W/m-K and the base temperature is 100°C. If the rod dissipates heat by convection from its tip determine the temperature distribution in the rod numerically.

Qn: 8 For the above problem determine the temperature profile across the length of the fin for different material: (a) steel, (b) gold and (c) silver.

Qn: 9 Solar radiation takes place on a concrete block such that net heat transfer is 0.2 MW/m^2. If the initial surface temperature is 30°C, determine the temperature distribution across the surface after 150 secs using the explicit finite difference scheme. Take the length of the domain as 250 mm and grid spacing as 50 mm.

Qn: 10 Solve the above problem using the implicit time marching scheme for an initial surface temperature of 25°C and compare the results with analytical solution.

QUIZ QUESTIONS

Qn: 1 The law governing the phenomena of conduction heat transfer is ____________________.

Qn: 2 Write the governing differential equation for one-dimensional steady state heat transfer with internal heat generation.

Qn: 3 Write the analytical solution for temperature distribution in one-dimensional steady state heat conduction without internal heat generation.

Qn: 4 Constant heat flux boundary condition is of the type:
(a) Dirichlet (b) Neumann (c) mixed (d) none

Qn: 5 Poison equation follows the type of differential equation:
(a) elliptical (b) parabolic (c) hyperbolic (d) none

Qn: 6 Lumped parameter analysis for transient heat conduction is valid for ____________________ condition.

Qn: 7 Extended surfaces used for enhancing heat dissipation are called ____________________

Qn: 8 Thermal diffusivity is a measure of heat absorbing capacity of material:
(a) true (b) false

Qn: 9 In the finite difference method, implicit methods converge faster as compared to explicit methods: (a) true (b) false.

Qn: 10 The two-dimensional steady state heat conduction equation without internal heat generation is known as Laplace equation: (a) true (b) false.

6 Modeling of Flow Problems

Flow problems find wide engineering applications in the area of atmospheric science, aerodynamics, cooling of electronic devices, bio-engineering fields etc. just to mention a few. These problems can be classified based on the variation of field variables, as was discussed for heat transfer problems. As temperature gradient in a domain decides the conduction heat transfer, so does the pressure gradient for flow of fluid in a flow domain. Flow problems can be classified as incompressible or compressible flow based on the variation of density of the fluid medium with pressure or pressure and temperature. Different type of flow fields are obtained for flow of fluid through conduits or internal flow and flow over solids. In reality, all flow problems are three-dimensional in nature, however, the significant variation of pressure gradient decides the flow field as one-dimensional, two-dimensional or three-dimensional. It is worth making clear the distinction between the terminologies, fluid medium and flow field. Fluid medium refers to the specific fluid under consideration whose velocity and pressure field are being computed, whereas the flow field refers to the resulting velocity and pressure fields obtained as a result of flow of the given fluid medium. The given fluid medium may be a single fluid such as water, air or it may be a mixture of gases such as exhaust gases from an internal combustion engine. The characteristic of the given fluid medium has to be well defined before proceeding to obtain the flow field. In most flow problems, especially in computational fluid dynamics problems, the main objective of simulation is to obtain fluid friction as a result of flow of fluid over some surfaces of the solids, or to compute the pumping power required to achieve the continuous flow of the fluid medium through the given conduit. In the case of flow analysis through turbomachines, the drag force and lift force exerted by the given shape of blades may be investigated. In flow problems, the conservation equations generally consist of mass conservation and momentum conservation. If heat transfer is also studied, then energy conservation equation also will be solved along with the flow equations. In the present scenario where good knowledge and awareness exist about computational fluid dynamics (CFD), many complex real-life problems can be solved using commercial software. In this chapter two types of flow problems are discussed with the aim of encouraging readers to develop their own computer program to gain self-confidence.

6.1 FLUID MECHANICS – FILLING OF WATER TANK

Analysis of flow problems requires the solution of the continuity and momentum equations for the given flow domain and such type of problems are generally studied under computational fluid dynamics. In reality, not all problems related to fluid flow involve the solution of continuity and momentum equations. A simple flow problem

DOI: 10.1201/9781003248071-6

that everyone comes across today is discussed in this section. Filling and discharge of water from a water tank is a common problem. The storage water tank size is decided based on the requirement of water for use in a day, and accordingly the size of pipe and capacity of pump are decided. However, the fluid mechanics behind such problems can be analyzed by modeling this problem using mass conservation and momentum equation in simpler terms in the following section. With the help of a mathematical model for such a problem, a simulation program can be developed using which the effect of various parameters can be analyzed.

6.1.1 Derivation of Mass and Momentum Conservation Equations

Figure 6.1 depicts a water tank of a certain capacity to hold water and discharge the same for utility. Water discharges through the outlet orifice pipe at the rate of $\dot{Q}_{out}$ with velocity, V. It is assumed that $h(t)$ is the height of water level in the tank at any given time, A_o and A_t are the cross-sectional area of the outlet orifice pipe and the water tank respectively. The main objective of the analysis is to determine how the velocity of exit water changes with a change in height of the water level in the tank. Hence, it is very clear that the problem under consideration is a transient problem. From the conservation of mass principle, one can write the following statement.

$$\begin{aligned}&\left(\text{Rate of change of mass of water in the tank}\right)= \\ &\qquad -\left(\text{Mass of water leaving the tank through the orifice pipe}\right)\end{aligned} \tag{6.1}$$

This was developed in basic thermodynamics for an open system as

$$\frac{dM_{CV}}{dt} = \dot{m}_{in} - \dot{m}_{out} \tag{6.2}$$

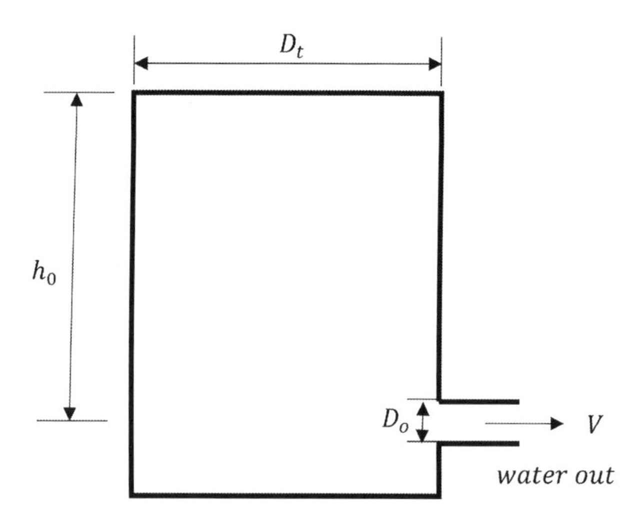

FIGURE 6.1 Schematic diagram for emptying tank problem.

In the above equation, $\dot{m}_{in} = 0$. After substituting the relevant variables in the above equation, one can get the equation as

$\rho \frac{dQ_t}{dt} = -\rho \dot{Q}_{out}$. As density, ρ is assumed constant, this equation can be further modified as

$$A_t \frac{dh(t)}{dt} = -A_o V \tag{6.3}$$

Equation (6.3) is the mass conservation equation. Now, the above equation can be solved once the value of velocity is known. It is known from basic fluid mechanics, that the velocity of water coming through the orifice, V is a function of water level, *h(t)* at any given time. The relation between velocity and height of water can be obtained from the momentum equation. When the Euler equation is simplified for an incompressible fluid, then the equation is reduced to Bernoulli's equation as written below.

$\frac{p}{\rho g} + \frac{V^2}{2g} + z = C$. Expressing the pressure term in terms of height of fluid, *h(t)* this equation is reduced to the following expression.

$h + \frac{V^2}{2g} + z = C$, where $p = \rho g h$. Substituting this equation between the datum level considered at the center of the orifice pipe at the exit of the tank and at any height, *h(t)* of water level in the tank, the following expression is obtained. It is assumed that the water at height *h(t)* does not have any velocity and is assumed to be zero. Now, assuming the water coming out of the orifice is the reference height, and also the water tank is vertical, we can express the fluid velocity as

$$V = C_d \sqrt{2gh} \tag{6.4}$$

where C_d is the coefficient of discharge of the orifice pipe. Substituting Equation (6.4) in Equation (6.3), the expression for height of water in the tank, *h(t)* is given as

$$\frac{dh(t)}{dt} = -C1\sqrt{h(t)} \tag{6.5}$$

where $C1 = \frac{A_o C_d}{A_t}\sqrt{2g}$

Equation (6.5) can be solved either for *h(t)* as a function of velocity or vice versa, in order to understand how the discharge from the tank varies according to the variation in the height of water.

6.1.2 Boundary Conditions and Initial Conditions

It can be observed that Equation (6.5) is a first order linear differential equation with time as the independent variable, and it requires an initial condition to solve the equation. As there is no spatial derivative in the above equation, no spatial boundary conditions need to be specified. The initial condition can be assumed as

$$@t = 0,\ h(t) = h_0$$

Using this above initial condition, Equation (6.5) can be solved analytically as follows.

6.1.3 Solution Using Analytical Method

$$\frac{dh(t)}{dt} = -C1\sqrt{h(t)}$$

$$\int_{h_0}^{h} \frac{dh}{\sqrt{h}} = -C1\int_{0}^{t} dt$$

$$2\left(\sqrt{h} - \sqrt{h_0}\right) = -C1(t - 0)$$

$$h(t) = \left(\sqrt{h_0} - \frac{1}{2}C1t\right)^2 \tag{6.6}$$

where h_0 is the height of water in the tank at time=0. Equation (6.6) is the analytical solution which computes the variation of height of water level with time as the water gets discharged through the orifice with area A_0. For problems involving emptying of tank, it is required to find out the total time taken to empty the tank. It is clear that when the tank is emptied, then $h(t)$ must be equal to zero. That is Equation (6.6) becomes

$\left(\sqrt{h_0} - \frac{1}{2}C1t\right)^2 = 0$ which gets simplified for the time, t as

$$t = \frac{2\sqrt{h_0}}{C1} \tag{6.7}$$

6.1.4 Computational Algorithm and Computer Program

Readers can find the above example in many textbooks on the basics of ordinary differential equations. The computational algorithm for this problem is straightforward as it requires only the initial height of the tank and other geometrical details. There are two types of solutions available for this problem, one is the computation of variation of height and the other one is calculation of time taken to empty the tank. The following is the computational algorithm.

(i) Read initial water level in the tank, ratio of orifice and tank diameters, coefficient of discharge of orifice.

(ii) Compute total time taken to empty the tank using Equation (6.7).
(iii) Assume time step and calculate number of time iterations using total time and time step.
(iv) Compute all the required constants to compute *h(t)* using Equation (6.6).
(v) Initialize the time level for t=0.
(v) Start the time do loop to compute *h(t)*
   (a) Compute *h(t)* using Equation (6.6) for the present time.
   (b) Compute velocity, *V* using Equation (6.4).
   (c) Go to step (a) until total number of time steps are completed
   (d) End the time loop
(vi) Write the output results, *t Vs h(t)* and *h(t) Vs V.*
(vii) End the program.

A computer program, ***Wtank_emptying.for*** in FORTRAN has been developed to study the variation of height of water level and velocity of water through the orifice with time.

6.1.5 Simulation Parameters and Discussion of Results

The purpose of modeling and simulation of the present problem is to understand how the level of water varies with time as the water gets discharged out from the tank. From the basics of fluid mechanics, it is understood that velocity of water is related to the square root of height of water level in the tank, thus a non-linear variation is expected for the variation of velocity with the height of water level. The simulation parameters for this problem can be identified as the height of water tank and diameter ratio. With an increase in height of the initial water level in the tank, the volume of water to be discharged also increases, and hence for any given diameter ratio, the time taken to empty the tank increases. When diameter ratio $\frac{D_o}{D_t}$ decreases, the time taken for emptying the tank also increases because the amount of water that can be discharged through the orifice increases. Figure 6.2 shows the simulation results obtained with the variation of the two simulation parameters. The variation of velocity of water through the orifice with height of water level is depicted in Figure 6.2(a) for initial heights of water, 2 m, 3 m and 4 m. As the velocity varies by the square root of height of water level in the tank, the variation is non-linear. It should be noticed that velocity variation obtained using the above three initial water levels almost coincide with each other, because of the selected number, which varies just by one meter. When the water level reaches zero, the velocity also reaches zero value as seen from the above figure. The variation of water level in the tank with time as water gets discharged through orifices of different diameter ratios is shown in Figures 6.2(b) to 6.2(d). Height of water level decreases following a non-liner path from the initial level of water, h_0 which is maximum and reaches zero value after some time. The time at which the height of water level becomes zero indicates the time taken to empty the tank from the initial level of water at h_0. For any given initial water level, the time taken for emptying the tank decreases with increases in the diameter ratio, that is, with increases in orifice diameter. Comparing all these three figures, one can easily notice that as the height of initial water level increases

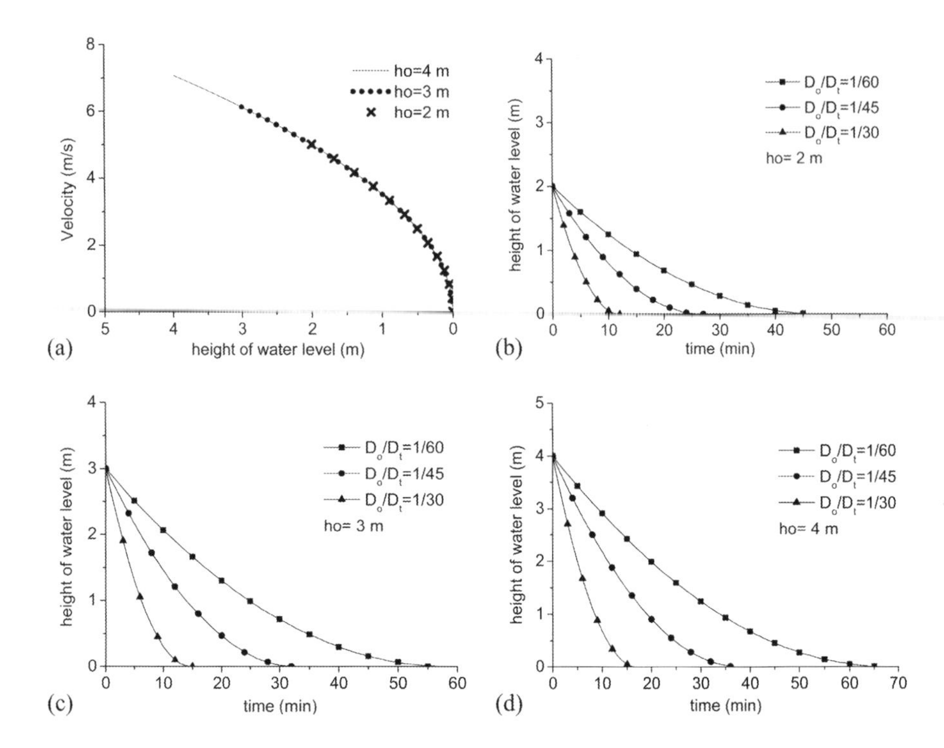

FIGURE 6.2 (a) Variation of velocity with height of tank, variation of height of water with time for (b) h_0=2 m, (c) $_0$ h=3 m and (d) h_0=4 m.

from 2 m to 4 m, the time taken for emptying the water tank increases for any given diameter ratio. The time taken to empty the tank is maximum for h_0=4 m with $\frac{1}{60}$ diameter ratio and minimum for h_0=2 m with $\frac{1}{30}$ diameter ratio.

6.2 TWO-DIMENSIONAL FLOW PROBLEMS – STOKES FLOW

The most common flow problems encountered in fluid dynamics applications are two-dimensional in nature. As was discussed earlier, the solution of flow problems involves the prediction of field variables consisting of velocity and pressure. This involves the solution of continuity and momentum equations for the flow field. The problem considered for numerical modeling is described in the following section.

6.2.1 Description of Problem

In modeling flow problems, the combination of continuity and momentum equations are called Navier-Stokes equations which were already derived. The details of the numerical solution of these equations will be discussed in Chapter 7. However, the simplified form of Navier-Stokes equations can be considered. The momentum equation describes the balancing of rate of change of momentum of fluid in a control volume to the net forces acting on the fluid medium. In general, force due to pressure and viscosity of the fluid are considered as the main contributing forces acting on the fluid, which need to be balanced with the change of fluid momentum. It has to be noted that not all type of flow problems requires the solution of complete Navier-Stokes equations. When the effect of viscous forces is ignored, then the simplification of Navier-Stokes equations gives rise to Euler equations. These equations may be used for the solution of field variables for both compressible and incompressible flow problems. In high speed flows, Euler equations play an important role in the formation of flow field as the inertial force exerted by the fluid medium will be much higher than that of the viscous force. When gases with small viscosity flow at high velocity, solution of Euler equations will provide the required solution without much error. Many flow problems studied in aerodynamic applications resemble the fluid momentum conservation expressed by the Euler equations. In contrast to the above situation, when fluids with high viscosity flow at low velocity, then the effect of viscous forces becomes more dominant compared to the force due to pressure, and fluid momentum becomes very small by virtue of low velocity flow. Study of Stokes flow finds applications in microfluidics, lab-on-chip etc. In this section, the solution of two-dimensional Stokes flow will be discussed in detail.

6.2.2 Mathematical Modeling

In Stokes flow, only the force due to the viscosity of the fluid gets balanced with the force due to the pressure field of the flow field when the inertial force of the fluid is considered negligible. It is only the pressure force that drives the flow by overcoming the viscous force of the fluid. Fluid momentum equations always balance the rate of change of momentum of the fluid to the net forces acting on the fluid. Hence, in a strict sense, in the absence of inertial force, the Stokes flow does not have momentum equation. However, the approximate momentum equations can be written from the Navier-Stokes equations by neglecting the inertial terms. Thus, we get the following equations as the governing equations for two-dimensional Stokes flow.

Continuity equation

$$\frac{\partial u}{\partial x} + \frac{\partial v}{\partial y} = 0 \tag{6.8}$$

Momentum equations

x-momentum

$$\mu\left(\frac{\partial^2 u}{\partial x^2} + \frac{\partial^2 u}{\partial y^2}\right) - \frac{\partial p}{\partial x} = 0 \tag{6.9}$$

y-momentum equation

$$\mu\left(\frac{\partial^2 v}{\partial x^2}+\frac{\partial^2 v}{\partial y^2}\right)-\frac{\partial p}{\partial y}=0 \tag{6.10}$$

Equations (6.8) to (6.10) are the final equations for the two-dimensional Stokes flow problem for incompressible fluid. One can easily notice that there are three unknowns, u, v and p and three equations, and hence there exists a solution for these equations. However, pressure appears only in the momentum equation and contributes implicitly to the continuity equation. Hence, these three equations can be recast in a form wherein all the three variables are related in some way in every equation. Here, the concept of vorticity can be introduced to rewrite the equations in velocity-vorticity form. It is well known that vorticity is defined as

$$\omega=\nabla\times\vec{V} \tag{6.11}$$

Vorticity about each axis in a three-dimensional coordinate can be written by expanding the above equation as

$$\omega=\nabla\times\vec{V}=\begin{vmatrix}\hat{i} & \hat{j} & \hat{k}\\ \frac{\partial}{\partial x} & \frac{\partial}{\partial y} & \frac{\partial}{\partial z}\\ u & v & w\end{vmatrix}$$

$$=\hat{i}\left(\frac{\partial w}{\partial y}-\frac{\partial v}{\partial z}\right)+\hat{j}\left(\frac{\partial u}{\partial z}-\frac{\partial w}{\partial x}\right)+\hat{k}\left(\frac{\partial v}{\partial x}-\frac{\partial u}{\partial y}\right) \text{ in which}$$

$$\omega_x=\left(\frac{\partial w}{\partial y}-\frac{\partial v}{\partial z}\right);\omega_y=\left(\frac{\partial u}{\partial z}-\frac{\partial w}{\partial x}\right);\omega_z=\left(\frac{\partial v}{\partial x}-\frac{\partial u}{\partial y}\right) \tag{6.12}$$

Equation (6.12) can be simplified for two-dimensional flow field on x-y plane as

$$\omega=\left(\frac{\partial v}{\partial x}-\frac{\partial u}{\partial y}\right) \tag{6.13}$$

The main objective of transforming Equations (6.8) to (6.10) is to eliminate the pressure term from the momentum equations. This can be done by subtracting Equation (6.10) from Equation (6.9) after differentiating these two equations with respect to the other directions. That is,

Differentiating Equation (6.9) with respect to y, we get

$\mu\frac{\partial}{\partial y}\left(\frac{\partial^2 u}{\partial x^2}+\frac{\partial^2 u}{\partial y^2}\right)-\frac{\partial^2 p}{\partial x\partial y}=0$, which can be rewritten as

$$\mu \frac{\partial u}{\partial y}\left(\frac{\partial^2}{\partial x^2}+\frac{\partial^2}{\partial y^2}\right)-\frac{\partial^2 p}{\partial x \partial y}=0 \tag{6.14}$$

Similarly, differentiating Equation (6.10) with respect to x, we get

$\mu \frac{\partial}{\partial x}\left(\frac{\partial^2 v}{\partial x^2}+\frac{\partial^2 v}{\partial y^2}\right)-\frac{\partial^2 p}{\partial x \partial y}=0$, which can be rewritten as

$$\mu \frac{\partial v}{\partial x}\left(\frac{\partial^2}{\partial x^2}+\frac{\partial^2}{\partial y^2}\right)-\frac{\partial^2 p}{\partial x \partial y}=0 \tag{6.15}$$

Now, the operation of (Equation (6.15)– Equation (6.14)) gives the following expression.

$\mu \frac{\partial v}{\partial x}\left(\frac{\partial^2}{\partial x^2}+\frac{\partial^2}{\partial y^2}\right)-\mu \frac{\partial u}{\partial y}\left(\frac{\partial^2}{\partial x^2}+\frac{\partial^2}{\partial y^2}\right)-\frac{\partial^2 p}{\partial x \partial y}+\frac{\partial^2 p}{\partial x \partial y}=0$. After cancelling the derivative terms of pressure, the equation is simplified as

$\mu \frac{\partial v}{\partial x}\left(\frac{\partial^2}{\partial x^2}+\frac{\partial^2}{\partial y^2}\right)-\mu \frac{\partial u}{\partial y}\left(\frac{\partial^2}{\partial x^2}+\frac{\partial^2}{\partial y^2}\right)=0$, which can be rewritten as

$\mu\left(\frac{\partial v}{\partial x}-\frac{\partial u}{\partial y}\right)\left(\frac{\partial^2}{\partial x^2}+\frac{\partial^2}{\partial y^2}\right)=0$ and using the definition of vorticity from Equation (6.13), this equation can be written as

$\mu\left(\frac{\partial^2 \omega}{\partial x^2}+\frac{\partial^2 \omega}{\partial y^2}\right)=0$. The viscosity of the fluid medium cannot be equal to zero and hence this is reduced to the following expression.

$$\left(\frac{\partial^2 \omega}{\partial x^2}+\frac{\partial^2 \omega}{\partial y^2}\right)=0 \tag{6.16}$$

Equation (6.16) is called a vorticity transport equation. The vorticity transport equation expressed by Equation (6.16) is a single equation for the unknown vorticity, ω. However, the effect of the continuity equation which involves two velocity components, u and v, has to be involved in the final solution of the governing equations along with the above vorticity transport equation. Hence, a second equation relating velocity and vorticity can be obtained by taking the curl of the vorticity definition as follows:

We know that vorticity is defined as

$$\vec{\omega}=\nabla \times \vec{V} \tag{6.17}$$

By applying curl operation on Equation (6.17) we get

$$\nabla \times \vec{\omega}=\nabla \times\left(\nabla \times \vec{V}\right) \tag{6.18}$$

From vector calculus

$$\nabla \times \left(\nabla \times \vec{V}\right) = \nabla\left(\nabla \cdot \vec{V}\right) - \nabla^2 \vec{V} \tag{6.19}$$

Now imposing the continuity constraint, $\nabla \cdot \vec{V}$, given by Equation (6.19), we get

$$\nabla \times \left(\nabla \times \vec{V}\right) = -\nabla^2 \vec{V}$$

Substituting the definition of $\omega = \nabla \times \vec{V}$ in the above expression we get finally

$$\nabla^2 \vec{V} = -\nabla \times \vec{\omega} \tag{6.20}$$

This can be rewritten for a two-dimensional flow problem as follows

$$\nabla^2 \vec{V} = -\nabla \times \omega \tag{6.21}$$

Equation (6.21) is called a velocity Poisson equation in canonical form, which governs the kinematic aspects of the problem and establishes the relationship between velocity field and vorticity field. Now, expanding Equation (6.21) for the two-dimensional domain, we get the following equations.

$$\frac{\partial^2 u}{\partial x^2} + \frac{\partial^2 u}{\partial y^2} = -\frac{\partial \omega}{\partial y} \tag{6.22a}$$

$$\frac{\partial^2 v}{\partial x^2} + \frac{\partial^2 v}{\partial y^2} = \frac{\partial \omega}{\partial x} \tag{6.22b}$$

Equations (6.16) and (6.22) are the final governing equations that have to be solved for velocities u, v, and vorticity, ω. These three equations are linear in nature, and velocity Poisson equations are coupled to the vorticity equation indirectly through the vorticity boundary conditions expressed as

$$\omega_b = \nabla \times \vec{V} \tag{6.23}$$

It has to be appreciated that vorticity is a derived quantity and it is a boundary phenomenon; this means it is generated at the boundary due to the velocity gradients at the boundaries. In any flow problem, the given boundary conditions for the field variables have to be enforced correctly, then only the flow field in the domain will evolve according to the underlying physics of the problem. In the solution of governing equations for the present Stokes flow problem, velocity Poisson equations required a vorticity field to compute the load for the velocity field, whereas in order to solve the vorticity transport equation, vorticity boundary conditions have to be computed exactly using the velocity values. Hence, it is clear that the velocity Poisson equations are coupled to the vorticity transport equation through the vorticity boundary conditions and this can be achieved through an iterative procedure until both vorticity and velocity fields are converged. The computational algorithm for the solution of Stokes equations will be discussed in detail in the following section.

6.3 THREE-DIMENSIONAL STOKES FLOW

It has been discussed in the previous section that Stokes equations in velocity-vorticity form give rise to simple linear equations, and they can be easily solved using finite element method. In this section the derivation of a three-dimensional Stokes flow problem will be carried out in detail. While solving three-dimensional flow problems using any numerical technique, care must be taken to employ a suitable simultaneous equations solver which can handle a large number of equations. For example, a simple $11 \times 11 \times 11$ mesh with an 8-node brick element gives rise to 1331 simultaneous equations in the form $Ax=b$, which need to be solved to obtain the field variable. In the three-dimensional computational domain, the increase in mesh size leads to cubic variation in the final number of nodes in the domain. A $51 \times 51 \times 51$ computational mesh produces 132,651 equations for an 8-node brick element. Of course, there are efficient and powerful iterative solvers that have been developed to solve equations of the order of 500,000 as found in a lot of commercial software. As was pointed out many times in this textbook, the organization of this textbook always insists that users themselves develop computer code, and wherever possible, the methods of developing simple computer programs will be demonstrated through some examples problems. The Stokes flow problem is taken up with the aim of encouraging readers to develop their computer codes with the supporting sample programs given in this textbook in some of the chapters. In order to deal with three-dimensional flow problems with a large number of simultaneous equations, a new type of finite element algorithm developed by the author, called the global matrix-free finite element (GMFFE) algorithm, will also be discussed in this section. The computer program used to obtain the simulation results for the present three-dimensional Stokes flow problem, has been developed using the GMFFE algorithm. The main principle of this algorithm lies in avoiding the assembly of element-level matrices to produce global matrices. It is well known that during the assembly of element matrices in finite element programming, the resulting global matrices will be always diagonally dominant. The significant characteristic of these global matrices is that they contain only a fraction of the total $N \times N$ entries of such matrices, where N denotes the number of nodes in the computational domain. For this reason, there are many techniques to store only the non-zero entries of such global matrices. Initially the conservation equations for three-dimensional Stokes flow are derived as discussed in the following section.

6.3.1 Governing Equations for Three-Dimensional Stokes Flow

The governing equations for Stokes flow problems can be directly deducted from the generalized Navier-Stokes equations. However, the solution of Navier-Stokes equations themselves will be discussed in a separate chapter. Hence, in this section, once again the governing equations for Stokes flow will be derived using velocity-vorticity form for simplicity of obtaining a numerical solution using the finite element method. As was explained earlier, without fluid momentum, there is nothing like a momentum equation for Stokes flow, and it is not derived from the basic momentum conservation principle. When the viscous force offered by the fluid flow is more significant compared to the inertial force of the fluid, then the flow field is just balanced between the fluid viscous force and force due to the pressure field. Though the fluid momentum is assumed to be absent in such flow problems, while computing the velocity field in a lid-driven cavity flow, the top plate of the cavity is assumed to move with a velocity equal to unity. With this brief analysis, let us write the conservation equation for three-dimensional Stokes flow for incompressible flow as follows in vector form.

Continuity Equation

$$\nabla \cdot \vec{V} = 0 \tag{6.24}$$

Momentum Equation

$$\mu \nabla^2 \vec{V} - \nabla p = 0 \tag{6.25}$$

When Equation (6.25) is expanded for three-dimensional flow field, four equations will be obtained including the continuity equation expressed by Equation (6.24) and these equations have to be solved for the four field variables, u, v, w and p. As pressure, p does not appear in the continuity equation explicitly, the solution of these equations becomes mathematically involved before getting the numerical solution. At present, the discussion on this will be avoided by focusing on the conversion of the governing equations to velocity-vorticity form. The vorticity is defined as the curl of velocity field as expressed by Equation (6.17) and also curl of gradient of a scalar quantity becomes zero. Taking the curl of Equation (6.25) results in the following equation.

$\mu\left(\nabla \times \nabla^2 \vec{V}\right) - \nabla \times \nabla p = 0$. In this equation the second term becomes equal to zero, hence the fluid momentum equation becomes

$\mu\left(\nabla \times \nabla^2 \vec{V}\right) = 0$. It can be shown that the curl of the Laplacian of a vector is equal to the Laplacian of curl of that vector [1]. Hence, the momentum equation can be expressed as

$$\nabla^2 \left(\nabla \times \vec{V}\right) = 0 \tag{6.26}$$

$$\nabla^2 \left(\vec{\omega}\right) = 0 \tag{6.27}$$

Equation (6.27) is simply a Laplacian equation for the vorticity field. Now, the velocity field has to be connected to this vorticity field through some equations after satisfying the incompressibility constraint or the continuity condition expressed by

Equation (6.24). After repeating the derivations for the velocity Poisson equation as was discussed from Equations (6.17) to (6.21), we get the final form of the velocity Poisson equation as

$$\nabla^2\vec{V} = -\nabla \times \vec{\omega} \tag{6.28}$$

Expanding Equations (6.27) and (6.28) will give rise to the following set of governing equations for three-dimensional Stokes flow problem.

Vorticity Transport Equations

x-direction

$$\frac{\partial^2 \omega_x}{\partial x^2} + \frac{\partial^2 \omega_x}{\partial y^2} + \frac{\partial^2 \omega_x}{\partial z^2} = 0 \tag{6.29a}$$

y-direction

$$\frac{\partial^2 \omega_y}{\partial x^2} + \frac{\partial^2 \omega_y}{\partial y^2} + \frac{\partial^2 \omega_y}{\partial z^2} = 0 \tag{6.29b}$$

z-direction

$$\frac{\partial^2 \omega_z}{\partial x^2} + \frac{\partial^2 \omega_z}{\partial y^2} + \frac{\partial^2 \omega_z}{\partial z^2} = 0 \tag{6.29c}$$

Velocity Poisson Equations

x-direction

$$\frac{\partial^2 u}{\partial x^2} + \frac{\partial^2 u}{\partial y^2} + \frac{\partial^2 u}{\partial z^2} = \frac{\partial \omega_y}{\partial z} - \frac{\partial \omega_z}{\partial y} \tag{6.30a}$$

y-direction

$$\frac{\partial^2 v}{\partial x^2} + \frac{\partial^2 v}{\partial y^2} + \frac{\partial^2 v}{\partial z^2} = \frac{\partial \omega_z}{\partial x} - \frac{\partial \omega_x}{\partial z} \tag{6.30b}$$

z-direction

$$\frac{\partial^2 w}{\partial x^2} + \frac{\partial^2 w}{\partial y^2} + \frac{\partial^2 w}{\partial z^2} = \frac{\partial \omega_x}{\partial y} - \frac{\partial \omega_y}{\partial x} \tag{6.30c}$$

Equations (6.29a) to (6.29c) are the vorticity transport equations which are simple Laplace equations for vorticities, ω_x, ω_y and ω_z in the x, y and z-directions respectively. The velocity Poisson equations in the x, y and z-directions are given by Equations (6.30a) to (6.30c) for u, v and w velocities respectively. These six equations have to be solved for the following boundary conditions.

Boundary Conditions

Velocity field: x-y plane for z=1, u=1, v=w=0 and for z=0, u=v=w=0
y-z plane for x=0 and x=1, u=v=w=0
x-z plane for y=0 and y=1, u=v=w=0

Vorticity field: $\omega_b = \nabla \times V_b$ on x-y plane for z=0 and z=1, y-z plane for x=0 and x=1 and x-z plane for y=0 and y=1.

Vorticity is generated at the boundaries of the domain on which no-slip boundary conditions are imposed and they get diffused into the interior of the domain due to viscous and inertial forces. In the present problem, in the absence of inertial forces, it is the viscous force that helps to diffuse the vorticity field into the interior of the domain. Vorticity transport and velocity Poisson equations are not coupled and both of them are linear equations, which can be solved easily. However, these two sets of equations are coupled implicitly through the vorticity boundary conditions which depend on the velocity field, and the velocity field itself can be obtained only when the vorticity loading is computed. Thus, the coupling of these equations through vorticity boundary conditions has to be considered while developing the computational algorithm for the solution of these equations. It can be noticed that in three-dimensional flow problems, the solution of the velocity-vorticity form of equations produces three vorticity components and three velocity components; in total six field variables have to be computed. When the velocity-pressure form of equations is employed for three-dimensional flow problems, then only four field variables need to be computed, and this makes the velocity-pressure form of fluid momentum equations more compact compared to the velocity-vorticity form of equations. Nonetheless, the pressure field does not appear in the continuity equation directly, and its implicit presence poses difficulty in solving these types of equations. Now, the solution of the governing equations for three-dimensional Stokes flow problem using the finite element method is detailed in the following section.

6.3.2 Finite Element Solution Procedure

The governing equations for the three-dimensional Stokes flow problem expressed by Equations (6.29a) to (6.29c) and (6.30a) to (6.30c) can be solved by the application of Galerkin's weighted residual finite element method. Consider the implementation of the finite element method to Equation (6.29a) as expressed below.

$\int_\Omega [N]^T \left(\frac{\partial^2 \omega_x}{\partial x^2} + \frac{\partial^2 \omega_x}{\partial y^2} + \frac{\partial^2 \omega_x}{\partial z^2} \right) dxdydz = 0$. Integrating all the three diffusion terms independently using Galerkin's method as was discussed in Chapter 4, we get

$$-\int_\Omega \left(\frac{\partial [N]^T}{\partial x} \frac{\partial [N]}{\partial x} + \frac{\partial [N]^T}{\partial y} \frac{\partial [N]}{\partial y} + \frac{\partial [N]^T}{\partial z} \frac{\partial [N]}{\partial z} \right) dxdydz \{\omega_x\} + \int_\Gamma [N]^T \frac{\partial \omega_x}{\partial x} d\Gamma + \int_\Gamma [N]^T \frac{\partial \omega_x}{\partial y} d\Gamma + \int_\Gamma [N]^T \frac{\partial \omega_x}{\partial z} d\Gamma = 0 \quad (6.31)$$

In the present Stokes flow problem, only Dirichlet boundary conditions will be considered for the velocity and vorticity fields. Hence, the second term in Equation (6.31) will not contribute to the formation of finite element matrices. Eight node isoparametric elements are used to discretize the computational domain for the simulation of the three-dimensional Stokes flow problem. The shape functions for 8-node isoparametric elements were already discussed in Section 4.5 of Chapter 4. In continuation of the discussion for 8-node isoparametric elements, the integration of the equation given by Equation (6.31) can be discussed. It has to be remembered, integration of Equation (6.31) gives rise to the stiffness matrix. Hence, the relation between the global derivatives and local derivatives can be expressed as

$$\begin{Bmatrix} \frac{\partial N_i}{\partial x} \\ \frac{\partial N_i}{\partial y} \\ \frac{\partial N_i}{\partial z} \end{Bmatrix} = \begin{bmatrix} \frac{\partial \xi}{\partial x} & \frac{\partial \eta}{\partial x} & \frac{\partial \zeta}{\partial x} \\ \frac{\partial \xi}{\partial y} & \frac{\partial \eta}{\partial y} & \frac{\partial \zeta}{\partial y} \\ \frac{\partial \xi}{\partial z} & \frac{\partial \eta}{\partial z} & \frac{\partial \zeta}{\partial z} \end{bmatrix} \begin{Bmatrix} \frac{\partial N_i}{\partial \xi} \\ \frac{\partial N_i}{\partial \eta} \\ \frac{\partial N_i}{\partial \zeta} \end{Bmatrix} \tag{6.32}$$

where $\begin{bmatrix} \frac{\partial \xi}{\partial x} & \frac{\partial \eta}{\partial x} & \frac{\partial \zeta}{\partial x} \\ \frac{\partial \xi}{\partial y} & \frac{\partial \eta}{\partial y} & \frac{\partial \zeta}{\partial y} \\ \frac{\partial \xi}{\partial z} & \frac{\partial \eta}{\partial z} & \frac{\partial \zeta}{\partial z} \end{bmatrix} = J^{-1}$, the inverse of Jacobian matrix. It was already discussed that the Jacobian matrix is employed to achieve the required coordinate transformation between the global and local coordinates. Once, the Jacobian, J of the 3×3 matrix is estimated, the inverse of the Jacobian can be evaluated as

$$J^{-1} = \frac{1}{J} \begin{bmatrix} J_{22} * J_{33} - J_{32} * J_{23} & -J_{12} * J_{33} + J_{32} * J_{13} & J_{12} * J_{23} - J_{22} * J_{13} \\ -J_{21} * J_{33} + J_{31} * J_{23} & J_{11} * J_{33} - J_{31} * J_{13} & -J_{11} * J_{23} + J_{21} * J_{13} \\ J_{21} * J_{32} - J_{31} * J_{22} & -J_{11} * J_{32} + J_{31} * J_{12} & J_{11} * J_{22} + J_{21} * J_{12} \end{bmatrix} \tag{6.33}$$

After substituting Equation (6.33) in Equation (6.32), we get the following expression.

$$\begin{Bmatrix} \frac{\partial N_i}{\partial x} \\ \frac{\partial N_i}{\partial y} \\ \frac{\partial N_i}{\partial z} \end{Bmatrix} = \frac{1}{J} \begin{bmatrix} J_{22} * J_{33} - J_{32} * J_{23} & -J_{12} * J_{33} + J_{32} * J_{13} & J_{12} * J_{23} - J_{22} * J_{13} \\ -J_{21} * J_{33} + J_{31} * J_{23} & J_{11} * J_{33} - J_{31} * J_{13} & -J_{11} * J_{23} + J_{21} * J_{13} \\ J_{21} * J_{32} - J_{31} * J_{22} & -J_{11} * J_{32} + J_{31} * J_{12} & J_{11} * J_{22} + J_{21} * J_{12} \end{bmatrix} \begin{Bmatrix} \frac{\partial N_i}{\partial \xi} \\ \frac{\partial N_i}{\partial \eta} \\ \frac{\partial N_i}{\partial \zeta} \end{Bmatrix} \tag{6.34}$$

Let J^{-1} represent the inverse of the Jacobian, that is Equation (6.33), then Equation (6.34) can be rewritten as

$$\begin{Bmatrix} \frac{\partial N_i}{\partial x} \\ \frac{\partial N_i}{\partial y} \\ \frac{\partial N_i}{\partial z} \end{Bmatrix} = \begin{bmatrix} J_{11}^{-1} & J_{12}^{-1} & J_{13}^{-1} \\ J_{21}^{-1} & J_{22}^{-1} & J_{23}^{-1} \\ J_{31}^{-1} & J_{32}^{-1} & J_{33}^{-1} \end{bmatrix} \begin{Bmatrix} \frac{\partial N_i}{\partial \xi} \\ \frac{\partial N_i}{\partial \eta} \\ \frac{\partial N_i}{\partial \zeta} \end{Bmatrix} \tag{6.35}$$

For example, let us determine the global derivative of node 1 of the 8-node isoparametric element with respect to x, y and z-axes.

$$\frac{\partial N_1}{\partial x} = J_{11}^{-1} * \frac{\partial N_1}{\partial \xi} + J_{12}^{-1} * \frac{\partial N_1}{\partial \eta} + J_{13}^{-1} * \frac{\partial N_1}{\partial \zeta} \tag{6.36a}$$

$$\frac{\partial N_1}{\partial y} = J_{21}^{-1} * \frac{\partial N_1}{\partial \xi} + J_{22}^{-1} * \frac{\partial N_1}{\partial \eta} + J_{23}^{-1} * \frac{\partial N_1}{\partial \zeta} \tag{6.36b}$$

$$\frac{\partial N_1}{\partial z} = J_{31}^{-1} * \frac{\partial N_1}{\partial \xi} + J_{32}^{-1} * \frac{\partial N_1}{\partial \eta} + J_{33}^{-1} * \frac{\partial N_1}{\partial \zeta} \tag{6.36c}$$

Similarly, the global derivatives of other nodes of the 8-node isoparametric element can be determined. This procedure can be easily programmed to determine the above expressions, and the integration of these global derivatives are carried out using the Gaussian quadrature formula and hence the integration is reduced to simple summation in the program. Now, the stiffness matrix given by Equation (6.31) can be written for the x-momentum Equation (6.29a) as

$$\left[K_{ij}\right]\left\{\omega_{x,j}\right\} = 0 \tag{6.37a}$$

where

$$\left[K_{ij}\right] = \int_{\Omega} \left(\frac{\partial [N]^T}{\partial x} \frac{\partial [N]}{\partial x} + \frac{\partial [N]^T}{\partial y} \frac{\partial [N]}{\partial y} + \frac{\partial [N]^T}{\partial z} \frac{\partial [N]}{\partial z} \right) dxdydz$$

Repeating the procedure, the vorticity transport equations, (6.29b) and (6.29c) in y and z-directions can be expressed as

$$\left[K_{ij}\right]\left\{\omega_{y,j}\right\} = 0 \tag{6.37b}$$

$$\left[K_{ij}\right]\left\{\omega_{z,j}\right\} = 0 \tag{6.37c}$$

The velocity Poisson equations given by Equations (6.30a) to (6.30c) have to be solved using the finite element method. Application of Galerkin's weighted residual method to Equation (6.30a) will result in the following equation.

$\int_{\Omega}[N]^T\left(\frac{\partial^2 u}{\partial x^2}+\frac{\partial^2 u}{\partial y^2}+\frac{\partial^2 u}{\partial z^2}-\frac{\partial \omega_y}{\partial z}+\frac{\partial \omega_z}{\partial y}\right)dxdydz = 0$, which can be simplified further as

$$\int_{\Omega}[N]^T\left(\frac{\partial^2 u}{\partial x^2}+\frac{\partial^2 u}{\partial y^2}+\frac{\partial^2 u}{\partial z^2}\right)dxdydz - \int_{\Omega}[N]^T\left(\frac{\partial \omega_y}{\partial z}-\frac{\partial \omega_z}{\partial y}\right)dxdydz = 0$$

$$-\int_{\Omega}\left(\frac{\partial[N]^T}{\partial x}\frac{\partial[N]}{\partial x}+\frac{\partial[N]^T}{\partial y}\frac{\partial[N]}{\partial y}+\frac{\partial[N]^T}{\partial z}\frac{\partial[N]}{\partial z}\right)dxdydz\{u\}$$

$$+\int_{\Gamma}[N]^T\frac{\partial u}{\partial x}d\Gamma+\int_{\Gamma}[N]^T\frac{\partial u}{\partial y}d\Gamma+\int_{\Gamma}[N]^T\frac{\partial u}{\partial z}d\Gamma-\int_{\Omega}[N]^T\frac{\partial[N]}{\partial z}dxdydz\{\omega_y\}$$

$$+\int_{\Omega}[N]^T\frac{\partial[N]}{\partial y}dxdydz\{\omega_z\} = 0$$

Using suitable notations, the above equation can be expressed as

$$\left[K_{ij}\right]\left\{u_j\right\} = \left[Y_{ij}\right]\left\{\omega_{z,j}\right\} - \left[Z_{ij}\right]\left\{\omega_{y,j}\right\} \tag{6.38a}$$

Similarly, the finite element formulation for Equations (6.30b) and (6.30c) can be written as

$$\left[K_{ij}\right]\left\{v_j\right\} = \left[Z_{ij}\right]\left\{\omega_{x,j}\right\} - \left[X_{ij}\right]\left\{\omega_{z,j}\right\} \tag{6.38b}$$

$$\left[K_{ij}\right]\left\{v_j\right\} = \left[X_{ij}\right]\left\{\omega_{y,j}\right\} - \left[Y_{ij}\right]\left\{\omega_{x,j}\right\} \tag{6.38c}$$

where

$$\left[K_{ij}\right] = \int_{\Omega}\left(\frac{\partial[N]^T}{\partial x}\frac{\partial[N]}{\partial x}+\frac{\partial[N]^T}{\partial y}\frac{\partial[N]}{\partial y}+\frac{\partial[N]^T}{\partial z}\frac{\partial[N]}{\partial z}\right)dxdydz$$

$$\left[X_{ij}\right] = \int_{\Omega}[N]^T\frac{\partial[N]}{\partial x}dxdydz$$

$$\left[Y_{ij}\right]=\int_{\Omega}\left[N\right]^{T}\frac{\partial\left[N\right]}{\partial y}dxdydz$$

$$\left[Z_{ij}\right]=\int_{\Omega}\left[N\right]^{T}\frac{\partial\left[N\right]}{\partial z}dxdydz$$

The final equations expressed by Equations (6.37a) to (6.37c) and (6.38a) to (6.38c) have to be solved for the field variables, ω_x, ω_y, ω_z, u, v and w for the three-dimensional Stokes problem. These six equations are linear and they are coupled only through vorticity boundary conditions. Hence, an iterative computational procedure is followed to obtain the field variables. Figure 6.3 shows the computational procedure in the form of a flow chart. With the help of input parameters, the boundary conditions for the velocity field are enforced, in this problem, the top plate is assumed to move with unit velocity. All the stiffness and gradient matrices are evaluated using the Gaussian quadrature formula for the 8-node isoparametric elements. At the beginning of computations, the vorticity boundary values are assumed to be zero and hence the load vector for velocity Poisson equations can be computed. To start with, the velocity field is computed, as its boundary conditions and the load vector are already computed. Using this velocity field, the vorticity boundary conditions are evaluated and enforced at the boundaries of the domain. This will help us to solve the vorticity field, however, this is not the final converged field because originally the vorticity boundary conditions were assumed to be zero. In the first iterations, the vorticity field is very weak because the velocity field is not converged due to the assumed zero boundary values for the vorticity fields. After checking for the convergence, the vorticity field and velocity field are updated for the next iteration, and after a few iterations, this will soon will give rise to converged velocity and vorticity fields. Using the data of velocity and vorticity, the required flow field plots can be obtained.

Three-dimensional flow problems can be easily modeled using commercial software, however, if one needs to write own computer code for simple Stokes flow problems, the conventional finite element procedure will result in difficulties in solving the final set of simultaneous equations. When a cubic computational domain is discretized using $51\times51\times51$ mesh with 8-node brick element, then 125000 elements and 132651 nodes have to be handled in the computations. In order to process the matrices of these huge sizes, a computing machine with higher RAM will be required. As was pointed out earlier, the finite element assembly procedure results in diagonally dominant matrices that have to be solved for the field variables. There are many algorithms that have been developed in order to store only the non-zero entries of the final assembled matrices instead of storing all the entries. One such algorithm called the global matrix-free algorithm (GMFFE) has been proposed by the author, which will enable the solution of three-dimensional problems on a personal computer. It is better to understand the conventional method of incorporating the Dirichlet boundary conditions in finite element procedure before getting into the details of the global matrix-free algorithm.

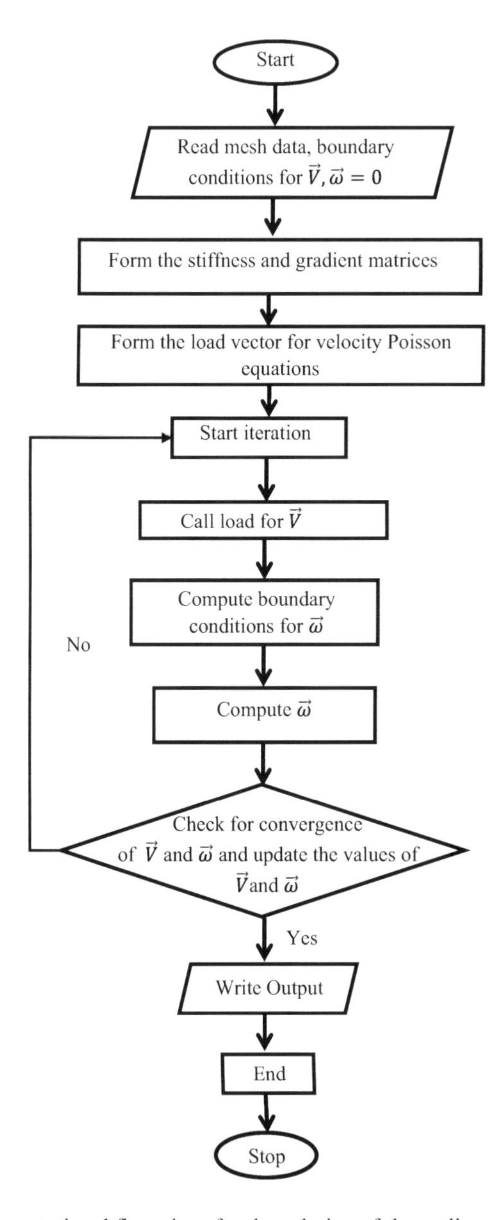

FIGURE 6.3 Computational flow chart for the solution of three-dimensional Stokes flow.

6.3.3 Enforcement of Dirichlet Boundary Conditions in Finite Element Solution Procedure

In the finite element solution procedure, after forming the global stiffness and load vector matrices, the final task is to modify these matrices to take care of the effect of boundary conditions. It is well understood that without the effect of boundary conditions, there will not be any change in a system under consideration in modeling. Only when the system interacts with its surroundings, there will be changes in the system due to the effect of interaction coming externally to the boundary of the domain. In problems where the boundary conditions are assumed to be static, that is, boundary conditions not changing with time, there will be definite values of the field variables on the boundaries of the domain, which need to be satisfied by the solution of the equations. In a given simulation scheme, the numerical results are obtained for the constraints imposed on the boundaries of the system, and hence, whenever the governing equations are derived or described, the corresponding initial and boundary conditions for the domain have to be provided. For the present Stokes flow problem, the simulation is focused on getting the results for the velocity field and vorticity field, hence, the values of all the three velocities and the three vorticities have to be enforced on the boundaries. Boundary conditions always mean in what way the field variables being computed by solving the given set of governing equations are related to the boundary constraints. For example, if heat flux is supplied on a boundary, then this heat flux is related to the gradient of the field variable temperature being computed in the simulation program. In the present Stokes flow problem, only Dirichlet boundary conditions are considered on the boundaries of the domain. In most numerical solution schemes, the final form of the simultaneous equations to be solved are modified to take care of the effect of boundary conditions, and this effect is called loading. The terminology loading is used as an analogy to an elastic medium being loaded with some force, and this will result in some deflection of the medium. One can view this as cause and effect; without cause, there is no net variation in the effect. Hence, the enforcement of boundary conditions to the final simultaneous equations provides enough loading on the flow field to make changes in the field variables. In the case of enforcement of Dirichlet boundary conditions, as the value of the field variables are known at the specified boundary nodes, the diagonal entry of the coefficient matrix is made equal to 1 and the value of the field variable at the respective node of the boundary is assigned as the load at the corresponding node.

6.3.3.1 Computational Steps to Incorporate Dirichlet Boundary Conditions

Let us consider the procedure of incorporating the Dirichlet boundary conditions in the final global matrices. The final form of the equations to be solved can be represented as $\left[K\right]\{\phi\}=\{f\}$ and this form can be expanded for 4×4 equations as

$$\begin{bmatrix} K_{11} & K_{12} & K_{13} & K_{14} \\ K_{21} & K_{22} & K_{23} & K_{24} \\ K_{31} & K_{32} & K_{33} & K_{34} \\ K_{41} & K_{42} & K_{43} & K_{44} \end{bmatrix} \begin{Bmatrix} \phi_1 \\ \phi_2 \\ \phi_3 \\ \phi_4 \end{Bmatrix} = \begin{Bmatrix} f_1 \\ f_2 \\ f_3 \\ f_4 \end{Bmatrix} \tag{6.39}$$

Let us assume that the above equations have to be modified for inclusion of Dirichlet boundary condition at node 3 and let us assume ϕ_3^* as the value known at the boundary node 3. Now, the equation corresponding to node 3, that is the third equation, needs to be modified, and also the effect of ϕ_3^* on other equations 1, 2, 3 and 4 on coefficients, K_{13}, $K_{23,}$ K_{33} and K_{43} has to be effected along with the corresponding changes in the load vectors, f_1, f_2, f_3 and f_4. Let us do it initially for the 3 column on the coefficients and the vectors.

$$f_1 = f_1 - K_{13} * \phi_3^* \tag{6.40a}$$

$$f_2 = f_2 - K_{23} * \phi_3^* \tag{6.40b}$$

$$f_3 = f_3 - K_{33} * \phi_3^* \tag{6.40c}$$

$$f_4 = f_4 - K_{43} * \phi_3^* \tag{6.40d}$$

As the coefficients K_{13}, $K_{23,}$ K_{33} and K_{43} are taken to the right hand side of Equation (6.39), their values in $[K]$ matrix takes zero values as

$$K_{13} = 0 \tag{6.41a}$$

$$K_{23} = 0 \tag{6.41b}$$

$$K_{33} = 0 \tag{6.41c}$$

$$K_{43} = 0 \tag{6.41d}$$

Now other coefficients in the equation corresponding to node 3 have to be made zero because the solution of the node is already obtained.

$$K_{31} = 0 \tag{6.42a}$$

$$K_{32} = 0 \tag{6.42b}$$

$$K_{33} = 0 \tag{6.42c}$$

$$K_{34} = 0 \tag{6.42d}$$

The equation corresponding to node 3 is modified as

$$K_{33} = 1.0 \quad \text{and} \quad f_3 = \phi_3^* \tag{6.43}$$

Equation (6.43) gives the final solution for the equation of node 3 in Equation (6.39). Now, because of the equalities expressed in Equation (6.43), the equalities given by Equations (6.40c), (6.41c) and (6.42c) become superfluous. However, the above processes are carried out using a computer code, and hence such repetitions cannot be avoided and such superfluous steps will not pose any difficulty later provided the corresponding equalities as expressed by Equation (6.43) are executed as the last process. In computer programming syntax, the last assigned equality step only will be taken as final and executed.

6.3.4 Global Matrix-Free Finite Element Algorithm

The application of the Galerkin's weighted residual method in the finite element procedure results in coefficient matrices for the field variables. Depending upon the type of governing equations to be solved, these coefficient matrices are algebraically added to produce a final set of simultaneous equations to be solved for the field variables. The coefficient matrix of these equations is generally sparse and diagonally dominant. Many algorithms have been developed to store only the non-zero entries of the coefficient matrix. Generally, these algorithms have been implemented on parallel computers especially, to solve three-dimensional problems. Since all the non-zero entries are stored in a compact vector form, parallel computation becomes very efficient to solve large-size problems. The present global matrix-free algorithm is capable of solving three-dimensional flow problems on a personal computer since it does not involve a global assembly procedure but keeps all the matrices and vectors at the element level. A comparison of the present algorithm with the bi-conjugate gradient method is discussed in the following section.

6.3.4.1 Matrix Storage Schemes for Large Size Problems and Solvers

The application of the finite element method gives rise to sparse matrices which are generally solved using bi-conjugate iterative solvers [2]. The large size sparse matrices are generally stored in compact vector storage form by storing only the non-zero entries of the matrices, which will save a lot of memory storage [3]. In the case of three-dimensional problems, which require finer meshes, the total number of non-zero entries also increases, thus increasing the size of the compact vector storage. Irons [4] proposed a novel solution scheme called an element-based frontal solver, in which a frontal width is computed from the contributions of variables with a given degree of freedom. However, this method is efficient only when the frontal width is minimum which is achieved by a suitable global element node numbering of the computational domain. Element-by-element solution scheme was developed by Hughes et al. [5] and this algorithm is based on the operator-splitting method. This method was found suitable for problems for which the capacitance matrix is diagonal, and this method was extended to structural and solid mechanics problems [6], which require conversion of elliptical equations into parabolic equations by introducing artificial time derivatives.

6.3.4.2 BICGSTAB and Element-by-Element Scheme for Parallel Computing

Conjugate gradient methods are highly attractive schemes to solve a large sparse system of equations because the solution scheme depends only on vectors obtained as products of coefficient matrices and vectors [7]. When these products are performed at the element level, a significant saving in computational time and effort can be achieved. Sheu et al. [8] implemented the BICGSTAB iterative solver in an element-by-element format to achieve computational efficiency in a parallel computation of three-dimensional Navier-Stokes equations using the finite element method. Thiagarajan and Aravamuthan [9] proposed a pre-conditioner for the conjugate gradient method along with an element-by-element solution scheme for parallel computation. Phoon [10] developed a generalized Jacobi (diagonal) preconditioning approach to implement the conjugate gradient iterative solver using an element-by-element strategy. The implementation of the element-by-element iterative solution procedure to solve large-scale problems on a personal computer is not straightforward though the scheme has been efficiently exploited in parallel computations [8–10]. The main reason for this restriction is the requirement of huge computer memory to store the large size global matrices. Even a compact vector storage scheme requires a memory space of 1,336,694 to store only the non-zero entries for a 3D flow problem with a mesh of size 31^3. Hence, the necessity of storing such huge size vectors restricts the use of the conjugate gradient iterative solvers such as the BICG iterative solvers [2] on personal computers.

6.3.4.3 Procedure to Implement Global Matrix-Free Finite Element Algorithm

The main idea behind the GMFFE algorithm is to avoid the assembly of matrices to obtain the final global matrices which are highly sparse in nature. The formation of sparse global matrices and the associated compact storage procedures can be avoided when all the coefficient matrices and the vectors are kept at the element itself, using an element-wise storage scheme. However, the conventional method of incorporating the Dirichlet boundary conditions cannot be implemented in this case because there is no global matrix. Hence, the Dirichlet boundary values have to be enforced at the element level itself. In a global matrix, the identification of elements completely vanishes after assembling the element-level matrices. It was understood from the previous section that while incorporating the Dirichlet boundary conditions, the load vector and the coefficient matrices get modified for the corresponding boundary nodes. When this procedure has to be executed at element level, then it is required to determine the number of elements contributing the Dirichlet value to a boundary node. That means it is required to find out, for a given boundary node, what the elements that share this node are, as one of the nodes for the element-nodal connectivity scheme. For example, in a three-dimensional computational domain, the boundary nodes are located on corners, edges and surfaces. Based on its location, a boundary node can be shared by certain number of elements as shown below.

Corner node – one element
Node on an edge – two elements
Node on a surface – four elements

Figure 6.4 represents a section of a horizontal boundary surface of a three-dimensional computational domain with the boundary shaded. Referring to the above

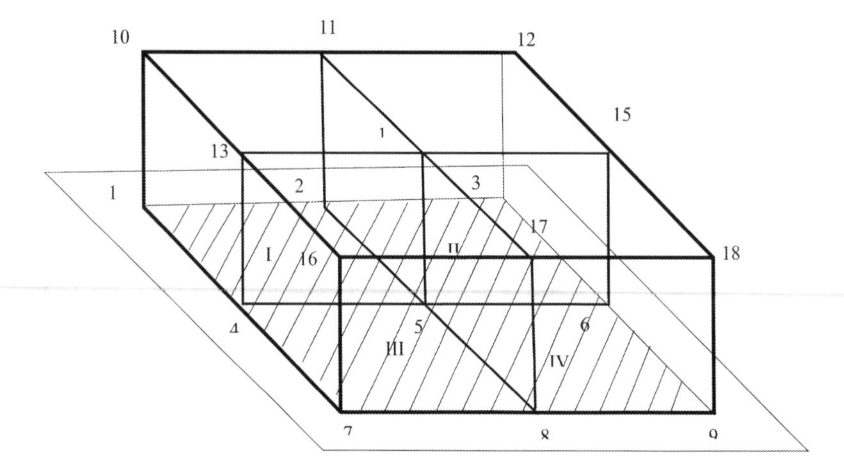

FIGURE 6.4 Schematic diagram of elements, I, II, III and IV on boundary and the respective contributing nodes.

figure, the element-nodal connectivity, nodes and their sharing elements can be expressed as

Element nodal connectivity

Element	Node numbers							
I	1	4	5	2	10	13	14	11
II	2	5	6	3	11	14	15	12
III	4	7	8	5	13	16	17	14
IV	5	8	9	6	14	17	18	15

Cornernodes–	Node	Elements
	1	I
	3	II
	9	IV
	7	III
Edge node:	4	I and III
	2	I and II
	8	III and IV
	6	II and IV
Surface node:	5	I, II, III, IV

Now, the elements that contain the boundary nodes have to be modified by making changes in the stiffness and load vector matrices to incorporate the effect of Dirichlet boundary conditions. This can be achieved by identifying the number of contributing elements to the boundary nodes, the respective element numbers and the respective rows of the elements that contain the boundary node. These data can be generated for all the boundary nodes by developing a subroutine for this purpose. Figure 6.5 explains the computational algorithm to generate the above data. In this algorithm, the outermost loop carries out the operations for all the boundary nodes.

Computer code 6.2 – Generation of data of elements surrounding the boundary nodes

Define matrices and constants

Ibu(nbn) – array with boundary nodes, nbn – array size for number of boundary conditions

nbc – total number of boundary nodes, nelem – total number of elements, nne – number of nodes per element

Inode(mel, nen) – array with element – nodal connectivity for all the elements, nen=8 for cubic element, mel – array size for elements

nelc(nbc,nen) – array with boundary nodes and corresponding number of contributing elements

nrow(nbc,nen) – array with boundary nodes and corresponding row numbers

nct(nbn) – array with total number elements surrounding a boundary node

Program

```
do ik = 1, nbc                          Do this for all boundary nodes
  i = ibu(ik)                           Identify the boundary node
  nc=0                                  Initialize the counter of elements
 **Scan all the elements to identify this boundary node**
  do j=1, nelem                         Do this for all the elements
**Scan a specific element for all its nodes**
   do ii=1, nne                         Do this for all the nodes of an element
    Inode = I (j, ii)                   Designate the node number
 if(Inode.eq.i) then                    Check the Inode is equal to the node i
  nc=nc+1                               Add to the element counter
  nelc(ik, nc)=j                        Store the number of elements in an array
  nrow(ik, nc)=ii                       Specify the row number
  endif                                 End the if statement
 enddo                                  End the scanning for the given element
 nct(ik)=nc                             Store the total number of elements surrounding node i
 enddo                                  End the scanning for all the elements
enddo                                   End the scanning for all the boundary nodes
```

FIGURE 6.5 Computer code for generating data of elements surrounding boundary nodes.

Each boundary node is assigned a variable name, then all the element-nodal connectivity data of all the elements are scanned to identify this boundary node. Then, that specific element is identified and from the element-nodal connectivity data its row number is also computed. When the scanning of the checking of each boundary node with the nodes of an element is performed, then the number of elements that have this particular boundary node are identified and are counted in a counter. Using the counter and connectivity data of nodes of an element, the row number of the node that matches with the given boundary node is also found out and stored in an array. Thus, by scanning all the elements, the total number of elements that have this boundary node is stored in another array.

The element level matrices are not assembled in the present algorithm, hence, the storage array for stiffness matrix consists of three arrays, for example, if the final global stiffness matrix is represented by *estffmku(mel,ii,jj)*, then the first array, '*mel*' stores the element number, '*ii*' stores the row number and '*jj*' stores the column number of the matrix. Similarly, the load vector matrix consists of two arrays, one for the elements and the second one for number of rows. In the conventional global assembly method, it will have only two arrays, one for the rows and the other for columns equal to the total number of nodes. In contrast, in GMFFE algorithm, the element array stores the total number of elements, row and column arrays, equal to nodes per element of the chosen finite element used to discretize the computational domain. Figure 6.6 shows the computer code to incorporate the Dirichlet boundary conditions in the global matrices. This algorithm works almost on the same principle of the general procedure to incorporate Dirichlet boundary conditions, with the only difference in calculating the fraction of the boundary value to be contributed to the elements surrounding any given boundary node. From the stored details of row and column of the boundary node appearing in the elements surrounding the boundary node, the exact rows and columns of those elements are modified to enforce the effect of Dirichlet boundary condition, and this process is repeated for all the boundary nodes. While developing this algorithm, sufficient care has been taken to produce the same effect as that of the global matrices by assembly of all the elements in the conventional practice. As the number of boundary nodes in a given computational domain, especially in three-dimensional domains, is very small compared to the total number of nodes, the computational effort for this task is insignificant. The process of generating information on the elements surrounding each boundary node is generated only once similar to the mesh generation data.

The procedure can be explained by considering the boundary node 4 located along the edge as shown in Figure 6.4. This node is surrounded by two elements, I and III. The coefficient matrices and the load vectors for elements I and III can be represented as

$$\left[K_{Ijk}\right]\left\{\phi_{Ij}\right\} = \left\{f_{Ij}\right\} \tag{6.44a}$$

$$\left[K_{IIIjk}\right]\left\{\phi_{IIIj}\right\} = \left\{f_{IIIj}\right\} \tag{6.44b}$$

Computer Code: 6.3 Incorporating Dirichlet boundary conditions in the final equations for global matrix-free finite element algorithm

Matrices and constants

lbu(nbn) – array with boundary nodes, nbn – array size for number of boundary conditions

nbc – total number of boundary nodes, nelem – total number of elements, nne – number of nodes per element; **lnode(mel, nen)** – array with element – nodal connectivity for all the elements, nen=8 for cubic element, mel – array size for elements; **nelc(nbc,nen)** – array with boundary nodes and corresponding number of contributing elements; **nrow(nbc,nen)** – array with boundary nodes and corresponding row numbers; **nct(nbn)** – array with total number elements surrounding a boundary node, **bu(nbn)** – value of boundary conditions; **glu(mel,nen)** – global load vector; **estffmku(mel,nen,nen)** – global stiffness matrix

```
do 21 i=1,nbc                                      Do this for all boundary nodes
      value=bu(i)
      nnc=nct(i)                                   Identify the number of elements for boundary node
      fract=(1.0/nnc)                              Compute the fraction for an element
      valuer=value*fract                           Compute the fraction of boundary value
do 22 k=1,nnc                                      Do this for all the elements for boundary node
      kel=nelc(i,k)                                Identify the column of element
      ii=nrow(i,k)                                 Identify the row of element
   do ij=1,nne                                     Do this for nodes of the element
      glu(kel,ij)=glu(kel,ij)-estffmku(kel,ij,ii)*value Modify the load vector at element level
      estffmku(kel,ii,ij)=0.0                      Make zero for the row entries of stiffness matrix
      estffmku(kel,ij,ii)=0.0                      Make zero for the column entries of stiffness matrix
      enddo                                        End this for all nodes of the element
      estffmku(kel,ii,ii)=1.0*fract          Assign the fraction of number of elements for boundary node
      glu(kel,ii)=valuer                     Assign the fraction of boundary value at the load vector
22     continue                                    Continue for all elements surrounding a boundary node
21    continue                                     Continue the process for all the boundary nodes
```

FIGURE 6.6 Computer program for incorporating Dirichlet boundary conditions using global matrix-free finite element algorithm.

where j and k refer to the row and column indices that vary from 1 to 8 for the selected 8-node brick element. The respective element-nodal connectivity for elements I and III is given as:

Element I: 1 **4** 5 2 10 13 14 11
Element III: **4** 7 8 5 13 16 17 14

The boundary node 4 appears in both the elements I and III but with different row numbers. It appears in the 2nd row in element I whereas it appears in the 1st row in element III. The algorithm for incorporating Dirichlet boundary conditions has been discussed by Reddy [11] by which the coefficient matrices and the load vectors of Equations (6.44a) and (6.44b) have to be modified for the known Dirichlet value at the boundary node 4. As the boundary node 4 is surrounded by two elements, I and III, then the Dirichlet value will be shared equally by these two elements, and accordingly the coefficient matrices and the load vectors of elements I and III have to be modified, that is in the present case, the contribution will be $\frac{1}{2}$ times the Dirichlet value at node 4 for elements I and III. This can be generalized that if ϕ_{bn} is the Dirichlet boundary value at a boundary node '*bn*', then each element surrounding this node gets a contribution equal to $\left(\frac{1}{nesbn}\right)\phi_{bn}$ where '*nesbn*' is the total number of elements surrounding the given boundary node. The above procedure is followed to modify all the element level coefficient matrices and the load vectors to incorporate the known Dirichlet values. When the implementation of this algorithm is completed for all the boundary nodes, then this will result in an element-wise system of simultaneous equations given as

$$\left[K_{ijk}\right]\left\{\phi_{ij}\right\} = \left\{f_{ij}\right\} \tag{6.45}$$

where i, j and k represent the element, row and column indices respectively.

For the solution of a large number of simultaneous equations, generally conjugate gradient iterative solvers are used, and these solvers make use of only the vectors obtained by the product of matrices and vectors for the solution procedure. In the GMFFE algorithm, all the computations are carried out at element level, and only vector level assembly is carried out to form the global vectors at the time of solution of simultaneous equations. This ensures the maximum size of a vector computed in the present algorithm is just equal to the total number of grid points irrespective of the global node numbering of the computational domain. In this aspect, the present approach is different from the frontal solver [4] whose efficiency depends upon the global node numbering of the elements in order to keep the front width minimum. In the GMFFE algorithm a simple conjugate gradient method without preconditioning developed by the author is used. The simplicity in implementing the present scheme on a personal computer is demonstrated by application to three-dimensional Stokes flow problems.

6.4 RESULTS FOR THREE-DIMENSIONAL STOKES FLOW

The GMFFE algorithm has been implemented to solve the three-dimensional Stokes flow equations as was discussed in the previous section. After reading the input data, the first step in the computation is to evaluate the number of elements surrounding each boundary node using the subroutine program given in Figure 6.6. Then, evaluation of matrices and load vector for velocity is carried out. The final converged solution is obtained by executing an iterative procedure to resolve the implicit coupling between vorticity transport equations and velocity Poisson equations through vorticity boundary conditions. The computational steps involved in the solution of three-dimensional Stokes flow are shown as a flow chart in Figure 6.7. The advantage of this algorithm is explained by carrying out computations using 11^3, 21^3, 31^3 and 51^3 meshes, and the storage space comparison with other established schemes is also compared.

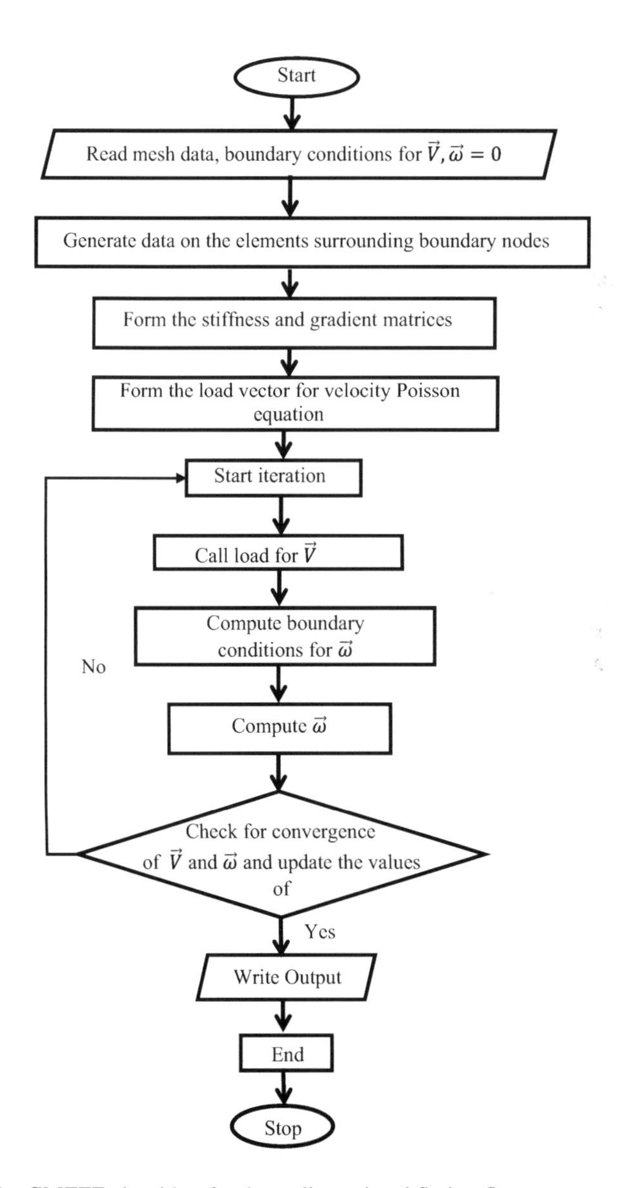

FIGURE 6.7 GMFFE algorithm for three-dimensional Stokes flow.

6.4.1 Comparison of Memory Storage of GMFFE Algorithm with Column Format Scheme

The computations for the solution of field variables of three-dimensional Stokes flow have been carried out on a Pentium-IV personal computer. In order to demonstrate the efficiency of the GMFFE algorithm to handle large-size problems on a personal computer, the details about the size of memory storage used in the present algorithm and the memory space required to store only the non-zero entries of the global coefficient matrix in vector format are computed for the meshes 11^3 (Mesh 1), 21^3 (Mesh 2), 31^3 (Mesh 3) and 51^3 (Mesh 4) and are compared with compact vector storage scheme as shown in Figure 6.8.

Figures 6.8(a) shows how the number of elements and nodes increase with refinement of meshes from 11^3 to 51^3. It can be noticed after 31^3 mesh, the number of nodes and elements increase very sharply, however the increase in number of nodes is very steep. Figure 6.8(b) depicts the comparison of storage of non-zero entries in column vector storage and global matrix-free finite element algorithm (GMFFE). As can be seen from this figure, the storage space in column vector scheme is smaller than that of GMFFE scheme because, the GMFFE algorithm stores the data in an array consisting of a number of elements, and nodes per element in row and column. It is worth analyzing the storage scheme of column vector method as a ratio of column vector storage scheme and number of nodes. The variation of this ratio with different mesh sizes is shown in Figure 6.8(c). With an increase in mesh size, this ratio increases as seen from the Figure 6.8(c). The most commonly used BICG iterative solvers [2], the non-zero entries in the coefficient matrices, are stored as a column vector and as seen from Figure 6.8(c), with refinement of mesh size, this storage size increases. The storage scheme proposed in the GMFFE algorithm can be compared with the column vector scheme. Figure 6.8(d) shows the ratio of storage space occupied by the GMFFE algorithm to the column vector scheme. It is very interesting to notice that as the number of nodes is increased by refining the mesh, this ratio decreases, thus indicating that the GMFFE algorithm is an efficient storage scheme for finer meshes.

This characteristic of the present algorithm has made it possible to execute the computer code for mesh 51^3 on a personal computer. Thus, the present scheme has proven to be highly efficient for storing and solving large number of equations obtained as a result of either multi-dimensional problems or mesh refinement. As a first attempt in implementing the global matrix-free finite element algorithm, a conjugate gradient iterative solver developed by the author without preconditioning has been employed in the present study. With this basic iterative solver, the convergence for the iterations could be achieved in a number of steps a little fewer than the total number of equations.

6.4.2 Flow Results for Three-Dimensional Stokes Flow Using 51^3 Mesh

6.4.2.1 Mesh Sensitivity and Validation Results

The computational domain for the three-dimensional Stokes flow problem is discretized using the transfinite interpolation technique (TFI) and initially three

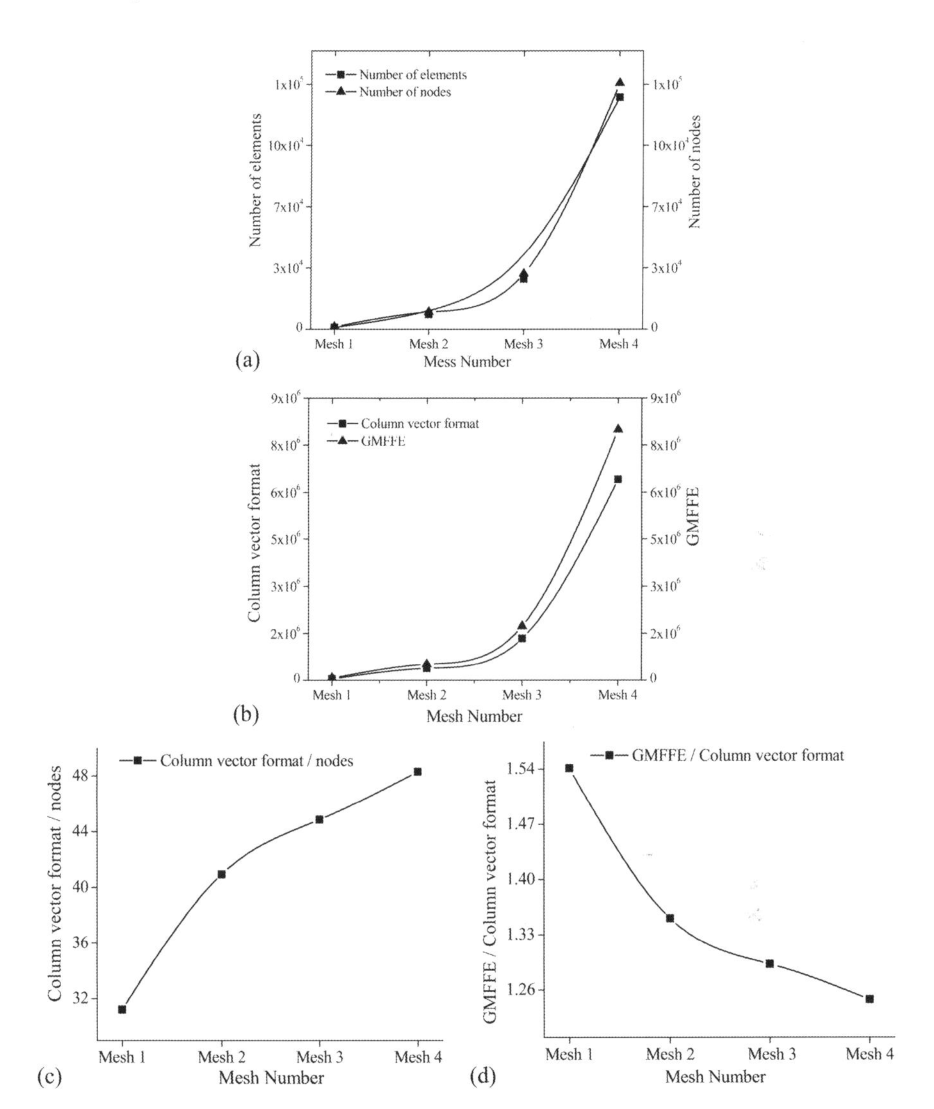

FIGURE 6.8 Comparison of memory storage of GMFFE algorithm and storage in column format.

meshes, 11^3, 21^3 and 31^3 have been generated for the purpose of mesh sensitivity study and to obtain validation results. The numerical simulation results have been obtained using 51^3 mesh as shown in Figure 6.9.

In order to implement the computational algorithm discussed in Figure 6.7, a computer program in FORTRAN, ***Stokes3D.for*** has been developed to obtain the simulation results. This computer program was executed on a Pentium-IV personal computer. Initially mesh sensitivity and validation results are obtained. The mesh sensitivity and validation results can be obtained from Ref. [12].

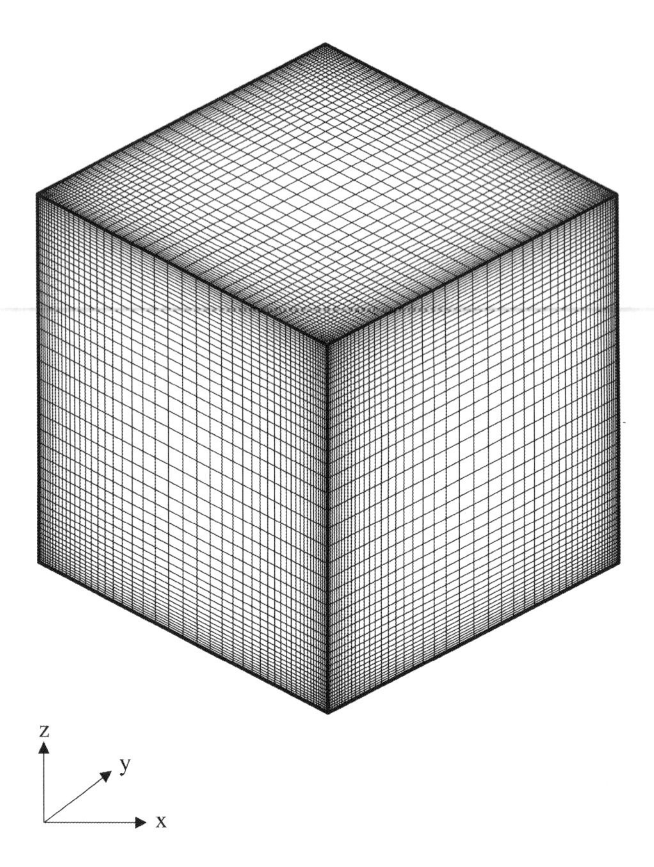

FIGURE 6.9 51^3 TFI Computational mesh for 3D Stokes flow.

6.4.2.2 Velocity Vectors Distribution

The simulation results for flow field have been obtained using the computer code after carrying out a mesh sensitivity study and validating the code. In any computer simulation program, especially for flow and heat transfer problems, it is required to test the computer program for mesh sensitivity and validation. Mesh sensitivity results confirm the correctness of implementation of the selected numerical method for discretizing the computational domain, in the present case, the finite element method, and when the chosen results converge with refinement of mesh, that shows the numerical results obtained for the solution of the governing equations is independent of the discretization of the computational domain. Validation results confirm that the computational algorithm and the computer code developed for the simulation program are capable of generating numerical results on par with the results obtained using some other numerical method and published in established literature. Thus, both these tests ensure that the computer code can be further implemented to simulate results as required to achieve the set objectives of the simulation program.

In the present three-dimensional Stokes flow problem, velocity and vorticity fields are obtained from simulation. Figure 6.10 shows the distribution of velocity vectors *u-w* on vertical plane *x-z*. The top lid of the cube moves with unit velocity in the *x*-direction from left to right. As the top lid moves from left to right, due to the assumption of no-slip boundary conditions, the fluid below the plate also moves from left to right and when it reaches the right wall, it turns back in the reverse direction, thus forming a re-circulatory flow pattern. A vortex structure is formed at the center of the flow field closer to the top moving plate and the velocity vectors are strong only near the top plate. This is because the fluid momentum is assumed to be theoretically zero because the convective terms are not considered in the Stokes equations and any fluid movement within the cubic cavity is achieved because of the pressure of the fluid and viscous forces. The central fluid core will get shifted if the inertial force of the fluid medium exceeds the viscous forces; however, in the present Stokes flow problem, only viscous forces are present, thus keeping the fluid vortex at the center of the cavity. As one moves from the top plate towards the bottom plate, the velocity vector field becomes very weak, reducing the velocity to zero at the boundaries. It is very interesting to notice the momentum boundary layer formation at the boundaries in the region varying from 0.5 to 0.75 in the *z*-direction on both sides of the plane. It has to be recalled that the velocity field is assumed to be zero on all the planes except at the top plate with u=1, v=w=0.

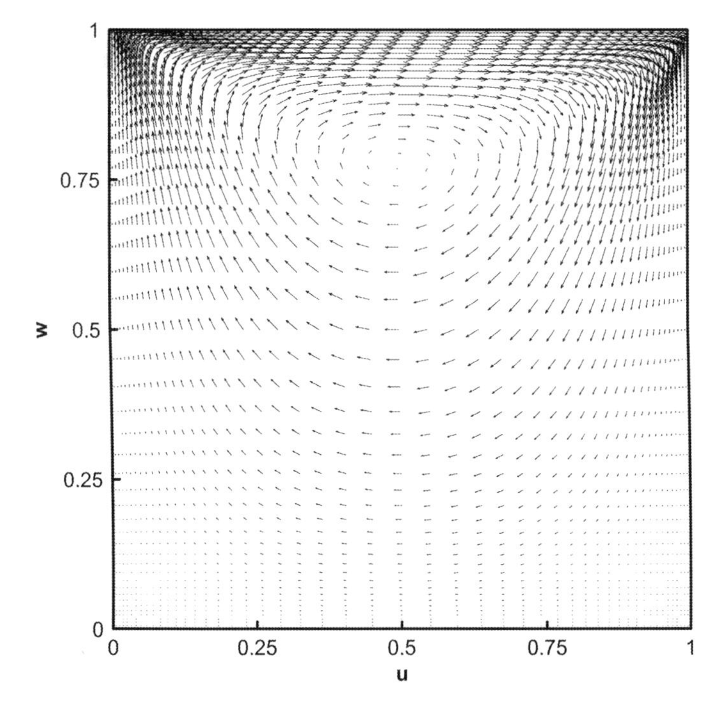

FIGURE 6.10 Velocity vector distribution on Y=0.5 plane.

It is essential to understand the nature of distribution of all the three vorticity components, ω_x, ω_y and ω_z to get the feel of vorticity transport within the flow domain. For this purpose, on the horizontal *x-y* plane at z=0.5, all the above vorticity components can be plotted on *x=0.5*, *y=0.5* and *z=0.5* planes. The definition of vorticity components in *x*, *y* and *z*-directions are described as below.

$$\omega_x = \left(\frac{\partial w}{\partial y} - \frac{\partial v}{\partial z}\right); \omega_y = \left(\frac{\partial u}{\partial z} - \frac{\partial w}{\partial x}\right) \text{ and } \omega_z = \left(\frac{\partial v}{\partial x} - \frac{\partial u}{\partial y}\right).$$

REFERENCES

1. Vector Calculus, P. C. *Maththews*, Springer, New York, 1998.
2. Press, W. H., Teukolshy, S. A., Vetterling, W. T. and Flannery, B. P. *Numerical Recipes in FORTRAN* – The art of scientific computing 2nd Edition, Cambridge University Press, Melbourne, Australia, 1992.
3. Saad, Y. *Iterative Methods for Sparse Linear Systems*, PWS Publishing Company, Boston, MA, 1996.
4. Irons, B. A frontal solution program for finite element analysis, *International Journal for Numerical Methods in Engineering*, 1970; 2: 5–32.
5. Hughes, T. J. R., Levit, I. and Winget, J. Element-by-element implicit algorithms for heat conduction, *Journal of Engineering Mechanics*, 1983; 109: 576–585.
6. Hughes, T. J. R., Levit, I. and Winget, J. An element-by-element solution algorithm for problems of structural and solid mechanics, *Computer Methods in Applied Mechanics and Engineering*, 1983; 36: 241–254.
7. Hestenes, M. R. and Stiefel, E. Methods of conjugate gradients for solving linear systems, *Journal of Research of the National Bureau of Standards*, 1952; 49: 409–436.
8. Sheu, T. W. H., Wang, M. M. T. and Tsai, S. F. Element-by-element parallel computation of incompressible Navier-Stokes equations in three dimensions, *SIAM Journal of Scientific Computing*, 2000; 21: 1387–1400.
9. Thiagarajan, G. and Aravamuthan, V. Parallelization strategies for element-by-element preconditioned conjugate gradient solver using high-performance Fortran for unstructured finite-element applications on Linux clusters, *Journal of Computing in Civil Engineering*, 2002; 16: 1–10.
10. Phoon, K. K. Iterative solution of large-scale consolidation and constrained finite element equations for 3D problems, in *International e-Conference on Modern Trends in Foundation Engineering: Geological Challenges and Solutions*, Indian Institute of Technology, Madras, India, January 26–30, 2004.
11. Reddy JN. *An Introduction to the Finite Element Method*, 2nd Edition, Mc-Graw Hill, Singapore, 1992. 12.
12. Murugesan, K., Lo, D. C. and Young, D. L. An efficient global matrix-free finite element algorithm for 3D flow problems, *Communications in Numerical Methods in Engineering*, 2005: 21: 107–118.

EXERCISE PROBLEMS

Qn: 1 A tank has a diameter of 30 cm and initially contains water up to a height of 50 cm. If water is drained from the tank through an outlet pipe of diameter 2.5 cm, calculate the time required to empty the tank.

Qn: 2 For the above problem plot the graph for time taken to empty the tank for height (a) 40 cm, (b) 50 cm and (c) 60 cm.

Qn: 3 Brine is fed into a tank at the rate of 0.03 m^3/s. The salt content in brine is 30 kg/m^3, while the tank initially has 3 m^3 of water. Outflow of liquid from the tank is at the rate of 0.015 m^3/s. Find the concentration of salt in the tank when tank has 5 m^3 of brine, given that the tank is well mixed.

Qn: 4 From the definition of vorticity field, compute the vorticity boundary condition along the bottom horizontal and right vertical walls of a square cavity. Develop the finite difference equations to compute the vorticity at the boundaries.

Qn: 5 For a two-dimensional flow field, derive the Stokes flow equations from Navier-Stokes equations in primitive variable form. Develop a computational algorithm for the solution of the velocity and pressure fields.

Qn: 6 In velocity-vorticity form of Stokes flow equations, there is no pressure term. Explain the methodology to compute pressure field from the velocity-vorticity fields obtained for a Stokes flow in a two dimensional flow domain.

Qn: 7 Generate computational mesh using TFI program for a rectangular cavity of size 2 × 1 (non-dimensional). Using the computer code Stokes3D.for, obtain results for Stokes flow in the rectangular cavity. Compare the velocity vectors and vorticity distributions obtained for the rectangular cavity with those results discussed for square cavity in this chapter. In which geometry is the vorticity distribution along the boundaries strong?

Qn: 8 Consider the finite element solution algorithm for a heat diffusion problem in three-dimensional geometry. Starting from a 11 × 11 × 11 mesh, compute the non-zero entries of the final global stiffness matrix by refining the mesh further. Develop some storage method to store only the non-zero entries of the stiffness matrix and their location in the matrix. Compare this method with the existing methods of storage of stiffness matrices.

Qn: 9 A global matrix-free finite element algorithm has been discussed in this chapter for efficient storage of non-zero entries of two-dimensional matrices obtained as a result of assembly of element level matrices. Develop this method further to apply this methodology for convective boundary condition. Draw a computational flow chart to implement the computational procedure for the convective boundary condition.

Qn: 10 Develop a methodology to implement the global matrix-free finite element algorithm for flux and adiabatic boundary conditions for a two-dimensional heat diffusion problem. Draw a flow chart to implement the computational procedure to incorporate the above boundary conditions in the main program.

QUIZ QUESTIONS

Qn: 1 Bernoulli's equation is nothing but the Euler equation when the fluid is assumed to be ________________________.

Qn: 2 In Stokes flow problems, ____ force is considered to be negligible.

Qn: 3 In velocity-vorticity form of Stokes flow equations, the momentum conservation equations are represented by the combination of (a) Helmholtz and Laplace equations, (b) Helmholtz and Poisson equations, (c) Laplace and Poisson equations or (d) none of the above.

Qn: 4 The number of governing equations for three-dimensional Stokes flow in primitive variable form of equations is ________________________ whereas it is ________________________ in velocity-vorticity form of equations.

Qn: 5 For the solution of large scale finite element equations, some of the efficient storage methods employed are ________________________ ________________________.

Qn: 6 In the solution of three-dimensional Stokes flow equations in velocity-vorticity form, the computation of vorticity boundary conditions on the right vertical surface involves the following velocity gradients: ________ ______________________________________.

Qn: 7 For accurate computation of vorticity values at the boundaries, the mesh near the boundaries has to be fine/coarse. Tick the right one.

Qn: 8 In the velocity vectors results obtained for three-dimensional Stokes flow, the fluid vortex is exactly located at the center of the plane y=0.5 and closer to the top moving lid because ______________________________.

Qn: 9 Vorticity distribution for Stokes flow depict symmetric distribution along the center line of the plane y=0.5 because ________________________ ________________________.

Qn: 10 Vorticity field is generated from the ________________________ ________________________ of a computational domain.

7 Navier-Stokes Equations

Flow problems are modelled to understand the effect of field variables such as velocity and pressure on the performance of a given system for fluid friction, heat transfer, species transport etc. The flow field obtained in a given situation depends on the type of geometry, characteristics of fluid, presence of other forces not related directly to the fluid medium, free flow or confined flow etc. The nature of flow field is determined by the type of variation of velocity and pressure in the system under consideration. Pressure differential maintained in a given flow domain decides the type of flow field variation. Fluid flow takes place mainly by virtue of pressure differential maintained in the system. When a fluid flows in a system, it flows with certain velocity and depending on the geometry, it may have either single or more than one velocity components. Change in velocity of a given lump of fluid results in change in momentum of the fluid which needs to be balanced for equilibrium conditions. When we talk about equilibrium conditions, one has to analyze whether in all real situations, equilibrium conditions are achieved. Let us consider the operation of a centrifugal pump used in a household for pumping water from a well and distributing to various utilities. Any pump is designed to run at a constant specific speed in most cases, to deliver the required discharge and pressure head. If one actually measures the discharge and pressure head in actual working conditions, it can be noticed that the pressure head and discharge do not remain constant for a given period of pumping of water. Of course, the manufacturer of the pump specifies the acceptable variations in these quantities when the pump runs at the design speed. When the system variables remain constant with the specified variations, then one can say that the pump lifts water at equilibrium conditions. A detailed study on pumps will indicate that in order to obtain the designed pressure head and discharge, the water that enters the inlet of the pump with nominal velocity and pressure undergoes many force balances within the casing of the pump.

7.1 MOMENTUM BALANCE OF FLUID IN A SYSTEM

Fluid mechanics basics has taught us the difference between fluid statics and fluid dynamics. Hydrostatic force plays a major role in fluid statics where the fluid is considered at rest without any velocity. Fluids with high density, especially incompressible fluids offer hydrostatic pressure during storage, which needs to be considered along with other forces. In fluid dynamics, the main focus is the study of fluid in motion, wherein the fluid may be compressible or incompressible. For the case of incompressible fluids, called liquids, this field of study is called hydrodynamics, in which the density of liquids is assumed to be constant in the given analysis. Fluid flow takes place in streams, sub-surface, through conduits, over surfaces or in open

DOI: 10.1201/9781003248071-7

atmospheres. Study of each of these flow problems can be treated as a separate field of study because each problem is defined by its own characteristics. In stream flows, the fluid is exposed to atmospheric conditions and in the case of sub-surface flow, the flow passage is formed by soil or earth structure. Flow of fluids through soil is studied as flow through porous media, which is governed by Darcy's law. When fluids flow through conduits, the resulting flow pattern is entirely different from the one formed in stream flow or sub-surface flow. Flow of fuels through injection nozzles in automobiles, water discharge through pipe lines, flow of exhaust gases from automobile and rocket engines exit etc. give rise to different types of flow patterns. Atmospheric science is another interesting field of study on fluid dynamics because wind flow pattern in a given locality contributes significantly to the resulting weather pattern in that locality. Experimental methods have been well established in fluid dynamics research over centuries to understand the mechanism and physics behind many fluid flow problems. However, in the past few decades, due to the availability of high-speed computers and improved knowledge on numerical methods, computational fluid dynamics (CFD) has occupied a huge space in industrial research. Aeronautics science is the area that made computational fluid dynamics an important tool for research, and this has helped scientists to understand the consequence of fluid flow over surfaces and contributed to better design of the present day aircraft. In most CFD programs, the fluid momentum equations are solved along with the continuity equation to simulate the flow field for a given problem. The commercially available CFD software packages have made it possible to analyze many three-dimensional systems with complex geometry. One can benefit more from the usage of such packages provided he/she has thorough knowledge of the physics behind the governing equations being solved in these kinds of software. Analysis of results obtained from these programs cannot be fruitful unless there is enough background to appreciate the results.

Fluid flow in a system is investigated with some specific purpose to achieve certain pre-determined goals. For example, in the design of the outside body structure of an automobile, the main focus is to achieve an aerodynamically efficient design in order to reduce power consumption by the vehicle to overcome the fluid friction at different speeds of the vehicle. When a CFD analysis is carried out for this problem, different shapes of profiles are considered and the resulting fluid friction is estimated. The velocity distribution over the surface of these profiles is simulated in order to compute the fluid friction. These velocity profiles are obtained by solving the momentum balance equations, which take care of all the forces acting on the fluid in the given computational domain. The fluid momentum balance equations are called Navier-Stokes equations, which consist of continuity equation also called mass conservation equation and momentum balance equation for each coordinate direction of the system. The momentum balance of fluid flow in a system can be easily understood by considering the example of the pumping of water using a centrifugal pump as shown in Figure 7.1. Water enters the pump through the suction pipe from the lower level of water. The very purpose of pumping water is to raise its pressure head so that it can be stored in a tank at a height such that it can be drawn for use by gravity. The pump is run by an electric motor, which gives mechanical energy

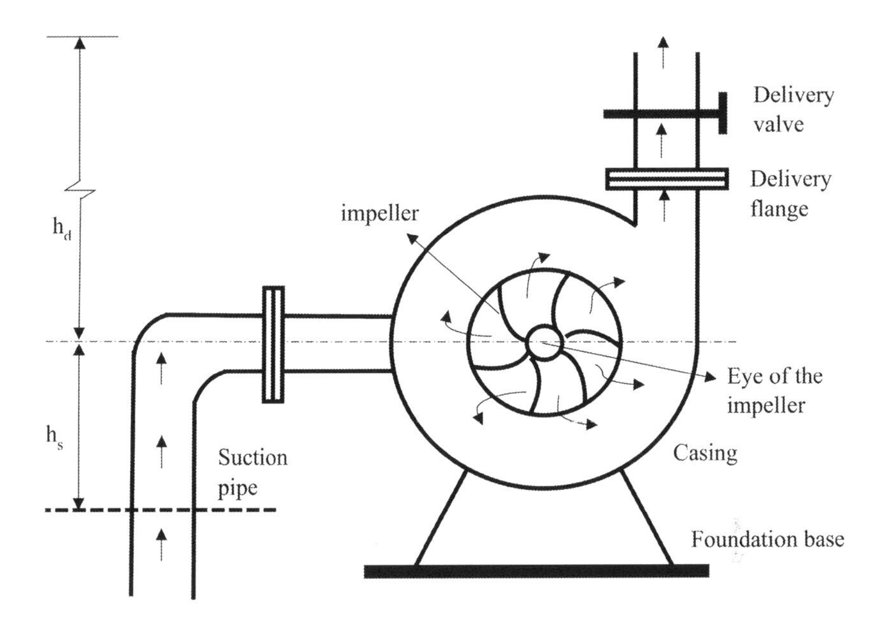

FIGURE 7.1 Schematic diagram of a centrifugal pump.

to the water entering through the suction pipe. In order to handle a certain quantity of water and increase its pressure head, the impeller blades are designed such that the shaft work from the electric motor is utilized maximum to achieve the target. As the water flows through the impeller, its kinetic energy increases and by allowing it to pass through a convolute casing of the pump, some of the kinetic energy is converted into pressure. With this increased pressure head, the water can easily reach the storage tank height through the discharge pipe. The momentum of water increases as it passes through the impeller of the pump and the pressure of water increases at the discharge pipe compared to its pressure at the suction pipe. When the same pump has to pump a liquid with viscosity higher than water, then the shaft work required to achieve the design parameters of pressure and discharge will also change due to increased fluid resistance offered by the liquid. Thus, there exists a definite balance of different types of forces within a pump so that the water is discharged at the design pressure and volume flow rate. When this force balance is achieved, then the system is said to be in equilibrium.

7.1.1 Fluid Dynamics

Fluid dynamics focuses on the study of variation of velocity and pressure for fluid flow in a given system. The fluid that flows has a certain mass on its own, and hence any change in velocity of the fluid will give rise to a change in momentum, and according to Newton's II law, this change in momentum of fluid must be equal to the net forces acting on the fluid. In order to understand various terms appearing in Navier-Stokes equations, it is essential to deliberate on different forces acting on fluid medium. Viscosity is an important physical property of a fluid that determines the fluid friction offered by a fluid in a flow field. Fluids in statics do not offer any frictional force though they possess viscosity, the fluid friction comes into effect only when the fluid flows in a system. From Newton's law of viscosity, one can easily appreciate that force due to viscosity of a fluid is directly proportional to velocity gradient. Hence, depending on the flow regime, type of flow and geometry, the variation of velocity in a given domain is determined. This means a fluid has to flow in a given fluid path after overcoming its own resistance for flow. If water has to flow in a pipe at a specified velocity and pressure, then some amount of work has to be done on water in the form of pump work to maintain the flow after overcoming the fluid friction. Now, the force due to fluid friction can be considered as one of the forces acting on the fluid in a flow field. Fluid flow takes place mainly because of a negative pressure gradient generated in the direction of flow and pressure is an inherent intensive property of a fluid that contributes to the mechanism behind fluid flow. As fluid flows over a surface, a force due to pressure of the fluid is exerted on the fluid itself. When velocity gradients of fluid are considered for the contribution of tangential and normal stresses of the fluid, then the normal stress is equated to the pressure of fluid according to Stokes law. Other than the forces due to fluid friction and fluid pressure, there may be body forces acting on the fluid.

The momentum of a fluid is obtained by the product of its mass and change in velocity, which is an important characteristic of any fluid flow system. Fluid systems are characterized by the nature of variation of velocity patterns in the domain. Flow over aerodynamic structure is deliberately designed such that the required lift force is maximum with minimum drag force. In the design of impeller blades in a pump, care is taken to create blade profiles that will produce the least amount of unwanted forces that have to be balanced to keep the pump in equilibrium condition, however, the profile should give rise to enough kinetic head. Thus, flow systems are designed with specific objectives to achieve desired flow pattern, which is the result of an ultimate force balance that is satisfied in each coordinate direction. Changes in momentum of fluid due to flow pattern need to be balanced by the net forces acting on the fluid. The net force is obtained as a result of constituent forces due to surface forces and body forces. Fluid friction and pressure give rise to surface forces whereas forces due to buoyancy or magnetic field contribute to the body forces. It has to be appreciated that during momentum balance of fluid in a system, the mass of fluid has to be conserved, of course, this has to be accounted for during mathematical modeling of the flow system. As momentum of fluid is linked with the mass of fluid, then either creation of mass of fluid or destruction of mass of fluid will result in an unrealistic system at the same time disturbing the momentum balance of the fluid.

7.2 NAVIER-STOKES EQUATIONS IN PRIMITIVE VARIABLES FORM

Any real problem is complex in nature due to geometric irregularity or multiple physical principles that govern the behavior of the system. Though experimental techniques are the most reliable and well-established methodology for analysis, it is not always possible to carry out experimental trials for all types of problems in engineering applications. In order to save time, effort and money involved in performing experiments for certain types of engineering problems, modeling has become a handy tool to analyze the behavior of the system under varying operating conditions without actually fabricating the system. With the availability of high-speed computers, modeling and simulation techniques are effectively used in all branches of science and engineering. This has resulted in the development of many kinds of commercial software for applied problems. Though software can do the required simulation analysis for a given problem, knowledge about the governing equations and solution methodology will help the engineers to have a better understanding of the physics underlying the problem. Generally, any CFD problem involves flow analysis along with heat transfer and mass transfer, especially in mechanical engineering-related problems.

CFD analysis mainly involves numerical solution of Navier-Stokes equations for a given geometry. Any solution methodology depends only upon the type of governing equations. The geometry of the physical system does not affect the solution methodology. As regards the Navier-Stokes equations, they represent the force balance of the fluid medium in a given flow situation with mass conservation. Whenever a fluid flows through a system, it comes across the system boundary, resulting in velocity and pressure gradients, which will in turn affect the force balance of the fluid medium. For example, when water flows through a turbine or a pump, the fluid undergoes a significant change in velocity and pressure as the fluid flows over the passages of the blades and the casing. The system performance is measured only under equilibrium conditions. Knowledge about the force balance of the fluid medium in the turbine or pump is highly essential in order to provide the required mechanical balance for the stable operation of the system. As the fluid passes through the system the conservation of mass principle also has to be satisfied for the flow situation.

7.2.1 Navier-Stokes Equations

The Navier-Stokes equations in primary variable form, that is, velocity and pressure are the most commonly used form of equations in CFD applications. It consists of a continuity equation that represents mass conservation within a control volume and momentum conservation equations in the respective coordinate directions. The detailed derivation of these equations were discussed in Chapter 2. For a two-dimensional field problem, these equations can be expressed as

Continuity Equation

$$\frac{\partial u}{\partial x}+\frac{\partial v}{\partial y}=0 \tag{7.1}$$

x-Momentum Equation

$$\frac{\partial u}{\partial t}+u\frac{\partial u}{\partial x}+v\frac{\partial u}{\partial y}=-\frac{\partial p}{\partial x}+\frac{1}{\mathrm{Re}}\left(\frac{\partial^2 u}{\partial x^2}+\frac{\partial^2 u}{\partial y^2}\right) \tag{7.2}$$

y-Momentum Equation

$$\frac{\partial v}{\partial t}+u\frac{\partial v}{\partial x}+v\frac{\partial v}{\partial y}=-\frac{\partial p}{\partial y}+\frac{1}{\mathrm{Re}}\left(\frac{\partial^2 v}{\partial x^2}+\frac{\partial^2 v}{\partial y^2}\right) \tag{7.3}$$

The main difficulty in the solution of the above equations is that the pressure does not appear explicitly in the continuity equation and it has to be satisfied implicitly by satisfying the incompressibility constraint. In the momentum equation, the continuity equation has to be satisfied, so that the mass conservation principle is not violated while dealing with the momentum balance. That means when fluid flow takes place through a control volume, once the mass conservation is satisfied, then there cannot be any generation or depletion of mass of the fluid within the control volume. If mass conservation is not satisfied due to some reason or other, then the momentum of fluid equivalent to the mass generated or depleted will cause momentum imbalance, and the resulting velocity and pressure fields will not be realistic according to the underlying physics. The solution of Navier-Stokes equations have to be viewed in two major steps, one is to propose a methodology that will resolve the issue of pressure term appearing only in the momentum equations and not in the continuity equation and the second one is the implementation of the numerical scheme such as finite difference, finite volume and finite element methods. It is the first step that distinguishes the solution methodology of Navier-Stokes equations from one method to another.

7.2.2 Application of Predictor-Corrector Method

Apart from taking care of the coupled and non-linear nature of the Navier-Stokes equations, the difficulty arising out of handling the pressure term also has to be considered before the application of the numerical solution procedure. The predictor-corrector scheme is one of the most commonly used methods to solve the Navier-Stokes equations in primitive variable form. This method solves the equations in two steps; in the first step, the momentum equations are solved without the pressure term to obtain pseudo velocities and then these velocities are corrected by including the effect of pressure after solving the pressure Poisson equation. The predictor-corrector method, also called Eulerian velocity correction method, is based on Helmholtz-Hodge decomposition principle as proposed by Chorin and Marsden [1]. This algorithm is an explicit algorithm, in which the velocity and pressure fields are computed by marching in time using the values of converged velocity at the previous time step. Using the algorithm proposed by Chorin and Marsden [1], a vector can be expressed as

$$V(x)=u(x)+\nabla\varphi(x) \tag{7.4}$$

by satisfying

$$\nabla\cdot u=0$$

and $\varphi(x)$ is a scalar.

The predictor-corrector method also called the Eulerian velocity correction method can be well understood by implementing this technique on the vector form of Navier-Stokes equations.

The continuity and momentum equations can be expressed in vector form as

$$\nabla \cdot V = 0 \tag{7.5a}$$

$$\frac{\partial V}{\partial t} + V \cdot \nabla V = -\nabla p + \frac{1}{\text{Re}} \nabla^2 V \tag{7.5b}$$

Discretizing the time derivative using the finite difference technique in the above equation, we get

$$\frac{V^{n+1} - V^n}{\Delta t} = -\nabla p^{n+1} - V^n \cdot \nabla V^n + \frac{1}{\text{Re}} \nabla^2 V^n \tag{7.5c}$$

A pseudo velocity which does not have the effect of scalar pressure quantity is defined as V^*.

By adding and subtracting V^* in Equation (7.5c) we get

$$\frac{V^{n+1} - V^*}{\Delta t} + \frac{V^* - V^n}{\Delta t} = -\nabla p^{n+1} - V^n \cdot \nabla V^n + \frac{1}{\text{Re}} \nabla^2 V^n \tag{7.5d}$$

Now Equation (7.5d) can be split into two parts using Equation (7.4) as

$$\frac{V^* - V^n}{\Delta t} = -V^n \cdot \nabla V^n + \frac{1}{\text{Re}} \nabla^2 V^n \tag{7.5e}$$

$$\frac{V^{n+1} - V^*}{\Delta t} = -\nabla p^{n+1} \tag{7.5f}$$

Simplification of Equation (8.5e) will produce the following equation.

$$V^* = V^n - \Delta t V^n \cdot \nabla V^n + \frac{\Delta t}{\text{Re}} \nabla^2 V^n \tag{7.6}$$

By taking divergence of Equation (7.5f) we get

$$\nabla \cdot \left(\frac{V^{n+1} - V^*}{\Delta t} \right) = -\nabla \cdot \nabla p^{n+1}$$

Since $\nabla \cdot V^{n+1} = 0$ as proposed by the Helmholtz-Hodge decomposition principle

$$\frac{\nabla \cdot V^{n+1}}{\Delta t} - \frac{\nabla \cdot V^*}{\Delta t} = -\nabla \cdot \nabla p^{n+1}$$

$$\nabla^2 p^{n+1} = \frac{1}{\Delta t} \nabla \cdot V^* \tag{7.7}$$

Equation (7.7) is called the pressure Poisson equation. The continuity equation at the current time level, *(n+1)* is satisfied while forming the pressure Poisson equation. It has to be noticed that using the predictor-corrector method, the field variables, velocities and pressure are computed using explicit time marching scheme. Hence, while computing the pressure field using the pressure Poisson equation, the pseudo velocities are only used, and the divergence free condition has to be enforced at the current time step. The corrected final velocity field at current time level, *(n+1)* is computed as

$$\frac{V^{n+1} - V^*}{\Delta t} = -\nabla p^{n+1}$$

Then using Equation (7.5f), the corrected velocity can be obtained as

$$V^{n+1} = V^* - \Delta t \Delta p^{n+1} \tag{7.8}$$

Now Equations (7.6) to (7.8) have to be solved to get the velocity and pressure fields. The computational algorithm can be explained as follows:

For the given time step, using the velocities at the previous time step

(i) Compute the fictitious velocities, u^* and v^* using Equation (7.6)
(ii) Solve the pressure Poisson equation using Equation (7.7) to obtain the pressure field
(iii) Correct the fictitious velocities using Equation (7.8)

The above procedure is repeated until steady state solution is achieved. Then go to the next time level. In order to solve these equations, the boundary conditions for velocities and pressure have to be specified.

7.2.3 Finite Element Solution Procedure

The governing equations of momentum balance as modified using the predictor-corrector method have been solved using Galerkin's finite element method as explained below. Equations (7.6) to (7.8) expressed in vector form need to be expanded for the application of the finite element method. Equation (7.6) can be expanded for u velocity in the x-direction momentum equation as

$$\frac{\partial u^*}{\partial t} = -\left[u\frac{\partial u}{\partial x} + v\frac{\partial u}{\partial y} \right]^n + \frac{1}{\mathrm{Re}}\left[\frac{\partial^2 u}{\partial x^2} + \frac{\partial^2 u}{\partial y^2} \right]^n \tag{7.9}$$

which can be further simplified as

$$u^* = u^n + \Delta t^n \left\{ -\left[u\frac{\partial u}{\partial x} + v\frac{\partial u}{\partial y} \right]^n + \frac{1}{\text{Re}}\left[\frac{\partial^2 u}{\partial x^2} + \frac{\partial^2 u}{\partial y^2} \right]^n \right\} \tag{7.9a}$$

Similarly, the momentum equation in y-direction can be expressed as

$$\frac{\partial v^*}{\partial t} = -\left[u\frac{\partial v}{\partial x} + v\frac{\partial v}{\partial y} \right]^n + \frac{1}{\text{Re}}\left[\frac{\partial^2 v}{\partial x^2} + \frac{\partial^2 v}{\partial y^2} \right]^n$$

which can be further simplified as

$$v^* = v^n + \Delta t^n \left\{ -\left[u\frac{\partial v}{\partial x} + v\frac{\partial v}{\partial y} \right]^n + \frac{1}{\text{Re}}\left[\frac{\partial^2 v}{\partial x^2} + \frac{\partial^2 v}{\partial y^2} \right]^n \right\} \tag{7.9b}$$

Solution of Equations (7.9a) and (7.9b) will produce results for the pseudo velocities, u^* and v^* in the computational domain. In order to affect the pressure field, the pressure Poisson equation has to be solved. Now, the pressure Poisson equation given by Equation (7.7) can be written as

$$\frac{\partial^2 p}{\partial x^2}^{n+1} + \frac{\partial^2 p}{\partial y^2}^{n+1} = \frac{1}{\Delta t^n}\left[\frac{\partial u^*}{\partial x} + \frac{\partial v^*}{\partial y} \right] \tag{7.10}$$

As the last step in the solution procedure, the pressure field obtained from the solution of the pressure Poisson equation has to be added to the pseudo velocities in order to obtain the final velocities, u and v. The corrected velocities can be obtained using Equation (7.8). The corrected u velocity in x-direction can be expressed as

$$\frac{\partial u}{\partial t} = -\frac{\partial p}{\partial x} + \frac{\partial u^*}{\partial t}$$

$$u^{n+1} = u^* - \Delta t^n \frac{\partial p}{\partial x}^{n+1} \tag{7.11a}$$

Similarly, the corrected v velocity in y-direction can be expressed as

$$\frac{\partial v}{\partial t} = -\frac{\partial p}{\partial y} + \frac{\partial v^*}{\partial t}$$

$$v^{n+1} = v^* - \Delta t^n \frac{\partial p}{\partial y}^{n+1} \tag{7.11b}$$

It has to be noticed that the pseudo velocities are obtained using Equations (7.9a) and (7.9b) with the known velocity fields at the previous time level, n. Then, the pressure

Poisson equation expressed by Equation (7.10) is solved to obtain the pressure field at time level *(n+1)* using the pseudo velocities as load vector. With the pressure field obtained, the u and v velocities are corrected to get the true velocities at the current time level, *(n+1)* using Equation (7.11).

The above equations can be solved using Galerkin's weighted residual finite element method as follows. Consider the x-momentum equation expressed by Equation (7.9a) for pseudo velocity, u^*. Application of Galerkin's method to Equation (7.9a) gives the following equation.

$$\int N^T \left\{ \begin{aligned} &(u^* - u^n) + \Delta t^n \left[u\frac{\partial u}{\partial x} + v\frac{\partial u}{\partial y} \right]^n \\ &-\Delta t^n \frac{1}{\text{Re}} \left[\frac{\partial^2 u}{\partial x^2} + \frac{\partial^2 u}{\partial y^2} \right]^n \end{aligned} \right\} dA = 0 \tag{7.12}$$

Before proceeding to the integration of various terms in the above expression, the element type that will be used to discretize the computational domain has to be decided. For the purpose of simplicity, linear triangular element is assumed to be used to discretize the flow domain. Of course, the finite element implementation for the solution of Navier-Stokes equations using bilinear quadrilateral isoparametric element will also be explained in the next section while solving the Navier-Stokes equations in velocity-vorticity form. The integration of derivatives and double derivatives using Galerkin's method for a linear triangular element have to be understood. The shape functions for a linear triangular element can be expressed as

$$N_i(x, y) = \frac{1}{2A}\left(a_i + b_i x + c_i y\right) \tag{7.13a}$$

$$N_j(x, y) = \frac{1}{2A}\left(a_j + b_j x + c_j y\right) \tag{7.13b}$$

$$N_k(x, y) = \frac{1}{2A}\left(a_k + b_k x + c_k y\right) \tag{7.13c}$$

where $a_i = x_j y_k - x_k y_j;\; b_i = y_j - y_k;\; c_i = x_k - x_j$

$$a_j = x_k y_i - x_i y_k;\; b_j = y_k - y_i;\; c_j = x_i - x_k$$

$$a_k = x_i y_j - x_j y_i;\; b_k = y_i - y_j;\; c_k = x_j - x_i \text{ and } 2A = \det\begin{vmatrix} 1 & x_i & y_i \\ 1 & x_j & y_j \\ 1 & x_k & y_k \end{vmatrix}$$

The gradient matrix for a linear triangular element for any variable ϕ can be expressed as [2]

$$\{g\} = \begin{bmatrix} \frac{\partial N_i}{\partial x} & \frac{\partial N_j}{\partial x} & \frac{\partial N_k}{\partial x} \\ \frac{\partial N_i}{\partial y} & \frac{\partial N_j}{\partial y} & \frac{\partial N_k}{\partial y} \end{bmatrix} \begin{Bmatrix} \phi_i \\ \phi_j \\ \phi_k \end{Bmatrix}$$

$$= \frac{1}{2A} \begin{bmatrix} b_i & b_j & b_k \\ c_i & c_j & c_k \end{bmatrix} \begin{Bmatrix} \phi_i \\ \phi_j \\ \phi_k \end{Bmatrix}$$

$$\{g\} = [B]\{\phi\} \tag{7.14}$$

Now, the integration of the diffusion term in a two-dimensional domain using a triangular element can be evaluated as

$$\int_{\Omega} [N]^T \left(\frac{\partial^2 \phi}{\partial x^2} + \frac{\partial^2 \phi}{\partial y^2} \right) dA = -\int_{\Omega} \left(\frac{\partial [N]^T}{\partial x} \frac{\partial [N]}{\partial x} + \frac{\partial [N]^T}{\partial y} \frac{\partial [N]}{\partial y} \right) dA \{\phi\}$$
$$+ \int_{\Gamma} [N]^T \frac{\partial \phi}{\partial n} d\Gamma$$

The second term on the right-hand side of the above equation contributes only for the boundary condition, hence it can be treated at the time of including the boundary conditions after forming the global matrices. Using Equation (7.14), the first term in the above equation can be expressed as

$$[K] = \int_{\Omega} [B]^T [D][B] dA \tag{7.15}$$

Equation (7.15) is called the stiffness matrix, with *[D]* indicating the thermo-physical properties associated with the diffusion equation. In all the example problems discussed in this section, only fluids with isotropic properties will be discussed. Hence, the *[D]* matrix can be simply replaced with the respective physical properties. With the help of area coordinates, the integration of shape functions of triangular elements can be expressed as

$$\int_{A} L_1^a L_2^b L_3^c dA = \frac{a!b!c!}{(a+b+c+2)!} 2A \tag{7.16}$$

Using Equation (7.16) the following shape function product commonly used in finite element formulation can be evaluated as

$$\int_A [N]^T [N] dA = \int_A \begin{bmatrix} N_i^2 & N_iN_j & N_iN_k \\ N_iN_j & N_j^2 & N_jN_k \\ N_iN_k & N_jN_k & N_k^2 \end{bmatrix} dA$$

$\int_A N_i^2 dA = \frac{2!0!0!}{(2+0+0+2)!} 2A = \frac{A}{6}$; $\int_A N_iN_j dA = \frac{1!1!0!}{(1+1+0+2)!} 2A = \frac{A}{12}$. Hence,

$$\int_A [N]^T [N] dA = \frac{A}{12} \begin{bmatrix} 2 & 1 & 1 \\ 1 & 2 & 1 \\ 1 & 1 & 2 \end{bmatrix} \tag{7.17}$$

Now, integration of Equation (7.12) can be performed for linear triangular elements as follows.

$$\int N^T \left\{ \begin{array}{c} (u^* - u^n) + \Delta t^n \left[u \frac{\partial u}{\partial x} + v \frac{\partial u}{\partial y} \right]^n \\ - \Delta t^n \frac{1}{\text{Re}} \left[\frac{\partial^2 u}{\partial x^2} + \frac{\partial^2 u}{\partial y^2} \right]^n \end{array} \right\} dA = 0$$

$$\begin{aligned} &\int_A [N]^T u^* dA - \int_A [N]^T u^n dA + \Delta t^n \int_A [N]^T u^n \frac{\partial u^n}{\partial x} dA \\ &+ \Delta t^n \int_A [N]^T v^n \frac{\partial u^n}{\partial y} dA - \frac{\Delta t^n}{\text{Re}} \int_A [N]^T \left(\frac{\partial^2 u}{\partial x^2} + \frac{\partial^2 u}{\partial y^2} \right) dA = 0 \end{aligned} \tag{7.18}$$

Let us evaluate every term in Equation (7.18) independently for linear triangular element. Let us consider the first term $\int_A [N]^T u^* dA = \int_A [N]^T [N] dA \{u^*\}$. Substituting Equation (7.17) in this expression, we get

$$\int_A [N]^T u^* dA = \frac{A}{12} \begin{bmatrix} 2 & 1 & 1 \\ 1 & 2 & 1 \\ 1 & 1 & 2 \end{bmatrix} \begin{Bmatrix} u_i^* \\ u_j^* \\ u_k^* \end{Bmatrix} \tag{7.19a}$$

Now the second term in Equation (7.18) is evaluated as

$$\int_A [N]^T u^n dA = \int_A [N]^T [N] dA \{u^n\}$$

$$\int_A [N]^T u^n dA = \frac{A}{12}\begin{bmatrix} 2 & 1 & 1 \\ 1 & 2 & 1 \\ 1 & 1 & 2 \end{bmatrix}\begin{Bmatrix} u_i^n \\ u_j^n \\ u_k^n \end{Bmatrix} \tag{7.19b}$$

The convective term with velocity u in Equation (7.18) is evaluated as $\int_A [N]^T u^n \frac{\partial u^n}{\partial x} dA = \int_A [N]^T [N]\{u^n\} \frac{\partial N}{\partial x}\{u^n\} dA$. Using Equations (7.14) and (7.17), this expression can be rewritten as

$$\int_A [N]^T u^n \frac{\partial u^n}{\partial x} dA = \frac{A}{12}\begin{bmatrix} 2 & 1 & 1 \\ 1 & 2 & 1 \\ 1 & 1 & 2 \end{bmatrix}\begin{Bmatrix} u_i^n \\ u_j^n \\ u_k^n \end{Bmatrix}\begin{bmatrix} b_i u_i & b_j u_j & b_k u_k \end{bmatrix} \tag{7.19c}$$

Similarly, the convective term with velocity v, that is the fourth term in Equation (7.18) is written as

$$\int_A [N]^T v^n \frac{\partial u^n}{\partial y} dA = \frac{A}{12}\begin{bmatrix} 2 & 1 & 1 \\ 1 & 2 & 1 \\ 1 & 1 & 2 \end{bmatrix}\begin{Bmatrix} v_i^n \\ v_j^n \\ v_k^n \end{Bmatrix}\begin{bmatrix} c_i u_i & c_j u_j & c_k u_k \end{bmatrix} \tag{7.19d}$$

The diffusion term that appears as the last term in Equation (7.18) can be integrated as

$$\int_A [N]^T \left(\frac{\partial^2 u^n}{\partial x^2} + \frac{\partial^2 u^n}{\partial y^2}\right) dA = \int_A \left(\frac{\partial [N]^T}{\partial x}\frac{\partial [N]}{\partial x} + \frac{\partial [N]^T}{\partial y}\frac{\partial [N]}{\partial y}\right) dA \{u^n\}$$

$$\int_A [N]^T \left(\frac{\partial^2 u^n}{\partial x^2} + \frac{\partial^2 u^n}{\partial y^2}\right) dA = \begin{bmatrix} b_i \\ b_j \\ b_k \end{bmatrix}\begin{bmatrix} b_i & b_j & b_k \end{bmatrix}\begin{Bmatrix} u_i^n \\ u_j^n \\ u_k^n \end{Bmatrix} \tag{7.19e}$$

After substituting all the terms in the x-momentum equation, the final expression for the fictitious velocity, u^* can be expressed as

$$\frac{A}{12}\begin{bmatrix} 2 & 1 & 1 \\ 1 & 2 & 1 \\ 1 & 1 & 2 \end{bmatrix}\begin{Bmatrix} u_i^* \\ u_j^* \\ u_k^* \end{Bmatrix} = \frac{A}{12}\begin{bmatrix} 2 & 1 & 1 \\ 1 & 2 & 1 \\ 1 & 1 & 2 \end{bmatrix}\begin{Bmatrix} u_i^n \\ u_j^n \\ u_k^n \end{Bmatrix}$$

$$-\frac{A\Delta t^n}{12}\begin{bmatrix} 2 & 1 & 1 \\ 1 & 2 & 1 \\ 1 & 1 & 2 \end{bmatrix}\begin{Bmatrix} u_i^n \\ u_j^n \\ u_k^n \end{Bmatrix}\begin{bmatrix} b_i u_i & b_j u_j & b_k u_k \end{bmatrix} -$$

$$\frac{A\Delta t^n}{12}\begin{bmatrix}2 & 1 & 1\\ 1 & 2 & 1\\ 1 & 1 & 2\end{bmatrix}\begin{Bmatrix}v_i^n\\ v_j^n\\ v_k^n\end{Bmatrix}\begin{bmatrix}c_iu_i & c_ju_j & c_ku_k\end{bmatrix}-$$

$$\frac{\Delta t^n}{4A\,\mathrm{Re}}\begin{bmatrix}b_i\\ b_j\\ b_k\end{bmatrix}\begin{bmatrix}b_i & b_j & b_k\end{bmatrix}\begin{Bmatrix}u_i^n\\ u_j^n\\ u_k^n\end{Bmatrix}-$$

$$\frac{\Delta t^n}{4A\,\mathrm{Re}}\begin{bmatrix}c_i\\ c_j\\ c_k\end{bmatrix}\begin{bmatrix}c_i & c_j & c_k\end{bmatrix}\begin{Bmatrix}u_i^n\\ u_j^n\\ u_k^n\end{Bmatrix} \tag{7.20}$$

The y-momentum equation expressed by Equation (7.9b) can be expressed for finite element solution as follows.

$$\int N^T\begin{Bmatrix}(v^*-v^n)+\Delta t^n\left[u\dfrac{\partial v}{\partial x}+v\dfrac{\partial v}{\partial y}\right]^n\\ -\Delta t^n\dfrac{1}{\mathrm{Re}}\left[\dfrac{\partial^2 v}{\partial x^2}+\dfrac{\partial^2 v}{\partial y^2}\right]^n\end{Bmatrix}dA=0 \tag{7.21}$$

Now, following the procedure adopted for the x-momentum equation, the integration of all the terms in the above equation will finally give rise to the following expression.

$$\frac{A}{12}\begin{bmatrix}2 & 1 & 1\\ 1 & 2 & 1\\ 1 & 1 & 2\end{bmatrix}\begin{Bmatrix}v_i^*\\ v_j^*\\ v_k^*\end{Bmatrix}=\frac{A}{12}\begin{bmatrix}2 & 1 & 1\\ 1 & 2 & 1\\ 1 & 1 & 2\end{bmatrix}\begin{Bmatrix}v_i^n\\ v_j^n\\ v_k^n\end{Bmatrix}$$

$$-\frac{A\Delta t^n}{12}\begin{bmatrix}2 & 1 & 1\\ 1 & 2 & 1\\ 1 & 1 & 2\end{bmatrix}\begin{Bmatrix}u_i^n\\ u_j^n\\ u_k^n\end{Bmatrix}\begin{bmatrix}b_iv_i & b_jv_j & b_kv_k\end{bmatrix}-$$

$$\frac{A\Delta t^n}{12}\begin{bmatrix}2 & 1 & 1\\ 1 & 2 & 1\\ 1 & 1 & 2\end{bmatrix}\begin{Bmatrix}v_i^n\\ v_j^n\\ v_k^n\end{Bmatrix}\begin{bmatrix}c_iv_i & c_jv_j & c_kv_k\end{bmatrix}-\frac{\Delta t^n}{4A\,\mathrm{Re}}\begin{bmatrix}b_i\\ b_j\\ b_k\end{bmatrix}\begin{bmatrix}b_i & b_j & b_k\end{bmatrix}\begin{Bmatrix}v_i^n\\ v_j^n\\ v_k^n\end{Bmatrix}-$$

$$\frac{\Delta t^n}{4A\,\mathrm{Re}}\begin{bmatrix}c_i\\ c_j\\ c_k\end{bmatrix}\begin{bmatrix}c_i & c_j & c_k\end{bmatrix}\begin{Bmatrix}v_i^n\\ v_j^n\\ v_k^n\end{Bmatrix} \tag{7.22}$$

Now, the pressure Poisson equation given by Equation (7.10) has to be solved using finite element method. Application of Galerkin's finite element method to Equation (7.10) gives the following equation.

$$\int_A [N]^T \left[\frac{\partial^2 p^{n+1}}{\partial x^2} + \frac{\partial^2 p^{n+1}}{\partial y^2} - \frac{1}{\Delta t^n} \left(\frac{\partial u^*}{\partial x} + \frac{\partial v^*}{\partial y} \right) \right] dA = 0 \tag{7.23}$$

$$= -\int_A \left(\frac{\partial [N]^T}{\partial x} \frac{\partial [N]}{\partial x} + \frac{\partial [N]^T}{\partial y} \frac{\partial [N]}{\partial y} \right) dA \{p^{n+1}\}$$

$$-\frac{1}{\Delta t^n} \int_A [N]^T \left(\frac{\partial [N]}{\partial x} \{u^*\} + \frac{\partial [N]}{\partial y} \{v^*\} \right) dA$$

Now making use of the expression for diffusion term, the first term in the above equation and the gradient matrix, the integration of the equation can be performed to give rise to the following expression.

$$= -\frac{1}{4A} \begin{bmatrix} b_i \\ b_j \\ b_k \end{bmatrix} \begin{bmatrix} b_i & b_j & b_k \end{bmatrix} \begin{Bmatrix} p_i^{n+1} \\ p_j^{n+1} \\ p_k^{n+1} \end{Bmatrix} - \frac{1}{4A} \begin{bmatrix} c_i \\ c_j \\ c_k \end{bmatrix} \begin{bmatrix} c_i & c_j & c_k \end{bmatrix} \begin{Bmatrix} p_i^{n+1} \\ p_j^{n+1} \\ p_k^{n+1} \end{Bmatrix} -$$

$$\frac{A}{3\Delta t^n} \begin{bmatrix} 1 \\ 1 \\ 1 \end{bmatrix} \begin{bmatrix} b_i & b_j & b_k \end{bmatrix} \begin{Bmatrix} u_i^* \\ u_j^* \\ u_k^* \end{Bmatrix} - \frac{A}{3\Delta t^n} \begin{bmatrix} 1 \\ 1 \\ 1 \end{bmatrix} \begin{bmatrix} c_i & c_j & c_k \end{bmatrix} \begin{Bmatrix} v_i^* \\ v_j^* \\ v_k^* \end{Bmatrix}$$

Finally, the integrated form of pressure Poisson equation can be written as

$$\frac{1}{4A} \begin{bmatrix} b^2{}_i + c^2{}_i & b_i b_j + c_i c_j & b_i b_k + c_i c_k \\ & b^2{}_j + c^2{}_j & b_j b_k + c_j c_k \\ SYM & & b^2{}_k + c^2{}_k \end{bmatrix}$$

$$= -\frac{A}{3\Delta t^n} \begin{bmatrix} 1 \\ 1 \\ 1 \end{bmatrix} \begin{bmatrix} b_i & b_j & b_k \end{bmatrix} \begin{Bmatrix} u_i^n \\ u_j^n \\ u_k^n \end{Bmatrix} - \frac{A}{3\Delta t^n} \begin{bmatrix} 1 \\ 1 \\ 1 \end{bmatrix} \begin{bmatrix} c_i & c_j & c_k \end{bmatrix} \begin{Bmatrix} v_i^n \\ v_j^n \\ v_k^n \end{Bmatrix} \tag{7.24}$$

In Equations (7.20), (7.22) and (7.24), the time step Δt^n is expressed with superscript 'n', which indicates a varying time step. That means, during the solution of Navier-Stokes equations, for the initial transients, smaller time steps will be used for numerical stability because the proposed method is an explicit method, which is stable at smaller time steps. However, after achieving initial convergence of the velocity and pressure fields, the time step for the computations can be increased gradually in order to save the computational time. For this purpose, a variable time step is preferred in such a solution procedure. In Equation (7.24), the integration of velocity derivatives involving $[N]^T$ using Equation (7.16) gives rise to a simple column vector with a coefficient of $\frac{A}{3}$.

Once the pseudo velocities and pressure field are computed, then the corrected velocity at the current time level, *(n+1)* has to be determined using Equation (7.11). The corrected *u* velocity is obtained as

$$u^{n+1} = u^* - \Delta t^n \frac{\partial p}{\partial x}^{n+1}$$

Application of Galerkin's finite element method to the above equation gives the following equation.

$$\int_A [N]^T \left(u^{n+1} - u^* + \Delta t^n \frac{\partial p^{n+1}}{\partial x} \right) dA = 0$$

$$\int_A [N]^T [N] dA \{u^{n+1}\} - \int_A [N]^T [N] dA \{u^*\} + \Delta t^n \int_A [N]^T \frac{\partial [N]}{\partial x} dA \{p^{n+1}\} = 0$$

After integration of the above terms, the final equation for the true *u* velocity can be expressed as

$$\frac{A}{12}\begin{bmatrix} 2 & 1 & 1 \\ 1 & 2 & 1 \\ 1 & 1 & 2 \end{bmatrix} \begin{Bmatrix} u_i^{n+1} \\ u_j^{n+1} \\ u_k^{n+1} \end{Bmatrix} = \frac{A}{12}\begin{bmatrix} 2 & 1 & 1 \\ 1 & 2 & 1 \\ 1 & 1 & 2 \end{bmatrix} \begin{Bmatrix} u_i^* \\ u_j^* \\ u_k^* \end{Bmatrix} - \frac{A\Delta t^n}{3} \begin{bmatrix} 1 \\ 1 \\ 1 \end{bmatrix} \begin{bmatrix} b_i & b_j & b_k \end{bmatrix} \begin{Bmatrix} p_i^{n+1} \\ p_j^{n+1} \\ p_k^{n+1} \end{Bmatrix} \tag{7.25}$$

Similarly, the corrected *v* velocity at the time level, *(n+1)* can be expressed as

$$v^{n+1} = v^* - \Delta t^n \frac{\partial p}{\partial y}^{n+1}$$

Application of Galerkin's finite element method to the above equation gives the following equation.

$$\int_A [N]^T \left(v^{n+1} - v^* + \Delta t^n \frac{\partial p^{n+1}}{\partial y} \right) dA = 0$$

$$\int_A [N]^T [N] dA \{v^{n+1}\} - \int_A [N]^T [N] dA \{v^*\} + \Delta t^n \int_A [N]^T \frac{\partial [N]}{\partial y} dA \{p^{n+1}\} = 0$$

Integration of all the above terms gives the following expression for the true velocity, v.

$$\frac{A}{12}\begin{bmatrix} 2 & 1 & 1 \\ 1 & 2 & 1 \\ 1 & 1 & 2 \end{bmatrix}\begin{Bmatrix} v_i^{n+1} \\ v_j^{n+1} \\ v_k^{n+1} \end{Bmatrix} = \frac{A}{12}\begin{bmatrix} 2 & 1 & 1 \\ 1 & 2 & 1 \\ 1 & 1 & 2 \end{bmatrix}\begin{Bmatrix} v_i^{*} \\ v_j^{*} \\ v_k^{*} \end{Bmatrix} - \frac{A\Delta t^n}{3}\begin{bmatrix} 1 \\ 1 \\ 1 \end{bmatrix}\begin{bmatrix} c_i & c_j & c_k \end{bmatrix}\begin{Bmatrix} p_i^{n+1} \\ p_j^{n+1} \\ p_k^{n+1} \end{Bmatrix} \tag{7.26}$$

The Navier-Stokes equations in primitive variable form have been solved using predictor-corrector method in order to resolve the absence of pressure term in the continuity equation. This method involves the prediction of pseudo velocities without the effect of the pressure term in the momentum equations. Then the pressure Poisson equation is solved to obtain the pressure field at the current time level using the pseudo velocities as the load vector. Finally, the true velocity field is obtained by correcting the pseudo velocities with the pressure field. This process is repeated until a convergent solution is obtained for the velocity and pressure fields. Application of the finite element method to the pseudo velocity equations, pressure Poisson equations and corrected velocity expressions have produced the solution in matrix form for linear triangular elements. Equations (7.20) and (7.22) compute the pseudo velocities and the pressure field is obtained from Equation (7.24), and finally the corrected velocity field is obtained from Equations (7.25) and (7.26).

7.2.4 Computational Algorithm

This section details the procedure for developing the computational procedure to solve the final form of the finite element equations. The solution of Navier-Stokes equations using the predictor-corrector method involves the following solution algorithm.

I Prediction of pseudo velocities, u^* and v^* from Equations (7.20) and (7.22).

II Prediction of pressure field at current time level, $(n+1)$ by solving the pressure Poisson equation (7.24).

III Computing the true velocities, u and v by correcting the pseudo velocities, u^* and v^* using the pressure, $p^{(n+1)}$ using Equations (7.25) and (7.26).

In order to carry out the above computational procedure, different types of matrices have to be computed for the entire computational domain. With the help of a mesh generation program, the computational domain has to be discretized using a linear triangular element and a data file with nodal coordinates and element-nodal connectivity is generated for further computations. In general, finer mesh is employed at the boundaries of the computational domain in order to compute the velocity and pressure gradients accurately. Hence, in the mesh generation process, the size of the triangular elements or area of the elements will not be constant and they vary widely depending on the location of the elements. Once, the element-nodal connectivity and coordinates of all the nodes are stored, then it becomes easy to compute the area of the elements and other geometrical parameters. Following are some of the subroutines that will compute different matrices in the computer code.

7.2.4.1 Computer Program – Subroutines

Subroutine input – All the mesh data such as number of elements, number of nodes, *x*- and *y*- coordinates of nodes, element-nodal connectivity are given as input in this subroutine. The value of Re, time step and number of iterations are also read as input. The values of velocities and pressure at the boundaries and at time=0 are read in this subroutine.

Subroutine matrices – All the matrices and matrix products required for the Equations (7.20), (7.22), (7.24), (7.25) and (7.26) are computed. This is carried out over an outermost element do loop where the nodal connectivity of each element is read and the constants required for computing area, A and shape function derivatives, b_i, c_i etc. are evaluated. All the matrices are computed for every element. Only, stiffness and capacitance matrices are stored in arrays and all other matrix products can be performed directly in the program. The element level matrices are assembled to form the final global matrices of the form $Ax=b$. The required modifications for incorporating the boundary conditions are carried out in matrices A and b.

Subroutine initiate – The present computational algorithm employs an explicit time marching scheme and hence the field variables have to be stored in two time levels, '*n*' the previous time level and '*(n+1)*' the current time level. As the algorithm involves an iterative procedure for convergence, the field variables are required to be stored at the iterative level also using one more array. For example, *u1, v1, p1* are @ time level, '*n*', u_2, v_2, p_2 are @ time level, '*(n+1)*' and u_3, v_3, p_3 are @ iterative level. The boundary conditions are also enforced in this subroutine. All the boundary conditions of the field variables have to be updated at the end of every time level.

Subroutine pvelocity – The computation of velocities, u^* and v^* as expressed by Equations (7.20) and (7.22) are performed in this subroutine.

Subroutine pressure – The pressure Poisson equation given by Equation (7.24) is computed in this subroutine to obtain $p^{(n+1)}$.

Subroutine cvelocity – The corrected velocities, u and v as expressed by Equations (7.25) and (7.26) are evaluated in this subroutine.

Subroutine output - Once the converged solution is obtained for the field variables, all the results are stored in this subroutine. The format in which the output results are required has to be decided, and accordingly the output of the field variables is written in various output files using formatted syntax.

Subroutine solver – An iterative solver is used to solve the final simultaneous equations obtained in the form $Ax=b$.

Subroutine check – This subroutine checks the difference between two iterated values of velocity and pressure fields and quantifies the difference as an error. When this error is found to be less than or equal to the specified acceptable error for convergence, then those field variables of the iteration are treated as a converged solution.

Subroutine iterate – The computational algorithm is executed in this subroutine to solve the field variables at current time level, '$(n+1)$' using the values at time level 'n'. The outermost time loop is started and within this loop, an iterative loop is formed in order to achieve convergence of velocity and pressure fields. The computations are carried out in the following sequence: *call subroutine pvelocity, call subroutine pressure, call subroutine cvelocity, call check, call output.* When the field variables are converged then the results are written in the output files.

Figure 7.2 depicts the computational flow chart for the solution of two-dimensional Navier-Stokes equations using the predictor-corrector method. It has to be observed that though the computations are performed in transient sequence, the final objective is to evaluate the field variables at steady state conditions. As the predictor-corrector algorithm involves the explicit time marching scheme, the converged steady state solutions have to be obtained only by implementing the time marching scheme. In most of the computational fluid dynamics problems, the final flow field is required at steady state conditions because then only any heat transfer phenomenon associated with the fluid flow can be simulated. A lot of commercial software on computational fluid dynamics problems makes use of the primitive variable form of Navier-Stokes equations and hence, in this textbook, the discussion on solution procedure is restricted up to the point of computational algorithm, and it is hoped that with this explanation on the computational algorithm, the reader can easily understand the solution techniques employed in commercial CFD software. There are other types of formulations of Navier-Stokes equations that can be solved easily solved and one such formulation is discussed in the following section.

7.3 NAVIER-STOKES EQUATIONS IN VELOCITY-VORTICITY FORM

Momentum conservation equations in the primitive variable form are very commonly employed in the solution of CFD-related problems with and without heat and species transfer. As was discussed earlier in the previous sections, the pressure term which appears in the momentum equations is not present in the continuity equation and it contributes indirectly to the mass conservation relation. Hence, researchers have come up with different types of solution algorithms for the Navier-Stokes equations in primitive variable form. In CFD applications when the pressure field is not required compared to fluid friction and heat transfer, then scientists have resorted to a set of momentum equations without the pressure term. Stream function-velocity and stream function-vorticity are the other forms of Navier-Stokes equations. As it is well understood from basic fluid mechanics, the stream function does not contribute any engineering design parameters directly and it is used to visualize the flow pattern in a given flow domain. Though the solution of the above-mentioned two types of formulations is relatively easier than the primitive variable form of Navier-Stokes equations, the resulting field variables are not directly useful in any simulation analysis. Of course, it has to be remembered that any different type of formulation of Navier-Stokes equations have to be obtained only from the primitive variable form because they are the fundamental equations that represent the natural setting of mass

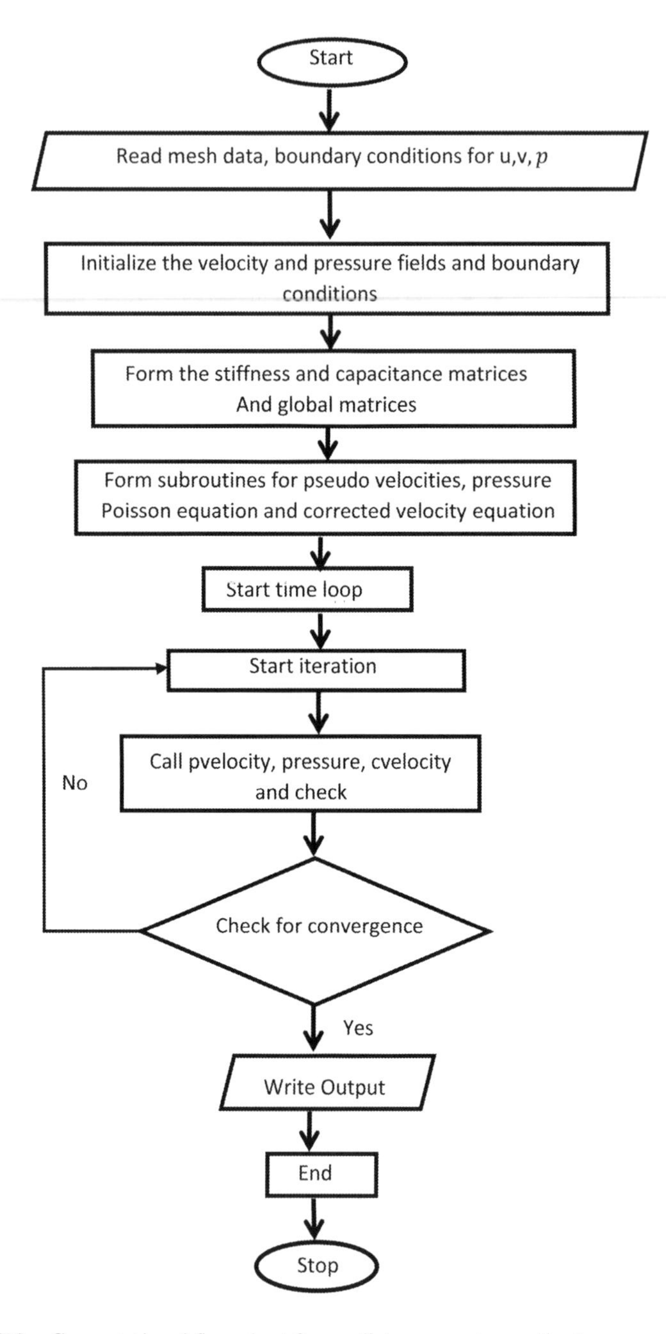

FIGURE 7.2 Computational flow chart for predictor-corrector method.

and momentum conservation principles in a given flow domain. Velocity-vorticity form of Navier-Stokes equations can be derived from the primitive variable form of Navier-Stokes equations to analyze problems wherein the main focus is only heat transfer and species transfer or other types of parameters that influence fluid flow. For example, the use of nanofluid as the working fluid in the domain and the effect of body forces such as magnetic force can be investigated easily in this type of equation as there is no need to employ a special type of solution algorithm to resolve the pressure term. In this set of equations, the pressure term does not appear and influence directly, thus making the solution algorithm straightforward. However, these equations cannot yield correct results unless the vorticity boundary conditions are enforced effectively on the boundaries. There exists coupling between the velocity and vorticity fields in these equations. The generalized derivation of these equations will be explained in detail in the following section.

7.3.1 Derivation of Velocity-Vorticity Equations as Generalized Formulation

The Navier-Stokes equations for laminar, incompressible Newtonian fluid flow in primitive variables form can be represented in vector form as

Continuity Equation

$$\nabla \cdot V = 0 \tag{7.27}$$

Momentum Equation

$$\frac{\partial \boldsymbol{V}}{\partial t} + (\boldsymbol{V} \cdot \nabla)\boldsymbol{V} = -\frac{1}{\rho}\nabla \boldsymbol{P} + \nu\left(\nabla^2 \boldsymbol{V}\right) + \boldsymbol{F}_b + \boldsymbol{F}_E \tag{7.28}$$

where $\boldsymbol{F}_b, \boldsymbol{F}_E$ respectively indicate buoyancy force due to thermal and species concentration and any external force exerted on the flow field, such as magnetic force, electromagnetic force etc. Here the coupling between the primitive variables, pressure and velocity imposes the main difficulty in obtaining the numerical solution for these equations. Generally, momentum equations are solved first, and during that process, the fluid incompressibility constraint given by Equation (7.27) has to be satisfied. That means the divergence free condition has to be satisfied in the flow field. Thus, the handling of the pressure term in this process creates a major challenge. The lack of an independent equation for the pressure field in the Navier-Stokes equations complicates the solution procedure. Though there are many standard numerical procedures to deal with these difficulties, it is always computationally demanding to solve Navier-Stokes equations in primitive variable form. For flow situations where pressure has a minor role to play, it is very beneficial to go for an alternative formulation to obtain a solution for flow field. One such formulation is the velocity-vorticity model of Navier-Stokes equations, which serves as a pressure-free formulation to represent vortex dynamics and kinematics of flow field. This formulation gives the advantage of avoiding the pressure term and directly computing the fluid vorticity generated as a result of fluid convection. Vorticity for a flow field can be defined in vector form as

$$\omega = \nabla \times V \tag{7.29}$$

Therefore, the vorticity transport equation can be obtained by taking curl of the fluid momentum equation expressed by Equation (7.28) as

$$\frac{\partial \omega}{\partial t} + \nabla \times \left(\boldsymbol{V} \cdot \nabla \boldsymbol{V} \right) = -\nabla \times \nabla \frac{\boldsymbol{P}}{\rho} + \nu \, \nabla^2 \left(\nabla \times \boldsymbol{V} \right) + \nabla \times \boldsymbol{F}_b + \nabla \times \boldsymbol{F}_M \tag{7.30}$$

Here $\nabla \times \nabla \frac{p}{\rho} = 0$, so the above vorticity transport equation can be rewritten without the pressure term as

$$\frac{\partial \omega}{\partial t} + \nabla \times \left(\boldsymbol{V} \cdot \nabla \boldsymbol{V} \right) = \nu \, \nabla^2 \left(\nabla \times \boldsymbol{V} \right) + \nabla \times \boldsymbol{F}_b + \nabla \times \boldsymbol{F}_M \tag{7.31}$$

Equation (7.31) can be further simplified using the vector relations [3]. Firstly, the second term in the left-hand side of the above equation can be expanded and simplified. Consider the vector relation

$$\begin{aligned} \left(V \cdot \nabla \right) V &= \frac{1}{2} \nabla \left(V \cdot V \right) - V \times \left(\nabla \times V \right) \\ &= \nabla \frac{V^2}{2} - \left(V \times \omega \right) \end{aligned}$$

Taking curl on both sides gives the second term of Equation (7.31) as follows.

$$\begin{aligned} \nabla \times \left(V \cdot \nabla \right) V &= \nabla \times \left[\nabla \frac{V^2}{2} - \left(V \times \omega \right) \right] \\ &= \nabla \times \nabla \frac{V^2}{2} - \nabla \times \left(V \times \omega \right) \\ &= \nabla \times \left(\omega \times V \right) \end{aligned}$$

Since $\nabla \times \nabla \frac{V^2}{2} = 0$, the right-hand side of the above equation can be expanded as

$$\nabla \times \left(\omega \times V \right) = \left(V \cdot \nabla \right) \omega - \left(\omega \cdot \nabla \right) V + \omega \left(\nabla \cdot V \right) - V \left(\nabla \cdot \omega \right) \tag{7.32}$$

Using the constraints that solenoidal vector field $\nabla \cdot \omega = 0$ and incompressibility $\nabla \cdot V = 0$ in the above equation, Equation (7.32) becomes

$$\nabla \times (\omega \times V) = (V \cdot \nabla)\omega - (\omega \cdot \nabla)V \tag{7.33}$$

Now, replacing Equation (7.33) for the second term of the Equation (7.31) gives

$$\frac{\partial \omega}{\partial t} + (V \cdot \nabla)\omega - (\omega \cdot \nabla)V = \upsilon \nabla^2 \omega + \nabla \times F_b + \nabla \times F_M \tag{7.34}$$

Equation (7.34) represents the vorticity transport equation for three-dimensional flow field. Using tensor and vector multiplication, the term $(\omega \cdot \nabla)V$ in Equation (7.34) can be shown to be equal to $(\nabla V) \cdot \omega$ and $(\nabla V) \cdot \omega = \ddot{\mathbf{S}} + \ddot{\mathbf{W}}$, where $\ddot{\mathbf{S}}$ is rate of strain tensor and $\ddot{\mathbf{W}}$ is angular rotation. One part of $(\nabla V) \cdot \omega$ increases the vorticity by stretching the vortex line whereas the remaining part contributes to the angular turning of the vortex line. This aspect of stretching and turning is completely absent in a two-dimensional flow field. Hence, ignoring these terms for the two-dimensional flow field, Equation (7.34) can be recast as

$$\frac{\partial \omega}{\partial t} + (V \cdot \nabla)\omega = \upsilon \nabla^2 \omega + \nabla \times F_b + \nabla \times F_M \tag{7.35}$$

Equation (7.35) represents the vorticity transport equation for generalized flow field.

Body Force Terms

The right-hand side of Equation (7.35) consists of a vorticity diffusion term and body force and external force terms. The body force due to buoyancy force can be evaluated for two-dimensional flow field. The curl of buoyancy force term in Equation (7.35) for two-dimensional flow field can be represented as

$$\nabla \times F_b = \frac{\partial F_{by}}{\partial x}\hat{j} - \frac{\partial F_{bx}}{\partial y}\hat{i} \tag{7.36}$$

where $\hat{i}$ and $\hat{j}$ are unit vectors in x and y-directions respectively. As the buoyancy force acts in the direction of gravitational force, that is in the y-direction, the buoyancy force in the x-direction becomes zero. With this modification, Equation (7.36) is modified as

$$\nabla \times F_b = \frac{\partial F_{by}}{\partial x}\hat{j} \tag{7.37}$$

Similarly, the curl of external force for the two-dimensional flow field can be represented as

$$\nabla \times F_M = \frac{\partial F_{My}}{\partial x}\hat{j} - \frac{\partial F_{Mx}}{\partial y}\hat{i} \tag{7.38}$$

Substituting Equations (7.37) and (7.38) in Equation (7.35), we get

$$\frac{\partial \omega}{\partial t} + (V \cdot \nabla)\omega = \upsilon \nabla^2 \omega + \frac{\partial F_{by}}{\partial x}\hat{j} + \frac{\partial F_{My}}{\partial x}\hat{j} - \frac{\partial F_{Mx}}{\partial y}\hat{i} \tag{7.39}$$

Equation (7.39) is called a vorticity transport equation for a two-dimensional laminar incompressible Newtonian fluid flow in canonical form. The physical significance of the terms appearing in Equation (7.39) is explained below.

$\frac{\partial \omega}{\partial t}$ - represents the transient variation of vorticity in the control volume.

$(V \cdot \nabla)\omega$ - represents the vorticity transport due to advective bulk fluid motion.

$\upsilon \nabla^2 \omega$ - represents diffusion of vorticity in the flow domain.

$\frac{\partial F_{by}}{\partial x}$ - force due to buoyancy acting in the y-direction.

$\frac{\partial F_{My}}{\partial x}\hat{j}$ - external body force acting in the y-direction.

$\frac{\partial F_{Mx}}{\partial y}\hat{i}$ - external body force acting in the x-direction.

The vorticity transport Equation (7.39) consists of two unknowns, velocity vector V and vorticity ω. In order to solve the vorticity transport equation, one more equation has to be obtained and this equation also should ensure divergence free solution for the flow field. The second equation also should have the same two variables, velocity and vorticity, and hence, the definition of vorticity field in terms of velocity field is considered as the equation to obtain the second equation.

The vorticity in a flow field is defined as

$$\boldsymbol{\omega} = \nabla \times V \tag{7.40}$$

By applying curl operation on both sides of Equation (7.40), we get

$$\nabla \times \boldsymbol{\omega} = \nabla \times (\nabla \times V) \tag{7.41}$$

The right-hand side of Equation (7.41) can be expanded as

$$\nabla \times (\nabla \times V) = \nabla(\nabla \cdot V) - \nabla^2 V \tag{7.42}$$

After imposing the continuity constraint, $\nabla \cdot V = 0$, Equation (7.42) becomes

$\nabla \times (\nabla \times V) = -\nabla^2 V$ and with the definition of vorticity given by Equation (7.40), we can write

$\nabla \times \boldsymbol{\omega} = -\nabla^2 V$, which can be rewritten as

$$\nabla^2 V = -\nabla \times \omega \tag{7.43}$$

Equation (7.43) is called velocity Poisson equations in canonical form which governs the kinematics aspects of the flow field and establishes the relationship between velocity and vorticity in the flow field. Finally, the canonical form of vorticity transport and velocity Poisson equations expressed by Equations (7.39) and (7.43) respectively can be expanded in Cartesian coordinates as follows.

Vorticity Transport Equation

$$\frac{\partial \omega}{\partial t}+u\frac{\partial \omega}{\partial x}+v\frac{\partial \omega}{\partial y}=\upsilon\left[\frac{\partial^2 \omega}{\partial x^2}+\frac{\partial^2 \omega}{\partial y^2}\right]+\frac{\partial F_{by}}{\partial x}\hat{j}+\frac{\partial F_{My}}{\partial x}\hat{j}-\frac{\partial F_{Mx}}{\partial y}\hat{i} \tag{7.44}$$

Velocity Poisson Equations

$$\frac{\partial^2 u}{\partial x^2}+\frac{\partial^2 u}{\partial y^2}=-\frac{\partial \omega}{\partial y} \tag{7.45a}$$

$$\frac{\partial^2 v}{\partial x^2}+\frac{\partial^2 v}{\partial y^2}=\frac{\partial \omega}{\partial x} \tag{7.45b}$$

Equations (7.44) and (7.45) represent the velocity-vorticity form of Navier-Stokes equations for laminar incompressible Newtonian fluid flow in a flow domain represented by two-dimensional Cartesian co-ordinates. Vorticity generated at all the boundaries of the domain gets advected and diffused into the entire flow domain. Hence, it is very important to impose exact vorticity boundary values so that the solenoidality condition is also satisfied for the vorticity field. The fluid kinematics is enforced by the velocity Poisson equations, which also assures a divergence free velocity field. Some of the advantages of velocity-vorticity formulation over other formulations can be described below.

(i) It is free from the pressure term, hence, the numerical difficulties that arise from enforcing the pressure boundary condition and incompressibility constraint are avoided.
(ii) It can be used for both two- and three-dimensional flow problems unlike stream function-vorticity formulation in which the stream function is a scalar function and can be represented only in two-dimensional form.
(iii) For external flow problems, the implementation of the vorticity boundary condition at unbounded domain is much easier than the boundary conditions for pressure.
(iv) Non-inertial effects that come from translation and rotation of the reference frames can be easily handled through initial and boundary conditions which gives greater computational advantage.
(v) The computation of vorticity boundary conditions is straightforward because they are obtained by using normal derivative of velocity components whereas in stream function-vorticity formulation, the second order derivatives of the stream function are required to compute the vorticity boundary values. Such boundary conditions are computed with first, second and third order accuracy.

(vi) The velocity components are readily available when required, e.g., for the transient period in unsteady flow problems for solving energy and species concentration equations.

Apart from the benefits, there also exist some difficulties in using velocity-vorticity formulation as described below.

- Special care is required while handling domains with multiple connectivity to ensure the equivalence of all the formulations.
- Additional computational efforts are required for three-dimensional flow problems, because six field variables are involved in velocity-vorticity formulation instead of four field variables in primitive variables formulation.

7.3.2 Computation of Vorticity Boundary Conditions

In the solution of the primitive variable form of Navier-Stokes equations, the boundary conditions have to be enforced for velocities and pressure which are the fundamental variables being computed in the solution procedure. However, in the case of velocity-vorticity formulation, vorticity is a derived quantity that is a function of the velocity field. Computation of vorticity requires velocity gradients at the boundaries. It has to be noticed that both the vorticity transport equation and the velocity Poisson equations are coupled through the vorticity field and vorticity is a boundary phenomenon. If the vorticities at the boundaries are not computed accurately and enforced at the boundaries, the loading for the velocity Poisson equations become very weak and this will result in slow convergence of the flow fields. Hence, the main challenge in velocity-vorticity formulation lies in the computation of the vorticity boundary values.

Second order accurate finite difference expressions [4] have been used to compute the vorticity at the boundary nodes. The derivation of expression for vorticity at boundary nodes can be discussed in detail by referring to Figure 7.3. This method is discussed only for normal boundary walls where the computational grid points are normal to the boundary walls. Consider a domain ABCD with four boundary nodes, i, j, k and m shown in the above figure. The x-y coordinate directions are indicated below the domain in the figure. As vorticity is a function of velocity gradients, a detailed derivation for vorticity at any boundary node can be obtained using forward difference equation for the gradients. For any given boundary node, the wall normal has to be identified so that the respective velocity gradients at that boundary node can be computed. In a two-dimensional domain, one wall normal in each coordinate direction is parallel to the coordinate axis and one wall normal is in the opposite direction. For example, the wall normal for the boundary node, 'k' on side DA is parallel to the positive x-axis, whereas the wall normal for the boundary node, m on side BC is parallel to the negative x-axis. Similarly, the boundary point 'i' has positive wall normal in the y-direction, whereas the wall normal to the boundary node, 'j' is parallel to the negative y-axis. Let us derive the expression for the boundary vorticity for different nodes shown in the domain ABCD. Initially, an expression for

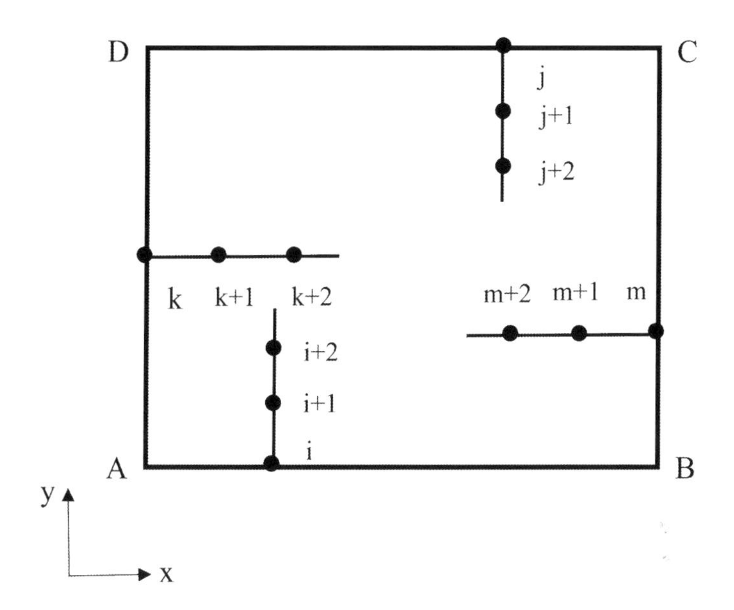

FIGURE 7.3 Computation of vorticity values at boundary nodes.

vorticity has to be developed. Consider the boundary node, 'i'. The vorticity at this boundary node can be expressed as

$\omega_i = \left(\frac{\partial v}{\partial x} - \frac{\partial u}{\partial y}\right)_i$. The wall normal to the boundary node, 'i' is parallel to the positive y-direction and velocity variation can be considered only in the y-direction as this node lies on the wall normal parallel to the x-axis. Hence, the contribution from $\frac{\partial v}{\partial x}$ term is zero for this boundary node. Now the vorticity definition gets modified as

$\omega_i = -\left(\frac{\partial u}{\partial y}\right)_i$. Likewise, for the boundary node, 'm', the vorticity is defined as

$\omega_m = \left(\frac{\partial v}{\partial x}\right)_m$. Now, the difference equation for vorticity at the boundary node can be derived by expanding the velocity and the vorticity using Taylor's series expansion scheme as follows.

Let us expand velocity 'v' for node, 'm' as

$$v_{m+1} = v_m + \left(\frac{dv}{dx}\right)_m \frac{\Delta x}{1!} + \left(\frac{d^2v}{dx^2}\right)_m \frac{\Delta x^2}{2!} + \left(\frac{d^3v}{dx^3}\right)_m \frac{\Delta x^3}{3!} + O(\Delta x^3) \quad (7.46)$$

After substituting $\omega_m = \left(\frac{\partial v}{\partial x}\right)_m$ in the above equation and neglecting higher order term, Equation (7.46) can be rewritten as

$$v_{m+1} = v_m + \omega_m \frac{\Delta x}{1!} + \left(\frac{d\omega}{dx}\right)_m \frac{\Delta x^2}{2!} + O(\Delta x^3) \tag{7.47}$$

Solving for $\left(\frac{d\omega}{dx}\right)_m$ from Equation (7.47), we get the following expression.

$$\left(\frac{d\omega}{dx}\right)_m = \left(v_{m+1} - v_m - \omega_m \Delta x\right)\frac{2}{\Delta x^2} \tag{7.48}$$

Similarly, let us expand the vorticity, ω_m about the point, '*m*' as

$$\omega_{m+1} = \omega_m + \left(\frac{d\omega}{dx}\right)_m \frac{\Delta x}{1!} + O(\Delta x^2) \tag{7.49}$$

Now, substituting Equation (7.48) for $\left(\frac{d\omega}{dx}\right)_m$ in Equation (7.49), the final expression for ω_m can be expressed as

$$\omega_m = \frac{2}{\Delta x}\left(v_{m+1} - v_m\right) - \omega_{m+1} \tag{7.50}$$

Equation (7.50) represents the generalized expression for vorticity at any boundary node in terms of contributing velocity values at the adjacent nodes and the vorticity at the adjacent node. The advantage of the above equation is that the vorticity at the boundary node is represented in terms of vorticity at the adjacent node. Using Equation (7.50), the vorticity at boundary nodes located on different boundary sides of the domain ABCD can be computed. The vorticity at boundary nodes, *i, j, k* and *m* shown in Figure 7.3 can be computed as follows.

7.3.2.1 Node i on Side AB – For Wall Normal Parallel to Positive y-Axis

$\omega_i = \left(\frac{\partial v}{\partial x} - \frac{\partial u}{\partial y}\right)_i$. As only Δy contributes, this expression is rewritten as

$$\omega_i = -\left(\frac{\partial u}{\partial y}\right)_i$$

$$\omega_i = \frac{2}{\Delta y}\left(u_i - u_{i+1}\right) - \omega_{i+1} \tag{7.51a}$$

7.3.2.2 Node j on Side CD – For Wall Normal Parallel to Negative y-Axis

$\omega_j = \left(\frac{\partial v}{\partial x} - \frac{\partial u}{\partial y}\right)_j$. As only Δy contributes, this expression is rewritten as

$$\omega_j = (-) - \left(\frac{\partial u}{\partial y}\right)_j = \left(\frac{\partial u}{\partial y}\right)_j$$

$$\omega_j = \frac{2}{\Delta y}\left(u_{j+1} - u_j\right) - \omega_{j+1} \tag{7.51b}$$

7.3.2.3 Node k on Side DA – For Wall Normal Parallel to Positive x-Axis

$\omega_k = \left(\frac{\partial v}{\partial x} - \frac{\partial u}{\partial y}\right)_k$. As only Δx contributes, this expression is rewritten as

$$\omega_k = \left(\frac{\partial v}{\partial x}\right)_k$$

$$\omega_k = \frac{2}{\Delta x}\left(v_{k+1} - v_k\right) - \omega_{k+1} \tag{7.51c}$$

7.3.2.4 Node m on Side BC – For Wall Normal Parallel to Negative x-Axis

$\omega_m = \left(\frac{\partial v}{\partial x} - \frac{\partial u}{\partial y}\right)_m$. As only Δx contributes, this expression is rewritten as

$$\omega_m = (-)\left(\frac{\partial v}{\partial x}\right)_m$$

$$\omega_k = \frac{2}{\Delta x}\left(v_k - v_{k+1}\right) - \omega_{k+1} \tag{7.51d}$$

where k is the index of boundary node and k+1 is the index of the node adjacent to the boundary node. Using the Equations (7.51a) to (7.51d), the vorticity values at all the boundary nodes of the computational domain can be evaluated. Generally, a separate subroutine is written to carry out these computations. In order to evaluate the vorticity values at the boundary nodes, the velocities on the boundary nodes and nodes adjacent to the boundary nodes in the normal direction have to be identified during the grid generation process, and these data have to be given as input to the subroutine along with the values of respective velocities at these nodes. Once these vorticity values at the boundary nodes are computed, they are stored in an array so that they can be used for the solution of the vorticity transport equation.

7.3.3 Solution Using Finite Element Method

The governing equations expressed by Equations (7.44) and (7.45) have to be solved for the velocity and vorticity fields in a given computational domain. In order to make these equations applicable to any size of computational domain and any type of fluid that satisfies incompressibility principle, these governing equations have to be made non-dimensional using scaling factors. After the application of the finite element solution procedure, the equations will be solved with a computer code for a specific problem. Quadrilateral isoparametric elements are assumed to be used to discretize the computational domain while applying the finite element method for the solution of the governing equations in velocity-vorticity form. In this section, an example problem on simple lid-driven square cavity will be considered to demonstrate the ability of velocity-vorticity formulation for predicting the flow fields. Hence, the body force terms appearing in Equation (7.44) will be assumed to be equal to zero in the present discussion. Of course, another example with buoyancy force also will be considered later. In a lid-driven square cavity flow problem, it is assumed that the cavity is of side equal to L and the top lid moves with a constant velocity equal to U_∞ in the positive x-direction. The governing equations in velocity-vorticity form for two-dimensional incompressible flow can be written as

Vorticity Transport Equation

$$\frac{\partial \omega}{\partial t}+u\frac{\partial \omega}{\partial x}+v\frac{\partial \omega}{\partial y}=\upsilon\left[\frac{\partial^2 \omega}{\partial x^2}+\frac{\partial^2 \omega}{\partial y^2}\right] \tag{7.52}$$

Velocity Poisson Equations

$$\frac{\partial^2 u}{\partial x^2}+\frac{\partial^2 u}{\partial y^2}=-\frac{\partial \omega}{\partial y} \tag{7.53a}$$

$$\frac{\partial^2 v}{\partial x^2}+\frac{\partial^2 v}{\partial y^2}=\frac{\partial \omega}{\partial x} \tag{7.53b}$$

The scaling factors that will be used to make the above equations non-dimensional, are defined as follows.
$X=\frac{x}{L}, Y=\frac{y}{L}, U=\frac{u}{U_\infty}, V=\frac{v}{U_\infty}, \Omega=\frac{\omega L}{U_\infty}, \tau=\frac{tU_\infty}{L}$. From these scaling factors, the actual variables can be expressed as
$x=XL, y=YL, u=UU_\infty, v=VU_\infty, \omega=\frac{\Omega U_\infty}{L}, t=\frac{\tau L}{U_\infty}$. Substitution of these expressions in Equation (7.52) will yield the following expression.

$$\frac{\partial \Omega}{\partial \tau}\frac{U_\infty}{L}\frac{U_\infty}{L}+U\frac{U_\infty}{L}\frac{\partial \Omega}{\partial X}\frac{U_\infty}{L}+V\frac{U_\infty}{L}\frac{\partial \Omega}{\partial Y}\frac{U_\infty}{L}=\upsilon\left[\frac{\partial^2 \Omega}{\partial x^2}\frac{1}{L^2}\frac{U_\infty}{L}+\frac{\partial^2 \omega}{\partial y^2}\frac{1}{L^2}\frac{U_\infty}{L}\right]$$

After simplification, the vorticity transport equation can be written as

$$\frac{\partial \Omega}{\partial \tau}+U\frac{\partial \Omega}{\partial X}+V\frac{\partial \Omega}{\partial Y}=\frac{1}{\mathrm{Re}}\left[\frac{\partial^2 \Omega}{\partial X^2}+\frac{\partial^2 \Omega}{\partial Y^2}\right] \tag{7.54}$$

Similarly, the velocity Poisson equations take the following form with scaling factors.

$$\frac{\partial^2 U}{\partial X^2}\frac{U_\infty}{L^2}+\frac{\partial^2 U}{\partial Y^2}\frac{U_\infty}{L^2}=-\frac{\partial \Omega}{\partial Y}\frac{U_\infty}{LL}$$

$$\frac{\partial^2 V}{\partial X^2}\frac{U_\infty}{L^2}+\frac{\partial^2 V}{\partial Y^2}\frac{U_\infty}{L^2}=\frac{\partial \Omega}{\partial X}\frac{U_\infty}{LL}$$

Finally, the velocity Poisson equations can be written as

$$\frac{\partial^2 U}{\partial X^2}+\frac{\partial^2 U}{\partial Y^2}=-\frac{\partial \Omega}{\partial Y} \tag{7.55a}$$

$$\frac{\partial^2 V}{\partial X^2}+\frac{\partial^2 V}{\partial Y^2}=\frac{\partial \Omega}{\partial X} \tag{7.55b}$$

7.3.4 Finite Element Formulation of Vorticity Transport Equation

In the application of Galerkin's weighted residual finite element method, let us assume an approximate solution for the dependent variable $\Omega \approx \sum_{i=1}^{nne} N_i \Omega_i$ where N_i is the shape function, i is node index and nne is number of nodes per element. The approximate solution gives rise to a residue, as shown below.

$$\frac{\partial \Omega}{\partial \tau}+U\frac{\partial \Omega}{\partial X}+\mathrm{V}\frac{\partial \Omega}{\partial Y}-\frac{1}{\mathrm{Re}}\left[\frac{\partial^2 \Omega}{\partial X^2}+\frac{\partial^2 \Omega}{\partial Y^2}\right]=R$$

The residue R has to be minimized so as to get the results with an acceptable specified error limit. This can be achieved by multiplying the residual R with a weighting function w and the integral of the product is required to be zero in every element of the computational domain. That is

$$\int_A w_i R dA = 0.$$

In Galerkin's weighted residual method the weighting function w is the same as the shape function used to approximate the solution. Hence, the above equation becomes

$$\int_A N_i^T R dA = 0$$

where

$$R=\frac{\partial \Omega}{\partial \tau}+U\frac{\partial \Omega}{\partial X}+V\frac{\partial \Omega}{\partial Y}-\frac{1}{\mathrm{Re}}\left[\frac{\partial^2 \Omega}{\partial X^2}+\frac{\partial^2 \Omega}{\partial Y^2}\right] \tag{7.56}$$

For any given element in the computational domain, the following equation has to be satisfied.

$$\int N^T \left[\frac{\partial \Omega}{\partial \tau} + U\frac{\partial \Omega}{\partial X} + V\frac{\partial \Omega}{\partial Y} - \frac{1}{\mathrm{Re}}\left[\frac{\partial^2 \Omega}{\partial X^2} + \frac{\partial^2 \Omega}{\partial Y^2} \right]\right] dA = 0 \tag{7.57}$$

Let

$$\Omega(X,Y) = \left[N\right]\{\Omega\}, U(X,Y) = \left[N\right]\{U\}, V(X,Y) = \left[N\right]\{V\}$$

The time derivative in Equation (7.56) can be approximated using first order accurate Taylor series expansion scheme as

$\frac{\partial \Omega}{\partial \tau} \approx \frac{\Omega^{n+1} - \Omega^n}{\Delta \tau}$ for a time step of $\Delta\tau$.

Using mean value theorem, the time derivative between the time levels 'n' and '$n+1$' can be written as

$$\{\Omega\} = (1-\Theta)\{\Omega\}^n + \Theta\{\Omega\}^{n+1} \tag{7.58}$$

where Θ is the weighting parameter. For Θ=0 and 1 respectively, explicit and implicit time marching solutions will be obtained. When Θ=0.5 a second order accurate semi-implicit time marching solutions can be obtained. In the present solution procedure, Θ=0.5 is employed.

The first term in the left-hand side of Equation (7.57) can be written as (7.59)

$$\int N^T \frac{\partial \Omega}{\partial \tau} dA = \int N^T N \frac{\{\Omega\}^{n+1} - \{\Omega\}^n}{\Delta \tau} dA \tag{7.59}$$

The integration of the advection term in Equation (7.57) takes the form

$$\int N^T U \frac{\partial \Omega}{\partial X} dA = \int N^T \left(N_K U_K{}^n\right) \frac{\partial N}{\partial X} dA \{\Omega\} \tag{7.60a}$$

Similarly,

$$\int N^T V \frac{\partial \Omega}{\partial Y} dA = \int N^T \left(N_K V_K{}^n\right) \frac{\partial N}{\partial Y} dA \{\Omega\} \tag{7.60b}$$

where $\left(N_K U_K{}^n\right) = \sum_{k=1}^{nne} N_K U_K{}^n$ and $\left(N_K V_K{}^n\right) = \sum_{k=1}^{nne} N_K V_K{}^n$

The diffusion term in Equation (7.57) can be integrated using the product rule as

$$\int N^T \frac{\partial^2 \Omega}{\partial X^2} dA = \int \frac{\partial}{\partial X}\left(N^T \frac{\partial \Omega}{\partial X} \right) dA$$

which gives

$$\int \frac{\partial}{\partial X}\left(N^T \frac{\partial \Omega}{\partial X}\right) dA = \int N^T \frac{\partial^2 \Omega}{\partial X^2} dA + \int \frac{\partial N^T}{\partial X}\frac{\partial \Omega}{\partial X} dA$$

Rearranging the above equation

$$\int N^T \frac{\partial^2 \Omega}{\partial X^2} dA = \int \frac{\partial}{\partial x}\left(N^T \frac{\partial \Omega}{\partial X}\right) dA - \int \frac{\partial N^T}{\partial X}\frac{\partial \Omega}{\partial X} dA$$

By applying Green's theorem, the first integral in the right-hand side of the above equation can be replaced as

$\int \frac{\partial}{\partial X}\left(N^T \frac{\partial \Omega}{\partial X}\right) dA = \int N^T \frac{\partial \Omega}{\partial \Gamma} d\Gamma$, which will contribute only for the boundaries.

Hence, the diffusion term becomes

$$\int N^T \frac{\partial^2 \Omega}{\partial X^2} dA = \int N^T \frac{\partial \Omega}{\partial \Gamma} d\Gamma - \int \frac{\partial N^T}{\partial X}\frac{\partial \Omega}{\partial X} dA$$

As $\int N^T \frac{\partial \Omega}{\partial \Gamma} d\Gamma$ is applicable only at the boundary of the computational domain, its contribution will be considered only at the time of incorporating the boundary conditions in the final global matrices. In the present example problem, only Dirichlet boundary conditions are assumed for velocity and vorticity fields, hence, this term will not have any contributions in the final matrices.

Hence

$$\int N^T \frac{\partial^2 \Omega}{\partial X^2} dA = -\int \frac{\partial N^T}{\partial X}\frac{\partial N}{\partial X} dA\{\Omega\} \tag{7.61a}$$

Similarly

$$\int N^T \frac{\partial^2 \Omega}{\partial Y^2} dA = -\int \frac{\partial N^T}{\partial Y}\frac{\partial N}{\partial Y} dA\{\Omega\} \tag{7.61b}$$

Substituting Equations (7.59) to (7.61) in Equation (7.57), we get

$$\int N^T N \frac{\{\Omega\}^{n+1} - \{\Omega\}^n}{\Delta t} dA + \int N^T \left(N_K U_K{}^n\right)\frac{\partial N}{\partial X} dA\{\Omega\} + \int N^T \left(N_K V_K{}^n\right)\frac{\partial N}{\partial Y} dA\{\Omega\}$$

$$-\frac{1}{\text{Re}}\left[-\int \frac{\partial N^T}{\partial X}\frac{\partial N}{\partial X} dA - \int \frac{\partial N^T}{\partial Y}\frac{\partial N}{\partial Y} dA\right]\{\Omega\} = 0$$

That is

$$\left[C_{ij}\right]\frac{\{\Omega\}^{n+1} - \{\Omega\}^n}{\Delta t} + \left[G_{ij}\right]\{\Omega\} + \frac{1}{\text{Re}}\left[K_{ij}\right]\{\Omega\} = 0 \tag{7.62}$$

After substituting Equation (7.58) in the above equation and rearranging the terms corresponding to time levels '*n*' and '*n+1*', we get

$$\left[C_{ij}\right]\left\{\Omega_i\right\}^{n+1} + \Theta\Delta t\left[G_{ij}\right]\left\{\Omega_i\right\}^{n+1} + \Theta\Delta t\vartheta\left[K_{ij}\right]\left\{\Omega_i\right\}^{n+1}$$

$$= \left[C_{ij}\right]\left\{\Omega_i\right\}^{n} - (1-\Theta)\Delta t\left[G_{ij}\right]\left\{\Omega_i\right\}^{n} - (1-\Theta)\Delta t\frac{1}{\text{Re}}\left[K_{ij}\right]\left\{\Omega_i\right\}^{n} \quad (7.63)$$

7.3.5 Finite Element Solution Procedure for Velocity Poisson Equations

Application of Galerkin's finite element solution procedure to Equation (7.55) gives the following equations.

$$\int N^T\left(\frac{\partial^2 U}{\partial X^2} + \frac{\partial^2 U}{\partial Y^2} + \frac{\partial \Omega}{\partial Y}\right)dA = 0$$

$$\int N^T\left(\frac{\partial^2 U}{\partial X^2} + \frac{\partial^2 U}{\partial Y^2}\right)dA + \int N^T\frac{\partial \Omega}{\partial Y}dA = 0$$

$$-\left[K_{ij}\right]\left\{U_i\right\} + \int N^T\frac{\partial \Omega}{\partial Y}dA = 0$$

$$-\left[K_{ij}\right]\left\{U_i\right\} + \left[Y_{ij}\right]\left\{\Omega_i\right\} = 0$$

$$\left[K_{ij}\right]\left\{U_i\right\} = \left[Y_{ij}\right]\left\{\Omega_i\right\} \quad (7.64a)$$

Similarly, the velocity Poisson equation in the *y*-direction can be integrated using Galerkin's finite element procedure as shown below.

$$\int N^T\left(\frac{\partial^2 V}{\partial X^2} + \frac{\partial^2 V}{\partial Y^2} - \frac{\partial \Omega}{\partial X}\right)dA = 0$$

Following the steps adopted for the integration of velocity Poisson in the x-direction, the final expression in matrix form can be written as

$$\left[K_{ij}\right]\left\{V_i\right\} = -\left[X_{ij}\right]\left\{\Omega_i\right\} \tag{7.64b}$$

where

$$\left[C_{ij}\right] = \int N^T N dA \quad \text{- capacitance matrix}$$

$$\left[K_{ij}\right] = \int \left[\frac{\partial N^T}{\partial X}\frac{\partial N}{\partial X} + \frac{\partial N^T}{\partial Y}\frac{\partial N}{\partial Y}\right] dA \quad \text{- stiffness matrix}$$

$$\left[X_{ij}\right] = \int N^T \frac{\partial N}{\partial X} dA \quad \text{- gradient matrix in } x\text{-direction}$$

$$\left[Y_{ij}\right] = \int N^T \frac{\partial N}{\partial Y} dA \quad \text{- gradient matrix in } y\text{-direction}$$

$$\left[G_{ij}\right] = \int N^T \left(N_k U_K{}^n\right)\frac{\partial N}{\partial X} dA + \int N^T \left(N_k V_K{}^n\right)\frac{\partial N}{\partial Y} dA \quad \text{- convection matrix}$$

Equations (7.63) and (7.64) are the final matrix form of the finite element solution for the velocity-vorticity form of Navier-Stokes equations.

7.3.6 Computational Algorithm

The solution of velocity-vorticity form of Navier-Stokes equations can be obtained by following the computational procedure shown in Figure 7.4. Though the equations are solved in transient procedure, only steady state solutions are obtained at the end of the computations. In most of the CFD problems, only the steady state solutions are required to evaluate the important flow parameters useful for the analysis of the problem. As seen from the flow chart, after forming the global level matrices for both the vorticity transport and velocity Poisson equations, the time iteration loop starts in order to obtain the converged solution at every time level. It has been pointed out earlier that vorticity transport equations and velocity Poisson equations are coupled through the vorticity boundary values. In the computational procedure, initially only the velocity field is initialized with the known velocities and the vorticity values at the boundaries are not known. As a first approximation, a small amount of vorticity values can be assigned as vorticity boundary values in order to initiate the computations. For solving the velocity Poisson equations, vorticity values are required. With the weak assigned boundary vorticity values, a weak loading of vorticity gradient can be computed for the solution of the velocity Poisson equations. Once the velocity field is obtained in the first iteration, the vorticity boundary conditions can be computed to solve the vorticity transport equations. Now, the vorticity field starts developing the flow domain, which will strengthen the loading required for the velocity Poisson equations, which in turn produce a strong velocity field, and

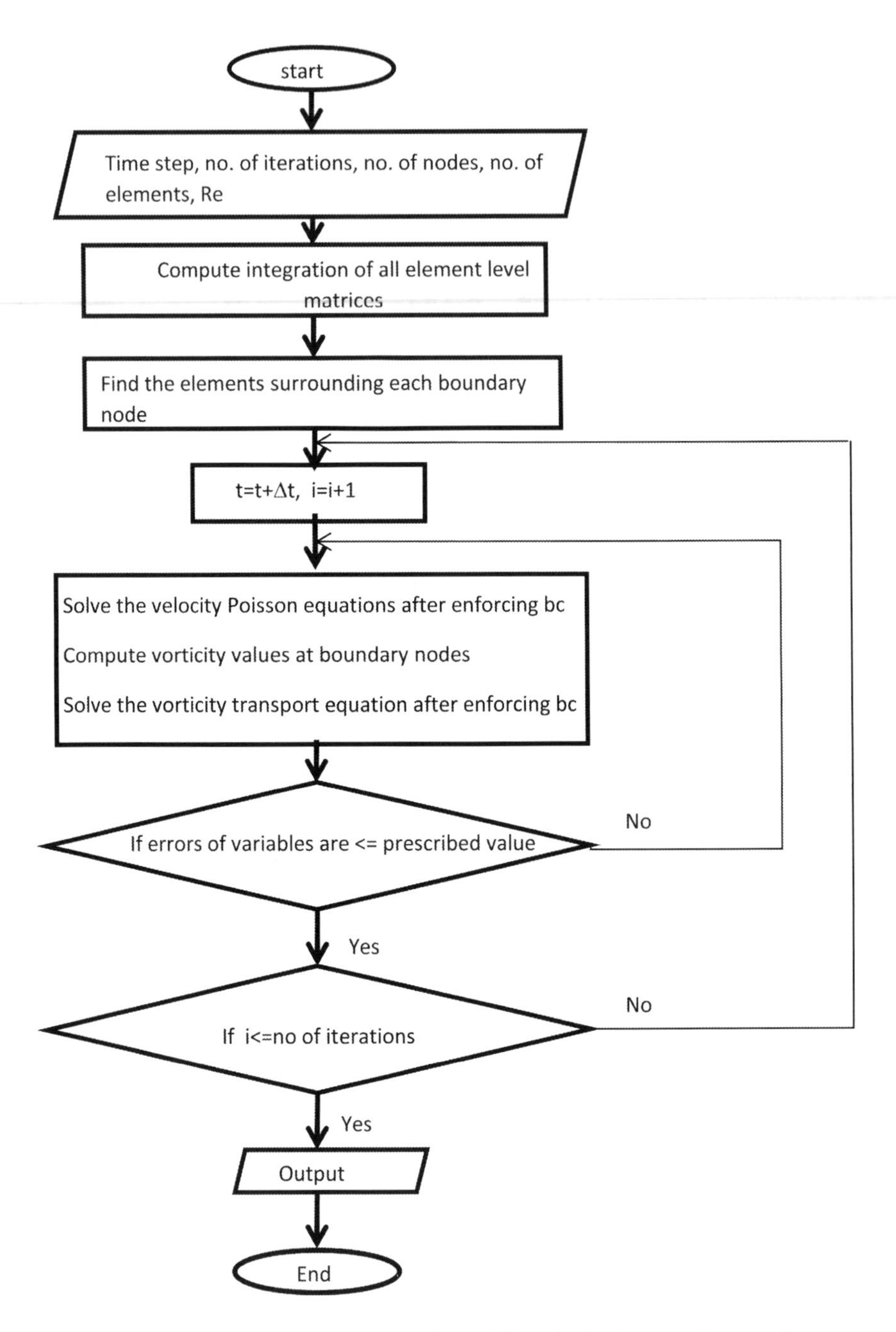

FIGURE 7.4 Computational flow chart for the solution of velocity-vorticity equations.

thus when this iteration between these equations is continued, a convergence state is reached after a few iterations at the given time level. This procedure will be repeated for all the time steps until steady state solution for the velocity and vorticity fields are obtained. The computational algorithm employed for the solution of velocity-vorticity equations is straightforward and simpler compared to the algorithms used for the primitive variable form of Navier-Stokes equations. However, the pressure field cannot be obtained directly from this computational algorithm as pressure is not a variable in this formulation. For CFD problems, where the main focus is on the study of flow field, fluid vortex formation, heat transfer and species transfer, the velocity-vorticity formulation can be easily employed. The computational procedure can be demonstrated by solving a lid-driven square cavity problem discussed in the following section.

7.3.7 Simulation of Lid-Driven Square Cavity Flow Problem

In the CFD literature it can be found that lid-driven square cavity flow problem is considered as a bench mark problem for validation of new numerical schemes or solution algorithms. The solution procedure explained for velocity-vorticity form of Navier-Stokes equations is explained by simulating flow parameters for a lid-driven square cavity flow problem at different Reynolds number (Re) values. Figure 7.5 shows the square cavity with boundary conditions specified for velocity and vorticity fields. It has to be noticed that the values of vorticity at the boundaries have to be computed using the procedure discussed already in the earlier section. Hence, the values of vorticity at the boundary specified as Ω_b have to be obtained using the computational procedure. Using the computational flow chart shown in Figure 7.4, a computer code in FORTRAN, ***velvor_2D.for*** is developed to obtain the simulation results.

7.3.8 Simulation Results

The characteristic of the fluid field developed in a lid-driven square cavity at Re=400 and 1000 can be well understood by making different plots of the flow parameters. Increase in Re value indicates the dominance of inertial force over the viscous force of the fluid in the flow domain. As the plate is moving from left to right at unit velocity, depending on the value of Re, the inertial force of the fluid dictates on the formation of the flow field. In order to understand the flow behavior in the square cavity, plots of velocity vector, streamline distribution, vorticity distribution, U-Y and X-V plots are made using the simulation results for Re=400 and 1000. A computational grid of size 31×31 mesh has been employed for Re=400 case, whereas 41×41 grid size has been employed for getting flow results at Re=1000. Figure 7.6 depicts the distribution of velocity vectors in the cavity for Re=400 and 1000. As the plate moves from left to right with unit velocity, the fluid sticking to the top wall also moves with the same velocity because of the assumption of no-slip boundary condition for the velocity field. The fluid is able to move freely until it reaches the

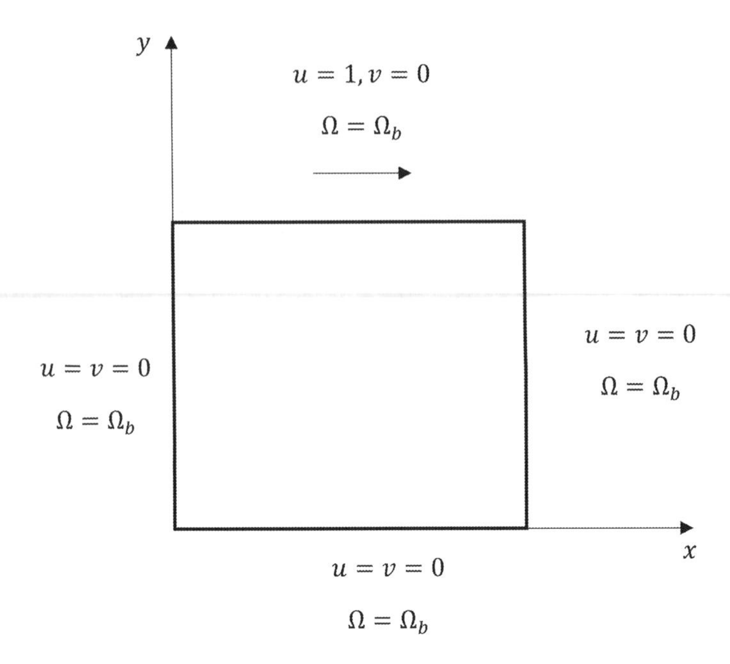

FIGURE 7.5 Schematic diagram for lid-driven square cavity problem.

right end wall, afterwards there will be change in direction, thus causing change in momentum of the fluid, which leads to the formation of a fluid vortex as seen from the above figure. A circulatory flow pattern is expected because the top lid moves continuously from left to right. The fluid core is formed at a point slightly off-set from the center of the cavity, towards the top right side of the cavity at Re=400. With an increase in Re value to 1000, due to increased fluid momentum, the fluid core gets shifted towards the center with a tendency to move downwards. It is very interesting to notice the formation of fluid boundary layer along the grid lines for Re=400 and 1000. The U-velocity has a higher value equal to unity as that of the moving top wall, compared to other walls, where its value is equal to zero due to no-slip boundary condition. The hydrodynamic boundary layer development over all the three walls except the top wall is clearly simulated by the program as seen from these figures. However, at both the bottom corners of the cavity, the velocity field looks very weak, however in order to satisfy mass conservation of the fluid medium, there must be fluid at these corners also, of course, with weak velocity field. The reason is that with these values of Re, the fluid medium could not transport momentum to the fluid in the corners of the cavity.

Figure 7.7 shows the streamline patterns in the cavity for Re=400 and 1000. Due to the weak momentum transport to the bottom corners of the cavity, separated fluid cells are formed at the corners and these cells are not able to be part of the main core of the fluid because of weak momentum transport. Vorticity distribution in the cavity are depicted for Re=400 and 1000 in Figure 7.8. Strong vorticity contours are observed at the cavity walls closer to the moving top wall, whereas the vorticity looks weak in the other regions of the cavity. With an increase in Re value from

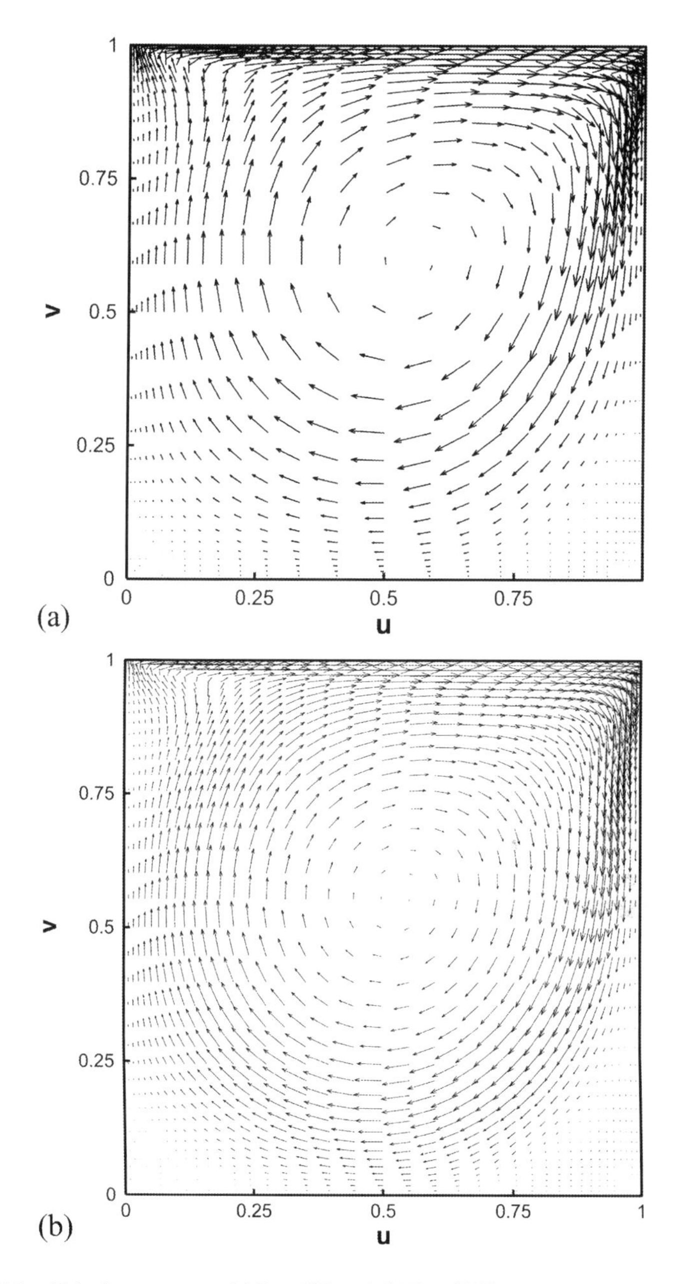

FIGURE 7.6 Velocity vectors at (a) Re=400 and (b) Re=1000.

400 to 1000, it is noticed that the vorticity contours engulf the entire cavity and the formation of a central fluid core is indicated by the near circular vorticity contours at the center of the cavity. The distribution of U velocity along the Y-direction at X=0.5 is an important flow characteristic that is used for validation purpose. Figure 7.9 shows such plots for Re=400 and 1000. It can be noticed the value of U velocity

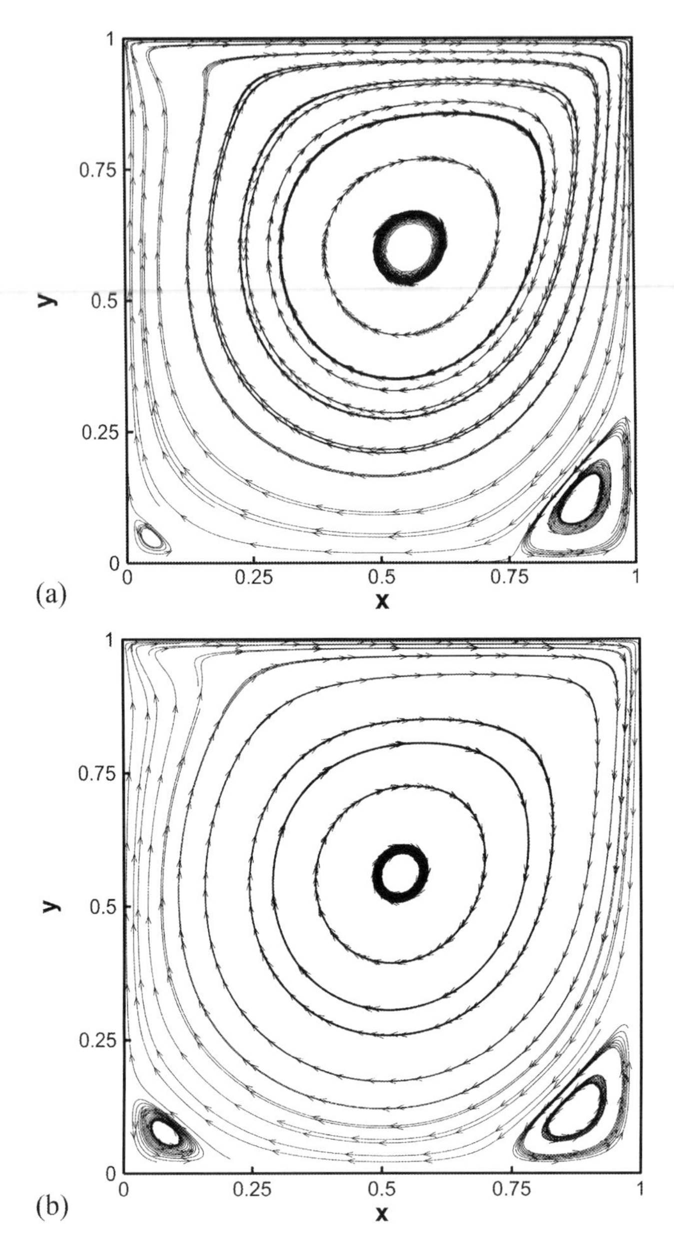

FIGURE 7.7 Streamline patterns for (a) Re=400 and (b) Re=1000.

varies from zero at Y=0 to 1 at Y=1. The velocity gradients formed at both Y=0 and Y=1 specify the type of flow pattern formed within the cavity for a given Re value. It can be observed that U velocity gradients in Y direction become steep at Y=0 and Y=1 with increases in Re value, and this increase indirectly indicates the increased momentum transport to the fluid closer to the wall at Y=0. It has to be remembered,

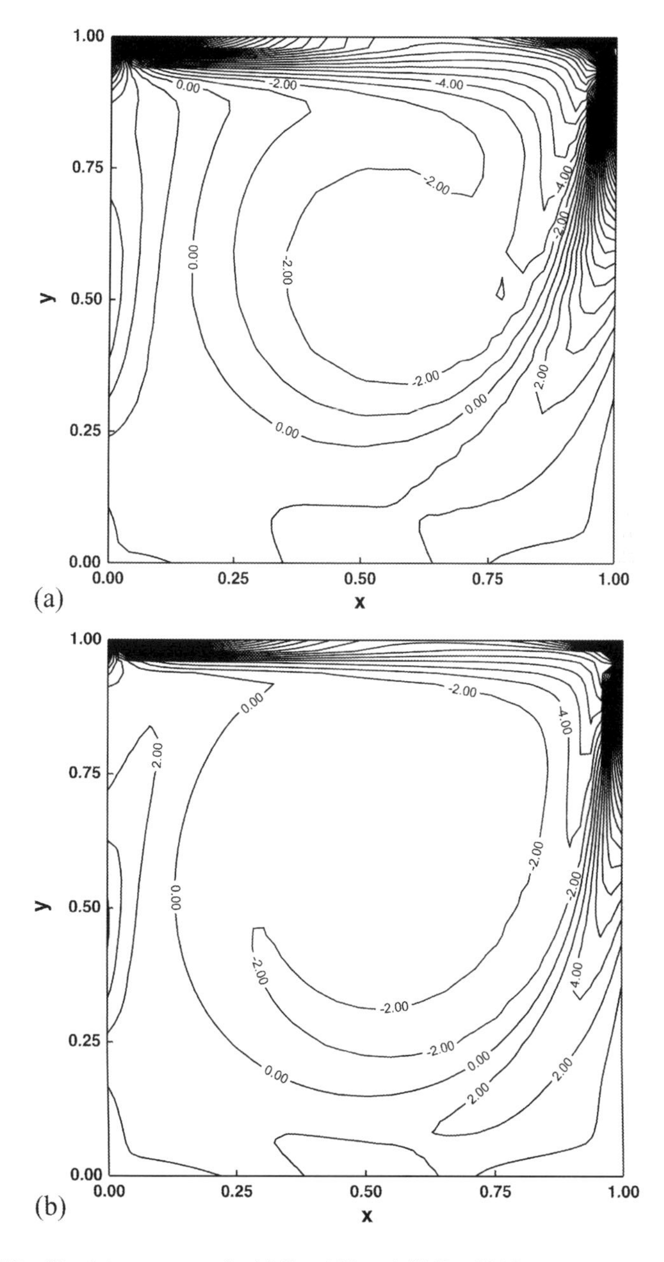

FIGURE 7.8 Vorticity contours for (a) Re=400 and (b) Re=1000.

only the fluid in contact with the top moving wall, has the highest momentum in the cavity and this momentum has to be transported to the entire fluid domain of the cavity after overcoming the viscous friction force offered by the specific fluid under consideration. The negative maximum value of U velocities in U-Y plot is an important characteristic of the flow field. The X-V plot drawn at Y=0.5 shown in Figure 7.10

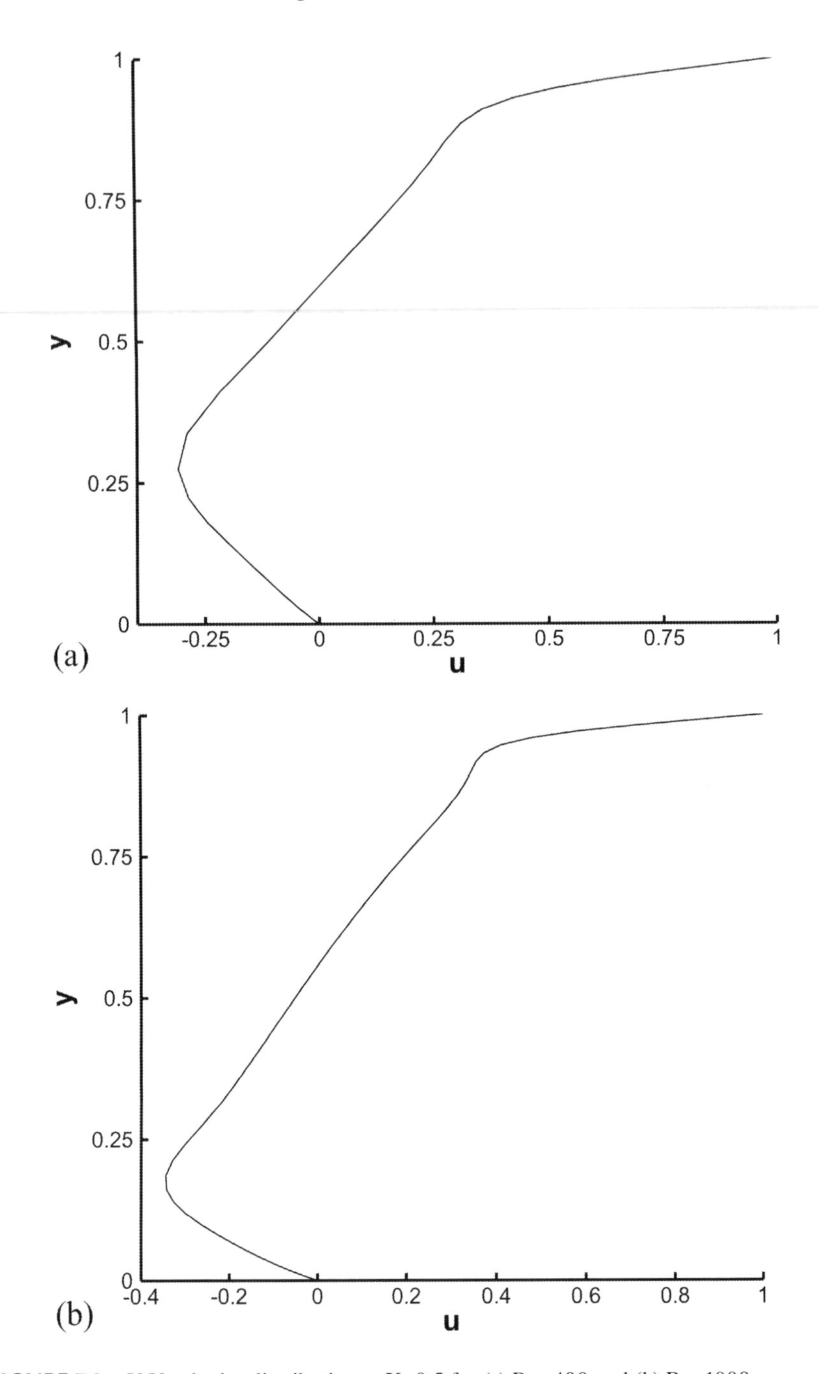

FIGURE 7.9 U-Y velocity distribution at X=0.5 for (a) Re=400 and (b) Re=1000.

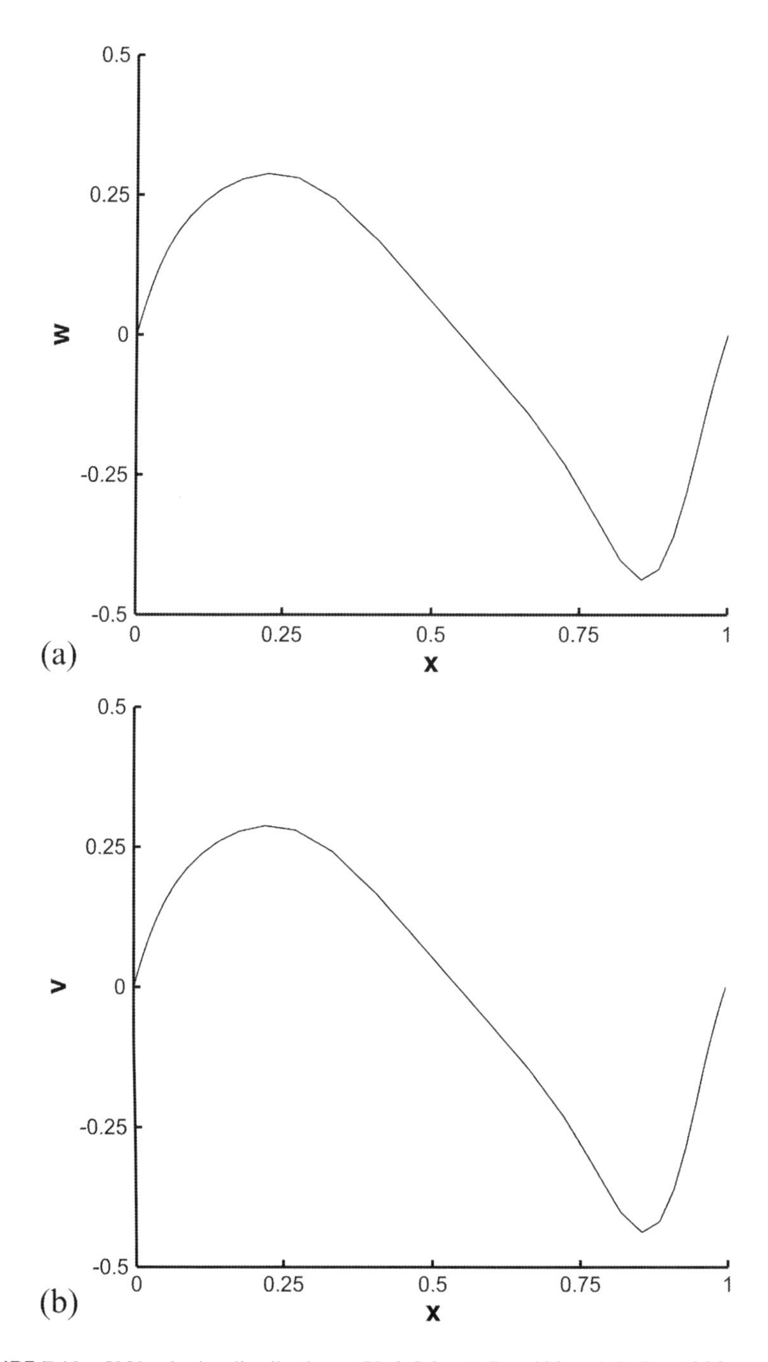

FIGURE 7.10 X-V velocity distribution at Y=0.5 for (a) Re=400 and (b) Re=1000.

indicates the change in direction of fluid vector due to the re-circulatory flow pattern formed between the two vertical walls of the cavity. At X=0 and X=1, the V velocity takes the value of zero because of the vertical walls. Due to the formation of the fluid core within the cavity, the fluid above the fluid core moves from left to right and the fluid below the fluid core moves from right to left, and this change in direction of flow field can be easily noticed in the above figure. The scale for V velocity in the y-axis has both positive and negative values with zero at the center.

7.3.9 Simulation of Natural Convection in a Square Cavity

In this example, fluid flow convection due to body force generated by temperature gradient within the cavity will be discussed. The movement of the top lid of the square cavity considered in the previous section provides the required momentum for the movement of the fluid in the cavity. There are flow problems in which the body force becomes the sole cause to garner momentum within the fluid for fluid flow. In the case of natural convection problems, whenever two sides of the domain are maintained at temperature differential, then there is a change in density of the fluid adjacent to the hot and cold walls. This difference in density of the fluid medium causes buoyancy force with the help of gravity and causes the fluid circulation, thus transporting heat from the hotter region to colder region. The density variation in the governing equation is employed with the help of Boussinesq approximation, which approximates the density variation only for the body force term and not in the flow field. When heat transport within the computational domain is considered, then the conservation equation for thermal energy has to be included in the governing equation. The energy equation derived in Chapter 2 is directly used in this section for two-dimensional flow domain. The governing equations for natural convection in velocity-vorticity form with only body force term can be written as [5, 6]

Vorticity Transport Equation

$$\frac{\partial \omega}{\partial t} + u\frac{\partial \omega}{\partial x} + v\frac{\partial \omega}{\partial y} = \upsilon\left[\frac{\partial^2 \omega}{\partial x^2} + \frac{\partial^2 \omega}{\partial y^2}\right] + \frac{\partial F_{by}}{\partial x}\hat{j} \tag{7.65}$$

Velocity Poisson Equations

$$\frac{\partial^2 u}{\partial x^2} + \frac{\partial^2 u}{\partial y^2} = -\frac{\partial \omega}{\partial y} \tag{7.66a}$$

$$\frac{\partial^2 v}{\partial x^2} + \frac{\partial^2 v}{\partial y^2} = \frac{\partial \omega}{\partial x} \tag{7.66b}$$

Energy Equation

$$\frac{\partial T}{\partial t} + u\frac{\partial T}{\partial x} + v\frac{\partial T}{\partial y} = \alpha\left[\frac{\partial^2 T}{\partial x^2} + \frac{\partial^2 T}{\partial y^2}\right] \tag{7.67}$$

The body force term in Equation (7.65) can be expressed as $\frac{\partial F_{by}}{\partial x}\hat{j} = \frac{\partial}{\partial x} g\beta_T\left(T - T_C\right)$, where β_T is the volumetric thermal expansion coefficient defined as

$\beta_T = -\frac{1}{\rho_0}\left(\frac{\partial \rho}{\partial T}\right)_p$ - the rate of change of density with temperature at constant pressure, T_c is the cold wall temperature and ρ_0 is the density at reference temperature, T_0. After the inclusion of body force term, Equation (7.65) can be written as

$$\frac{\partial \omega}{\partial t} + u\frac{\partial \omega}{\partial x} + v\frac{\partial \omega}{\partial y} = \upsilon\left[\frac{\partial^2 \omega}{\partial x^2} + \frac{\partial^2 \omega}{\partial y^2}\right] + \frac{\partial}{\partial x} g\beta(T - T_c) \quad (7.68)$$

Now, the governing equations (7.66a), (7.66b), (7.67) and (7.68) have to be solved for the four unknown field variables, u, v, ω and T. Generally, these equations are solved in non-dimensional form by making use of the following scaling parameters.

$$X = \frac{x}{L},\ Y = \frac{y}{L},\ U = \frac{uL}{a},\ V = \frac{vL}{a},\ \Omega = \frac{\omega L^2}{a},\ \tau = \frac{ta}{L^2},\ \theta = \frac{T - T_c}{T_h - T_c}.$$

It has to be noticed that thermal diffusivity, α has been used in all the above scaling parameters. This is because the temperature-associated body force term appearing in Equation (7.68) can be evaluated only when temperature, T is computed from the energy equation. That shows how the vorticity transport equation and the energy equation are coupled with each other to evolve the flow field, which in turn transports heat from the hot region to cold region. After implementing the scaling parameters, the non-dimensional form of the governing equations can be written as

Vorticity Transport Equation

$$\frac{\partial \Omega}{\partial \tau} + U\frac{\partial \Omega}{\partial X} + V\frac{\partial \Omega}{\partial Y} = \Pr\left[\frac{\partial^2 \Omega}{\partial X^2} + \frac{\partial^2 \Omega}{\partial Y^2}\right] + Ra\Pr\frac{\partial \theta}{\partial x} \quad (7.69)$$

Velocity Poisson Equations

$$\frac{\partial^2 U}{\partial X^2} + \frac{\partial^2 U}{\partial Y^2} = -\frac{\partial \Omega}{\partial Y} \quad (7.70a)$$

$$\frac{\partial^2 V}{\partial X^2} + \frac{\partial^2 V}{\partial Y^2} = \frac{\partial \Omega}{\partial X} \quad (7.70b)$$

Energy Equation

$$\frac{\partial \theta}{\partial \tau} + U\frac{\partial \theta}{\partial X} + V\frac{\partial \theta}{\partial Y} = \left[\frac{\partial^2 \theta}{\partial X^2} + \frac{\partial^2 \theta}{\partial Y^2}\right] \quad (7.71)$$

Equations (7.69) to (7.71) have to be solved for the velocity, vorticity and temperature fields within the computational domain. The Rayleigh number, Ra is related to Grashof number, Gr and Prandtl number, Pr by the following relation.

$Ra = Gr\,\mathrm{Pr}$, where $Gr = \dfrac{g\beta\Delta TL^3}{\upsilon^2}$, $\mathrm{Pr} = \dfrac{\nu}{\alpha}$. The Grashof number relates buoyancy force to the viscous force of the fluid whereas Prandtl number relates the momentum diffusivity to thermal diffusivity. The value of Rayleigh number indicates the strength of natural convection along a heated vertical wall.

7.3.9.1 Finite Element Solution Procedure

The Galerkin's weighted residual finite element method can be implemented on the lines of procedure employed for the solution of velocity-vorticity equations in the previous section. After the implementation of finite element procedure, the final matrix form of the governing equations can be expressed as

Vorticity Transport Equation

$$\begin{aligned}&\left[C_{ij}\right]\left\{\Omega_i\right\}^{n+1} + \Theta\Delta t\left[G_{ij}\right]\left\{\Omega_i\right\}^{n+1} + \mathrm{Pr}\times\Theta\Delta t\left[K_{ij}\right]\left\{\Omega_i\right\}^{n+1}\\ &= \left[C_{ij}\right]\left\{\Omega_i\right\}^{n} - (1-\Theta)\Delta t\left[G_{ij}\right]\left\{\Omega_i\right\}^{n} - \mathrm{Pr}(1-\Theta)\Delta t\left[K_{ij}\right]\left\{\Omega_i\right\}^{n}\\ &+\mathrm{RaPr}\Delta t\left[X_{ij}\right]\left\{\theta_i\right\}^{n}\end{aligned} \tag{7.72}$$

Velocity Poisson Equations

$$\left[K_{ij}\right]\{U\} = \left[Y_{ij}\right]\{\zeta\} \tag{7.73a}$$

$$\left[K_{ij}\right]\{V\} = -\left[X_{ij}\right]\{\zeta\} \tag{7.73b}$$

Energy Equation

$$\left[C_{ij}\right]\left\{\theta_i\right\}^{n+1} + \Theta\Delta t\left[K_{ij}\right]\left\{\theta_i\right\}^{n+1} = \left[C_{ij}\right]\left\{\theta_i\right\}^{n} - (1-\Theta)\Delta t\left[K_{ij}\right]\left\{\theta_i\right\}^{n} - \Delta t\left[G_{ij}\right]\left\{\theta_i\right\}^{n} \tag{7.74}$$

As Equations (7.72) and (7.74) are coupled through the temperature term, an iterative procedure has to be employed in order to resolve the coupling. For the simulation of results for natural convection in a differentially heated square cavity, a computer code in FORTRAN, ***velvor_NC.for*** has been developed and the results obtained are discussed briefly in the following section. Figure 7.11 shows the computational procedure for the solution of the Equations (7.72) to (7.74) to obtain simulation results.

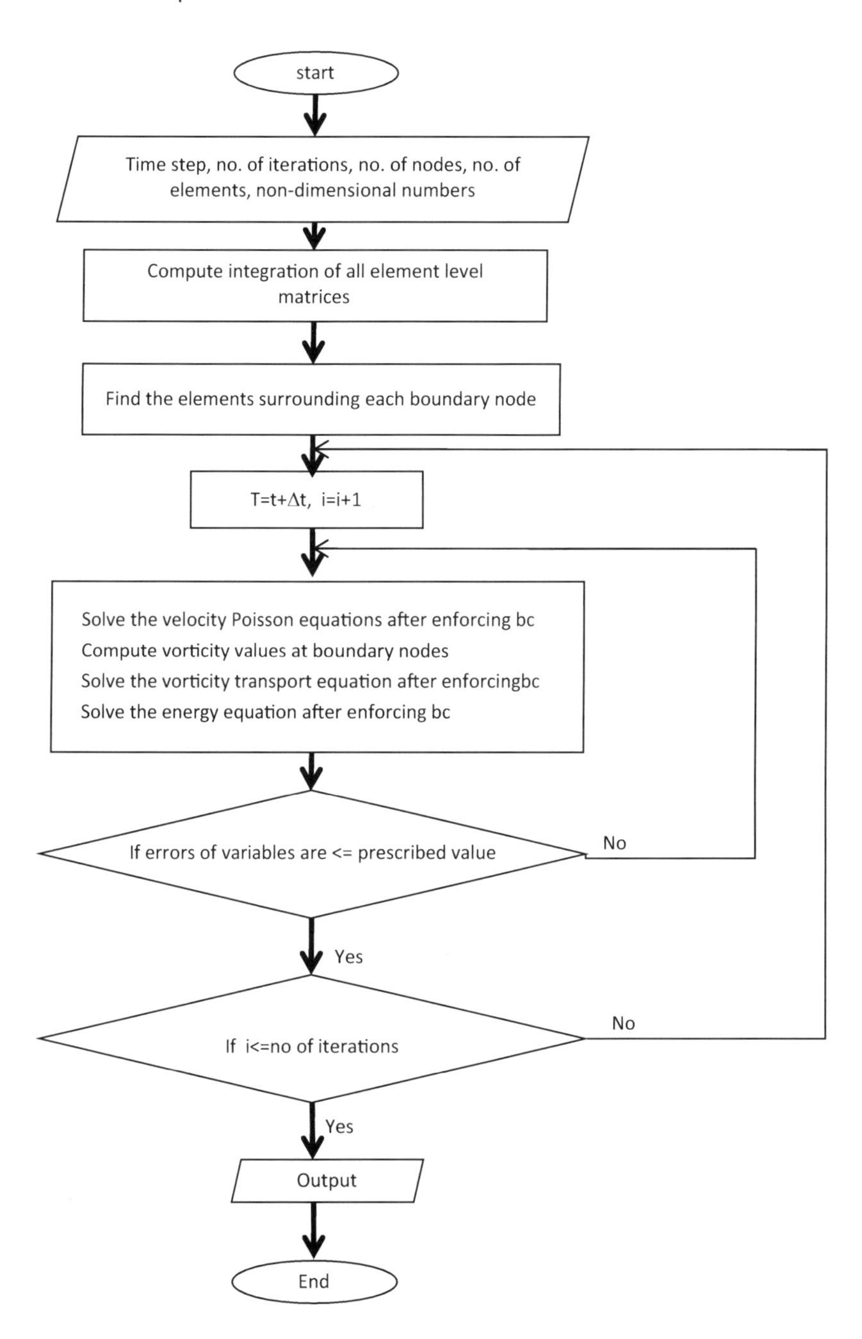

FIGURE 7.11 Flow chart of computational algorithm for natural convection problem.

7.3.9.2 Simulation Results for Natural Convection in a Differentially Heated Square Cavity

Figure 7.12 shows the schematic diagram of the differentially heated square cavity for which the simulation results have been obtained. As assumed in the previous lid-driven cavity problem, here none of the walls of the cavity is moving with certain velocity. The figure indicates a simple square enclosure with left side wall subjected to high temperature, θ_h and the right side wall at cold temperature, θ_c, whereas the other two horizontal walls are adiabatic for heat transfer. There is no momentum transfer from any moving fluid and hence, the fluid movement due to buoyancy force evolves very slowly. It is well known that when fluid is heated, the density decreases and hence the fluid adjacent to the left wall has lesser density compared to the fluid adjacent to the right cold wall. This difference in density causes the low-density fluid to move upwards due to buoyancy force and heavier fluid to move in the downward direction, thus creating fluid circulation within the cavity. This buoyancy force is very small, hence, the resulting fluid momentum is also weak at the beginning of the simulation. As the Rayleigh number is related to the natural convection phenomenon, an increase in Ra value increases the fluid

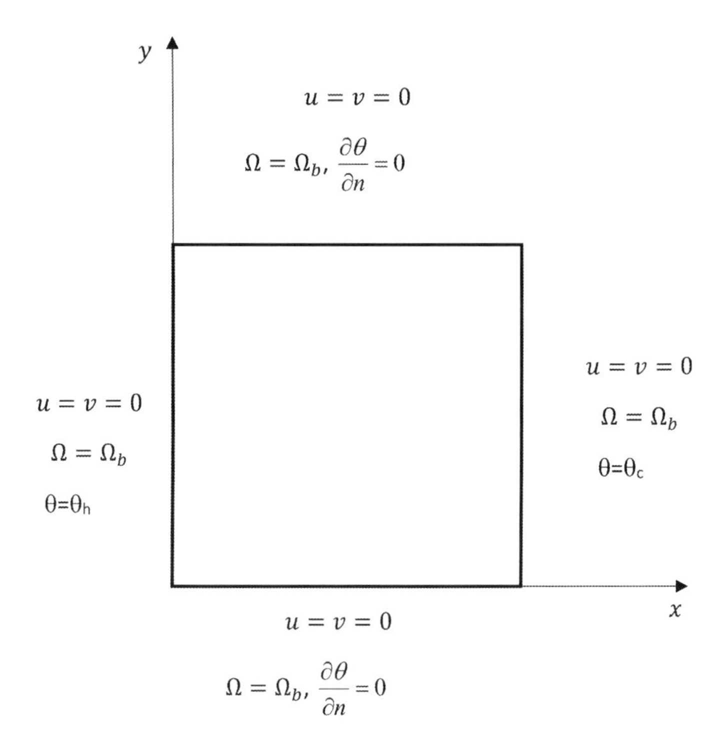

FIGURE 7.12 Schematic diagram for natural convection in a differentially heated square cavity.

momentum, thereby increasing the fluid convection. When Ra=10^3, the heat transfer takes place only by means of diffusion within the fluid in the cavity and when Ra increases beyond this value, a thin velocity boundary layer is observed to take place near the heated wall, thus setting up fluid momentum transport within the flow domain. Figure 7.13 shows the velocity vector distribution within the cavity for Ra=10^3, 10^4, 10^5 and 10^6. It can be noticed from this figure that for Ra=10^3, the velocity vectors form a circulatory fluid movement within the cavity and when this value increases further to 10^4, there is further development of the velocity boundary layer [5]. With further increases to 10^5 and 10^6, it is observed that the velocity boundary layers near the hot and cold walls become much stronger compared to the horizontal walls. The temperature contours within the cavity are depicted in Figure 7.14 for different Ra values. As the horizontal walls are adiabatic for heat transfer, the isothermal lines become normal to these walls for all the values of Ra. At Ra=10^3, the heat transfer seems to take place purely by diffusion and no

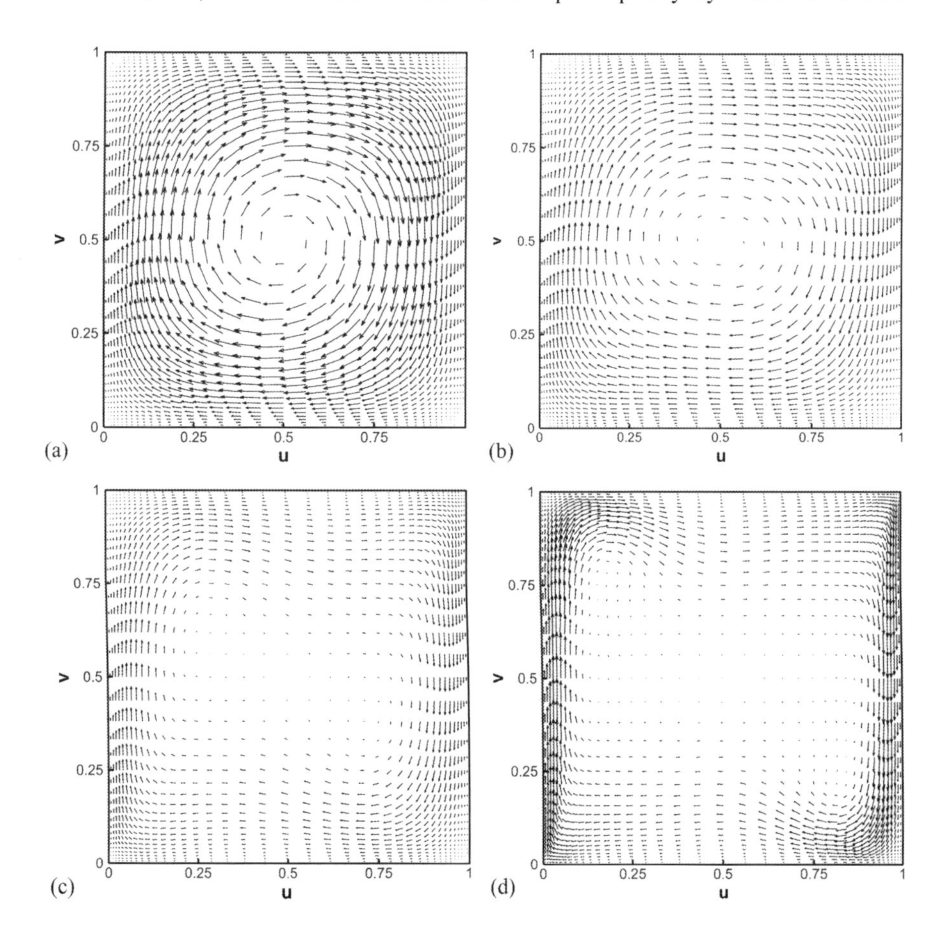

FIGURE 7.13 Velocity vectors for (a) Ra=10^3, (b) Ra=10^4, (c) Ra=10^5 and (d) Ra=10^6.

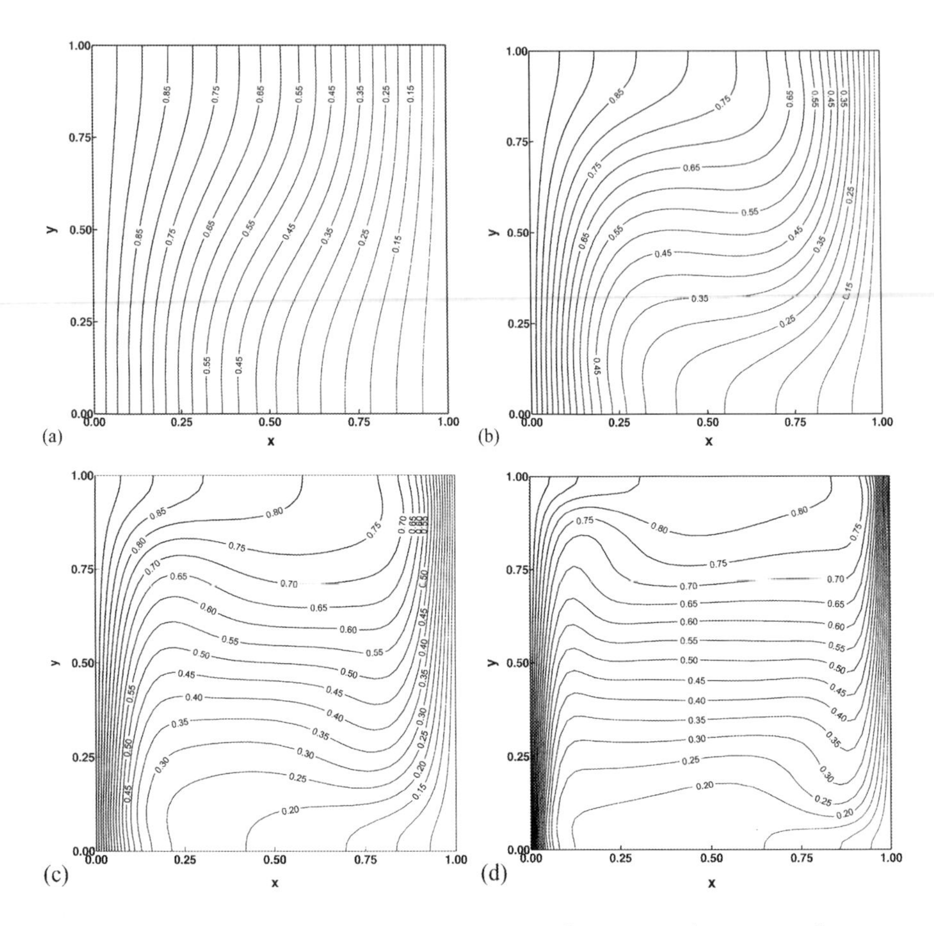

FIGURE 7.14 Temperature contours for (a) Ra=10^3, (b) Ra=10^4, (c) Ra=10^5 and (d) Ra=10^6.

observable thermal boundary layers are formed at this value. With an increase in the value of Rayleigh number, the thermal boundary layer starts developing near the vertical hot and cold walls as seen from Figures 7.14 (b) to 7.14(d). Temperature gradients near both the vertical walls increase, with increase in Ra value, thus promoting fluid convection within the cavity. The temperature distribution along the X-direction at Y=0.5 is shown in Figure 7.15 for different values of Ra. At Ra=10^3, the temperature distribution looks like a straight line with θ=1 at X=0 and θ=0 at X=1. With an increase in the value of Rayleigh number, the temperature gradient near the hot and cold vertical walls increases and this is due to fluid convection with increased buoyancy force. This encourages the formation of a strong thermal boundary layer at the cold and hot vertical walls, enhancing convective heat transfer.

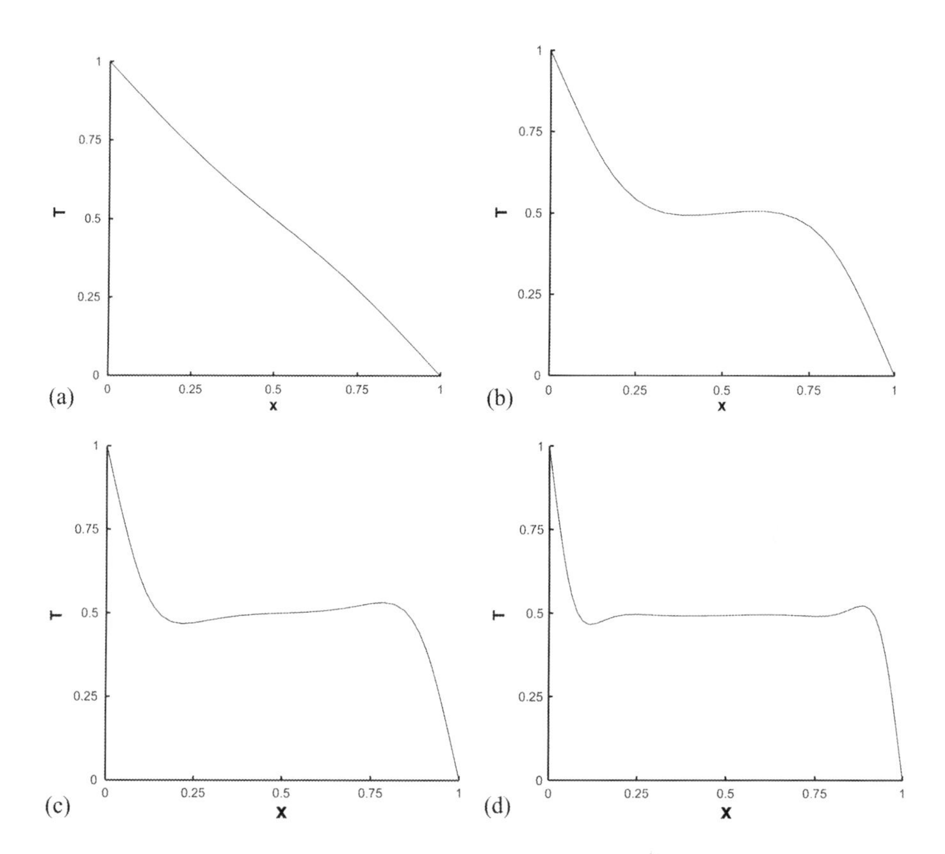

FIGURE 7.15 X-θ plot for (a) Ra=10^3, (b) Ra=10^4, (c) Ra=10^5 and (d) Ra=10^6.

REFERENCES

1. *A Mathematical Introduction to Fluid Mechanics*, edited by Alexandre Chorin and Jerrold E. Marsden, Springer-Verlag Publishing Company, Inc., New York, 1992.
2. *Applied Finite Element Analysis*, edited by Larry J. Segerlind, 2nd Edition, Wiley, Hoboken, NJ, 1984.
3. *Advanced Engineering Mathematics*, edited by Michael D. Greenberg, 2nd Edition, Prentice-Hall, Inc., Prentice, NJ, 1998.
4. Wong, K. L. and Baker, A. J. A 3D incompressible Navier-Stokes velocity-vorticity weak form of finite element algorithm, *International Journal of Numerical Methods in Fluids*, 2002; 38: 99–123.
5. Senthil Kumar, D. *Study of Double Diffusive Mixed Convection Problems*, PhD dissertation work, Mechanical & Industrial Engineering Department, Indian Institute of Technology Roorkee, Roorkee, 2009.
6. Nithish Reddy, P. *Study of Double Diffusive Convection in Enclosures*, PhD dissertation work, Mechanical & Industrial Engineering Department, Indian Institute of Technology Roorkee, Roorkee, 2016.

EXERCISE PROBLEMS

Qn: 1 Write down the Euler equation, Bernoulli's equation and Navier-Stokes equations for two-dimensional flow field. Distinguish the difference between these equations highlighting the assumptions made and limitations of these equations. Also suggest the method of solution of these equations to obtain the flow variables in a given computational domain.

Qn: 2 Express the Navier-Stokes equations in primitive variables form, velocity-vorticity form, stream-function-vorticity form. What are the advantages and limitations of these different forms of the momentum conservation equations? Explain the method of solution of these different forms of equations to obtain the flow field variables.

Qn: 3 Starting from the primitive variables form of Navier-Stokes equations, develop velocity-vorticity form of Navier-Stokes equations for three-dimensional flow field. Define the vorticity vector for x-, y- and z-directions. Also derive the velocity Poisson equations and finally write down the vorticity transport equations and velocity Poisson equations for all the three coordinates.

Qn: 4 For a three-dimensional computational domain, develop expressions for the computation of vorticity values at the boundaries of (i) x-z, (ii) z-x, (iii) y-z and (iv) x-y planes.

Qn: 5 Solve the three-dimensional velocity-vorticity form of Navier-Stokes equations using Galerkin's finite element method assuming 8-node brick elements for discretizing the computational domain. Write down the detailed computational algorithm for the solution of vorticity transport equations and velocity Poisson equations to obtain the six field variables, ω_x, ω_y, ω_z, u, v and w in the coordinate directions. Explain the iterative procedure involved if any.

Qn: 6 Develop a computational mesh of size 31 × 31 using the computer program ***Tfi_2D.for*** for a two-dimensional lid-driven cavity problem. Verify the mesh data and the node numbers for the boundaries for the computation of vorticity values at the boundaries. Assume the top lid of the cavity is moving with unit velocity from left to right in the x-direction. Then using the given computer code ***velvor_2D.for***, obtain flow results for Re=200. Plot the u-y values at x=0.5 and compare these results with available literature. Also plot the velocity vectors and vorticity contours on the x-y coordinate plane.

Qn: 7 Repeat Qn: 6 with different values of Re=300 and 400 for lid moving with unit velocity. Obtain velocity vectors after solving the equations using the computer code ***velvor_2D.for*** and plot the velocity vectors. With the help of the graphical software used for plotting the velocity vectors, identify the boundary layers developed over the left and right vertical walls of the cavity and compare these plots with the velocity vectors formed on the bottom wall of the cavity. Comment on the nature of velocity boundary layers developed on these three walls. Explain the physics behind the reason for these variations with the help of Navier-Stokes equations.

Qn: 8 For the solution of velocity-vorticity form of Navier-Stokes equations, the vorticity field on the boundary has to be enforced correctly in order to obtain the vorticity field according to the underlying physics. Then, with this vorticity field the velocity Poisson equations can be solved. In order to initiate flow field at the beginning of iterations in the solution procedure, small values of vorticity are assigned at the boundary points. Repeat **Qn: 6** by assuming different values for the vorticity at the boundaries as initial values. Study the effect of these initially assumed values on the convergence of flow field results.

Qn: 9 Consider natural convection in a vertical cavity of size 1×2 units. The left end vertical wall is kept at a temperature equal to unity whereas the right end vertical wall is kept at a temperature of zero. The dimension of the cavity and temperature values are represented as non-dimensional quantities in accordance with the formulation of the natural convection problem described using velocity-vorticity formulation. It is required to obtain velocity vectors, temperature and vorticity vectors in the cavity for $Ra=10^4$. Use the given computer code ***2dmeshnc.for*** to develop a TFI mesh of size 21×41 with required mesh data for the simulation. With the help of the given computer code ***ncbicg.for***, obtain simulation results to draw the velocity, temperature and vorticity contour plots.

Qn: 10 Repeat **Qn: 9** for a horizontal cavity of size 2×1 which is discretized using 41×21 mesh with the help of the TFI mesh generation computer code ***2dmeshnc.for.*** Execute the computer code ***ncbicg.for*** for $Ra=10^4$ to obtain velocity, temperature and vorticity vectors for the same boundary conditions as described in **Qn: 9**. Compare the development of velocity boundary layers over the hot and cold walls of the cavities. Comment on the evolution of temperature field along the hot and cold walls and also within the cavity. Which cavity gives rise to better natural convection and why?

QUIZ QUESTIONS

Qn: 1 In flow problems with heat transfer, in the analysis of velocity and thermal boundary layers, choose the correct statement from the following:
Velocity boundary layer precedes the thermal boundary layer.
Thermal boundary layer precedes the velocity boundary layer.

Qn: 2 In centrifugal pumps used to pump water, the ____________________ ____________________ force is more significant compared to viscous forces.

Qn: 3 The different terms in Navier-Stokes equations in a particular direction indicates force balance/energy balance in that direction. Tick the right one.

Qn: 4 Navier-Stokes equations have exact solutions: true/false. Tick the right one.

Qn: 5 The term that makes it complex to get a direct solution for Navier-Stokes equations is ____.

Qn: 6 The kinematics of the flow field in obtained using ____________________ ____________________equations in velocity-vorticity form of Navier-Stokes equations.

Qn: 7 In a lid-driven cavity flow problem where the lid moves from left to right, with increase in Reynolds number the fluid vortex formed moves towards (a) left, (b) right, (c) center, or (d) none of the above.

Qn: 8 In natural convection problems, ____________________ force is dominant compared to other forces in the fluid medium.

Qn: 9 For natural convection phenomenon in a square cavity with left wall subjected to hot temperature and right wall to cold temperature, the fluid rotation is observed in anti-clockwise direction/clockwise direction. Tick the right one.

Qn: 10 In the natural convection problem described in Qn: 8, with increase in Rayleigh number, the velocity vectors move (a) to the top side of the cavity, (b) to the bottom side of the cavity, (c) to the hot end and cold end of the cavity or (d) to the corners of the cavity.

Index

A

ADI, 204, 205
adiabatic boundary condition, 112–15, 123, 132, 135, 138, 207, 211, 221, 223
aerodynamic applications, 263
aeronautics science, 294
algebraic equations, 1, 8, 13, 23, 27, 30, 33, 35, 147, 239
analytical, 2, 8, 12, 23, 30–33, 35, 38, 39, 54, 55, 93, 106, 111, 120, 122, 128, 144–46, 148, 161, 221, 222, 255, 260
analytical methods, 23, 30, 32, 33, 38, 93
analytical solutions, 2, 12, 30, 32, 93, 144
area coordinates, 177, 303
atmospheric science, 257, 294

B

backward difference scheme, 104, 106, 199
BICG, 279, 286
BICG iterative solvers, 279, 286
BICGSTAB, 278
BICGSTAB iterative solver, 278
body forces, 67, 69, 76, 78, 79, 82, 87, 296, 313
Boolean, 182, 184, 191
boundary elements, 217
Boussinesq approximation, 336
Boyle's law, 8
buoyancy force, 17, 67, 68, 90, 313, 315, 322, 336, 340, 342

C

canonical form, 266, 316, 317
Cartesian coordinates, 44, 46
CFD, 311
Charles's law, 8
chemical engineering, 87
chemical reactors, 50
collocation method, 124
combined heat transfer, 53
compact vector storage, 278, 279
compressible flow, 257
compressible fluids, 73
computational algorithms, 13, 35, 172, 176
computational domain, 33, 60, 94–96, 108, 109, 111, 127, 132, 147, 155, 156, 159, 162, 163, 197, 217
computational fluid dynamics, 101, 102, 131, 257, 294, 311
computer programs, 2, 36, 102, 267
COMSOL, 35, 176
COMSOL multiphysics, 176
conduction heat transfer, 11, 41–45, 47, 50, 52, 53, 56, 58, 60, 61, 76, 87, 90, 96, 118, 126, 154, 194, 201, 210, 211, 217, 224, 225, 255, 257
conservation principle, 17–19, 38, 43, 44, 47, 65, 76, 88, 155, 156, 181, 198, 202, 206, 211, 223, 230, 234, 268, 298
continuum, 7, 25–28, 30, 37, 38, 43, 61, 65, 93–95, 147
continuum approach, 38
control volume, 10, 11, 14, 19, 28, 44, 47–50, 55, 62–64, 66, 68, 75–81, 86–88, 94, 95, 124–28, 130–39, 194, 198, 212, 230, 263, 297, 298, 316
convective boundary conditions, 141, 215, 220, 240, 245
convective heat transfer, 18, 23, 26, 41, 42, 53, 60, 61, 75, 76, 86, 87, 100, 105, 106, 126, 128, 154, 161, 206, 207, 213, 221–25, 236, 242, 248
convective heat transfer coefficient, 53, 87, 100, 105, 106, 161, 206, 207, 221–24, 242, 248
convergence, 25, 106, 141, 163, 221, 245, 274, 275, 285–87, 307, 310, 311, 318, 329, 345
cooling towers, 4, 29, 59
coordinate transformation, 164, 166, 168, 172, 175, 205, 243, 271
Crank-Nicholson, 146, 199, 202, 237
curl operation, 265, 316
cylindrical coordinates, 48–50, 139

D

Darcy' s equation, 228
Darcy' s law, 294
deflection of beam, 144
difference equations, 34, 95, 98–100, 111, 112, 114, 117, 118, 120, 123, 291
differential equations, 1, 7, 8, 13, 23, 24, 27–33, 35, 38, 50, 61, 62, 74, 93–95, 97, 99, 102, 107, 123, 147, 148, 154, 173, 176, 181, 227, 260

differential formulation, 7, 89
diffusion, 19, 23, 25, 37 51, 60, 61, 87 88, 91, 146, 200, 204, 205, 209, 228, 229, 232, 270, 291, 303, 305, 307, 315, 316, 324, 325
diffusion coefficient, 61, 87, 229
Dirichlet boundary, 51, 57, 90, 106, 112, 114, 117, 120, 121, 136, 138, 158–60 205–7, 233, 251, 254 271, 274–76, 280, 283 284, 325
diurnal cycle, 210, 226, 227, 232, 240, 244, 247
DoE, 12
drag coefficient, 100
drag force, 257, 296
drying, 18–20, 24, 87, 209, 210, 245, 253

E

eigen value, 20, 23, 31
Einstein, 1
electronic cooling, 54
element-by-element solution, 278, 279, 290
element-nodal connectivity, 156, 157, 194, 205, 280, 282, 283, 309, 310
element-wise storage, 279
emissivity, 53
energy conservation 14, 17, 18, 37, 38, 43, 44, 47, 75, 76, 88, 146,202, 206, 209, 211, 223, 230, 231, 234, 240, 242, 257
equations solver, 33, 106, 239, 267
Eulerian, 30, 298, 299
Eulerian velocity, 298, 299
exact solution, 93, 124, 152, 163
explicit scheme, 198–202, 204, 237

F

Fick's law, 88
finite difference method, 39, 94–97, 102, 103, 106, 118, 121–23, 126–28, 139, 143–45, 195, 200, 204, 237, 254, 255
finite element analysis, 154, 205, 207, 290, 343
finite element method, 39, 94, 124, 139, 147, 148, 150–54, 156, 159, 160, 162–64, 167, 172, 176, 180, 181, 194, 195, 205–7, 210, 213, 215, 220, 221, 223, 234, 239, 267, 268, 270, 273, 278, 279, 288, 291, 300, 302, 307–9, 321, 323, 338, 344
finite volume method, 95, 123–28, 130–33, 139, 141, 144–46, 151, 181
fin problem, 102, 107, 124, 126–28, 140–43, 150, 153, 158, 159, 163
flow fields, 60, 257, 318, 322
fluid dynamics, 23, 88, 101, 102, 131, 205, 257, 262, 293–95, 311
fluid friction, 7, 31, 60, 61, 82, 89, 257, 293, 294, 296, 311
fluid mechanics, 4, 10, 11, 16, 21, 22, 26–30, 32–35, 37, 41, 59, 89, 258, 259, 261, 293, 311, 343
FORTRAN, 35, 106, 120, 123, 128, 159, 186, 192, 224, 261, 287, 290, 329, 338
forward difference scheme, 199, 200
Fourier law, 11, 29, 41, 46, 49, 55, 86, 87, 211
Fourier number, 200–202
frontal solver, 278, 284

G

Galerkin, 124, 148, 151, 162, 199, 205, 278, 323, 338
Gaussian integration, 164, 168
Gaussian quadrature, 163, 180, 272, 274
global coordinates, 163, 164, 167, 173–75, 178, 180, 182, 183, 205
global matrices, 156–59, 240, 267, 276, 279, 303, 310, 325
global matrix-free, 267, 274, 278, 279, 286, 287, 291
GMFFE algorithm, 267, 284, 286, 287
Grashof number, 338
Green's theorem, 152, 153, 213, 235, 325

H

heat exchangers, 56, 59, 75
heat generation, 44, 46, 47, 54–58, 76, 81, 82, 87, 90, 129, 130, 133, 138, 145, 209–12, 214–18, 220, 221, 223–25, 254, 255
heat transfer, 27, 42, 80, 93, 96, 100, 128, 167, 186, 201, 203, 209, 211, 236, 242, 313
Helmholtz-Hodge, 298
Hooke' s law, 8
h-refinement, 163
hydrostatic force, 293

I

impeller blades, 295, 296
implicit scheme, 146, 198, 199, 201–4
incompressibility, 268, 298, 313, 314, 317, 321
incompressibility constraint, 268, 298, 313
initial conditions, 24, 231, 259
integral approach, 94, 147
integral formulation, 94, 147, 153, 180
internal energy, 44
interpolation technique, 156, 182
isoparametric element, 272
isoparametric formulation, 162–64, 167, 170, 172, 180, 205
non-isothermal diffusion coefficients, 232
iterative solver, 278, 279, 286, 287, 310

J

Jacobian matrix, 167, 175, 179, 207, 271

K

Kinematics, 314, 316, 317, 346
kinetic energy, 16, 18, 78, 295

L

Lagrangian, 62, 169
landmine detection, 210, 225, 226, 240, 242, 243
Laplace equation, 30, 55, 255
least square method, 148
lid-driven square cavity, 322, 329
lift force, 257, 296
load vectors, 242, 243, 277, 282, 284
logarithmic variation, 57
lumped parameter analysis, 26, 38, 255
lumped parameter model, 25–27, 37

M

Mach number, 60, 66
magnetic field, 296
magnetic force, 17, 75, 313
mass conservation equation, 65, 67, 259, 294
mass conservation principle, 17, 18, 65, 298
mass lumping method, 239
material derivative, 65, 66, 70
mathematical modeling, 1–3, 12, 14, 20, 25, 26, 28, 29, 31, 32, 35, 38, 41, 43, 59, 60, 75, 84, 87, 93, 296
MATLAB, 12, 35
matrix equations, 23, 105, 106, 111, 147, 162
mean value theorem, 195, 196, 237, 324
mesh generation, 33, 147, 151, 154, 156, 180–82, 184, 186, 188, 192, 206, 215, 282, 309, 345
mesh refinement, 163, 182, 286, 287
mesh sensitivity, 106, 107, 128, 129, 160, 161, 163, 203, 220–22, 224, 244, 287, 288
momentum balance, 16, 22, 24, 29, 67, 68, 71, 73, 75, 86, 294, 296, 298, 300
multi-block TFI, 188

N

natural boundary condition, 148, 153
natural convection, 336, 338–40, 345, 346
natural coordinates, 164, 167, 172–76, 180, 183, 187, 189, 190, 243
Navier-Stokes equations, 15, 32, 74, 101, 242, 262, 263, 267, 279, 290, 291, 293–95, 297–99, 302, 307, 309, 311, 313, 317, 318, 327, 329, 344–46
Neumann, 51, 52, 255
Newtonian fluid, 313, 316, 317
Newton' s law, 41, 72, 90, 296
Newton's second law, 66
non-linear elements, 205
non-linear equations, 24
nuclear rod, 254
nuclear waste, 19, 31, 46, 194, 203, 209
numerical approach, 33
numerical computations, 253
numerical procedures, 313
numerical simulation, 32, 163, 176, 200, 245, 253, 287
numerical stability, 200, 307
numerical technique, 13, 33, 35, 267
Nusselt number, 61, 100

O

one-dimensional, 11, 16, 29, 51, 54, 55, 60, 94–97, 99, 101, 102, 106, 107, 112, 128–30, 132, 145, 146, 149, 152–54, 156, 163, 164, 168, 169, 194, 197, 200, 204–7, 209–11, 213, 215, 220, 253, 255, 257
one-dimensional conduction, 11, 54, 55, 95, 204, 210
one-dimensional element, 168, 207
OpenFOAM, 176
operator-splitting method, 278

P

parabolic, 30, 255, 278
physical laws, 1–3, 7, 8, 13, 20, 22, 29, 30, 43
physical models, 3–5
physical phenomena, 12, 20, 29–31, 34, 198
piecewise approximation, 147, 148, 160, 162, 164, 167, 168, 178, 180
polynomial fitting, 101, 102
polynomial function, 139
porous materials, 24, 87
porous media, 294
porous solid, 228, 230, 253
post-processing, 101, 102, 242
Prandtl number, 338
preconditioning, 279, 284, 286
predictor-corrector method, 298
propagation problems, 21, 23, 24
pseudo velocities, 298, 300–302, 308, 309
psychrometric chart, 247

Q

quadratic elements, 169
quadratic variation, 150, 168, 178
quadrature, 163, 180, 272, 274
quadrilateral element, 172, 173, 175, 243
quadrilateral isoparametric, 321

R

radiation, 11, 17, 18, 37, 38, 41, 42, 53, 54, 76, 80, 82, 85, 90, 205, 210, 226–28, 232, 233, 236, 240, 243, 244, 250, 253, 255
Rayleigh number, 338, 340, 342, 346
Rayleigh Ritz, 207
rectangular elements, 172
relative humidity, 245, 248
Robin boundary, 53

S

semi-implicit scheme, 199, 202, 203, 205
shape function, 151, 152, 154, 155, 167–69, 175, 176, 195, 304, 310, 323
shear stress, 29, 31, 60, 61, 72, 78, 101
Sherwood number, 61
simulation analysis, 3, 12, 13, 297, 311
simulation process, 12
simulation program, 36, 101, 107, 111, 194, 209, 243, 244, 248, 258, 276, 288
simulations, 107, 201, 203, 242
solar, 18, 19, 37, 54, 87, 90, 205, 209, 210, 226–28, 232, 233, 236, 240, 242–44, 250, 253, 255
solar collector, 54, 90
solar energy, 87, 209, 210, 227
solar ponds, 19, 87, 209
sparse matrices, 278
sparse matrix, 218
species concentration, 18, 19, 61, 87, 88, 209, 313, 318
species conservation, 18, 38, 61, 87, 88
stability conditions, 199
Stefan-Boltzmann constant, 53
stiffness matrices, 214, 217, 218, 234, 243, 291
Stokes equations, 267, 289
Stokes flow, 262–64, 266–71, 274–76, 284–89, 291, 292
storage scheme, 279, 286, 287
stream-function-vorticity, 344
streamline, 329, 330
subparametric, 180
subroutines, 102, 243, 310
superparametric formulation, 180
syntax, 2, 35, 36, 242, 278, 310
system behavior, 1, 4, 20, 25, 31

T

tangential forces, 75
tangential stresses, 75, 78, 79, 83, 84
Taylor's series expansion, 97, 98, 107, 324
temperature contours, 122, 123, 251, 252
TFI, 156, 181, 182, 186, 188, 192, 206, 244, 286, 291, 344, 345
TFI meshing, 188
TFI technique, 181, 182, 186, 192
thermodynamics, 11, 17, 41, 42, 76, 88, 146, 201, 258
three-dimensional domain, 50, 153, 176, 178, 181, 192
total energy equation, 82, 83, 90
transfinite interpolation technique, 156
transient heat conduction, 24, 26, 31, 54, 146, 198, 202, 227, 231, 255
triangular elements, 170, 172, 177, 181, 182, 303, 304, 309
tridiagonal matrix, 104
two-dimensional, 16, 24, 29, 37, 54, 60, 107–9, 112, 117, 118, 120–22, 128, 131–33, 138, 139, 152, 153, 168, 170, 172, 181, 182, 186, 191, 192, 204, 206, 207, 225, 228, 230, 231, 235, 253, 255, 257, 262–64, 266, 291, 292, 297, 303, 311, 315–18, 322, 336, 344
two-dimensional applications, 128
two-dimensional computational domain, 109, 132, 253
two-dimensional conduction, 107, 118, 120
two-noded linear element, 167

U

unsteady, 10, 14, 19, 24, 37, 38, 43, 44, 47, 50, 53, 66, 89, 146, 206, 318
unsteady flow conditions, 66, 89
unstructured, 163, 181, 182, 205, 290
unstructured meshes, 181

V

validation results, 107, 128, 129, 161, 220, 221, 244, 286, 287, 288
variational formulation, 94, 147
velocity-pressure form, 270
velocity-vorticity, 101, 268, 270, 291, 292, 311, 313, 317, 318, 322, 327–29, 336, 338, 343–46
velocity-vorticity equations, 313, 328, 338
velocity-vorticity formulation, 317, 318, 322, 329, 345
viscous, 17, 67, 75, 83–85, 87, 263, 268, 270, 289, 329, 333, 338, 345

viscous dissipation, 83–85, 87
viscous forces, 67, 75, 263, 289, 345
viscous friction, 333
volume coordinates, 177, 178
vortices, 184–86, 192
vorticity, 101, 264–72, 274, 276, 285, 289, 290, 291, 292, 313–23, 325, 327, 329–31, 333, 336–38, 344, 345
vorticity boundary, 266, 267, 270, 274, 285, 291, 292, 313, 317, 318, 327
vorticity equation, 266
vorticity field, 266, 268, 270, 274, 276, 291, 292, 316–18, 327, 345

W

WBT, 247
weighted residual method, 94, 123, 124, 147, 148, 150–53, 162, 169, 176, 195, 273, 278, 323
weighting function, 94, 124, 148, 151, 152, 323
wet bulb temperature, 209, 245, 247
wind tunnel, 4, 6, 12

Y

Young's modulus, 72